MASTERING APPLIED ENGINEERING MECHANICS WITH CHATGPT

Mastering Applied Engineering Mechanics with ChatGPT

Statics-Dynamics-Vibrations

Mehrzad Tabatabaian

MERCURY LEARNING AND INFORMATION
Boston, Massachusetts

Mercury Learning and Information
121 High Street, 3rd Floor
Boston, MA 02110
info@merclearning.com

M. Tabatabaian. *Mastering Applied Engineering Mechanics with ChatGPT: Statics-Dynamics-Vibrations.*
ISBN: 978-1-50152-389-2

The publisher recognizes and respects all marks used by companies, manufacturers, and developers as a means to distinguish their products. All brand names and product names mentioned in this book are trademarks or service marks of their respective companies. Any omission or misuse (of any kind) of service marks or trademarks, etc. is not an attempt to infringe on the property of others.

Library of Congress Control Number: 2025937804

242526321   This book is printed on acid-free paper in the United States of America.

Our titles are available for adoption, license, or bulk purchase by institutions, corporations, etc. All of our titles are available in digital format at various digital vendors.

All companion files for this title are available by visiting the website for the book at *https://sciendo.com/book/ 9781501523892* and by clicking on the "COMPANION FILES" tab.

To my family, whose unwavering support made this publishing journey possible—and each milestone a celebration.

CONTENTS

PREFACE

Welcome to *Mastering Applied Engineering Mechanics with ChatGPT*!

The aim of this book is to revolutionize the way engineering students engage with their coursework and help professionals prepare for review and exams. Traditional textbooks often present information in a static and one-dimensional format, making it challenging for students to actively participate in the learning process.

In this book, a novel approach that uses ChatGPT, an advanced language model, is introduced to provide dynamic and interactive learning prompts. These prompts are designed to stimulate critical thinking, encourage problem-solving, and deepen the understanding of important engineering concepts. While the integration of AI in education is still an evolving area, educators and researchers are exploring how to effectively incorporate AI tools, such as ChatGPT, into the learning environment. This ongoing task aims to find the best ways for AI to complement traditional learning methods and enhance educational outcomes. This book contributes to that aim, supporting both students in their learning journey and educators in enhancing their teaching methods.

This book is divided into three comprehensive parts, each containing several chapters, with approximately ten prompts per chapter. In total, the book offers more than 300 prompts, all carefully designed to align with the standard curriculum of an engineering student. The chapters cover topics essential for building a strong foundation in engineering principles. Additionally, an AI-aided tutor is integrated through a custom ChatGPT app for each part, offering interactive support and tailored guidance for learners as they progress through the material.

This book is organized into the following parts:

- Part 1: Statics
- Part 2: Dynamics
- Part 3: Mechanical Vibrations

Learning should be an active and enjoyable process. *Mastering Applied Engineering Mechanics with ChatGPT* aims to empower engineering students to take control of their learning journey and develop the skills they need to succeed in their academic and professional endeavors.

This book is supported by three interactive GPT Apps—Statics Wizard Tutor, Dynamics Wizard Tutor, and Vibration Wizard Tutor—designed to complement their respective parts. Each App provides interactive prompt-type communications, instant feedback, and visualizations, offering readers a dynamic way to review concepts, solve problems, and strengthen understanding across the covered topics. All companion files for this title are available by visiting the website for the book at *https://sciendo.com/book/ 9781501523892* and by clicking on the "COMPANION FILES" tab.

Hopefully, you find this book exciting and transformative. Happy learning!

Mehrzad Tabatabaian, PhD, PEng
Vancouver, BC, Canada
September, 2025

INTRODUCTION

By keeping the prompts general and allowing the apps to customize the experience, this book provides a versatile and comprehensive learning tool that can be used by a wide range of students and professionals.

Purpose of the book: This book is designed to provide engineering students and professionals with a comprehensive understanding of statics, dynamics, and vibrations through a series of interactive prompts. The goal is to enhance problem-solving skills, reinforce theoretical concepts, and prepare readers for practical applications in mechanical engineering. Readers can also highlight any expression or prompt and use the Ask Microsoft Copilot™ feature (when connected) to explore related information.

How to use the prompts: Each prompt is crafted to challenge the reader's understanding of core principles within a given topic. Students and professionals are encouraged to attempt solving the prompts independently before referring to the solutions and explanations provided by ChatGPT (or any other preferred generative AI). The prompts are organized progressively, starting with foundational concepts and gradually building toward more complex applications.

Overview of statics as a subject: *Statics* is a fundamental branch of mechanics that focuses on analyzing forces and their effects on bodies in a state of equilibrium. It lays the groundwork for understanding how structures and components behave under various loading conditions, which is essential for the design and analysis of engineering projects. Important topics include force vectors, equilibrium of particles and rigid bodies, moments, couples, and the analysis of trusses, beams, and other structural elements. Mastery of statics is crucial for ensuring the stability and integrity of buildings, bridges, vehicles, machinery, and various engineering systems.

Overview of dynamics as a subject: *Dynamics* is a core branch of mechanics that deals with the study of forces and their effects on the motion of bodies. Unlike statics, which focuses on equilibrium conditions, dynamics examines bodies in motion and the forces causing such motion. This subject is crucial for understanding and predicting the behavior of moving systems, which is essential for designing and analyzing various engineering applications. Important topics include kinematics of particles and rigid bodies, kinetics involving Newton's second law, work-energy principles, impulse-momentum principles, and mechanical vibrations. Mastery of dynamics is vital for ensuring the proper functioning and safety of engineering systems such as vehicles, machinery, aerospace structures, and any applications where motion and forces are involved.

Overview of mechanical vibrations as a subject: *Mechanical vibrations* is a specialized branch of dynamics that focuses on the oscillatory motion of mechanical systems. It examines the causes, behaviors, and responses of systems subjected to periodic or transient forces. Vibrations can be advantageous in applications such as vibrating screens, ultrasonic cleaning devices, and precision instruments, or detrimental, leading to issues like structural fatigue, noise, and failure in machinery. This subject is critical for analyzing and designing systems to minimize unwanted vibrations and enhance performance. Important topics include free and forced vibrations, damping mechanisms, multi-degree-of-freedom systems, vibration isolation, and modal analysis. Mastery of mechanical vibrations is essential for engineers to ensure the durability, reliability, and efficiency of systems such as vehicles, machinery, bridges, and aerospace structures.

Features of *Mastering Applied Engineering Mechanics with ChatGPT: Statics-Dynamics-Vibrations*:

1. Comprehensive coverage

   • Explore essential topics across statics, dynamics, and mechanical vibrations

   • Structured into 30 chapters with 300 carefully crafted prompts for in-depth understanding

2. Interactive learning approach

   - Engage actively with prompts designed to stimulate critical thinking

   - Develop problem-solving skills through hands-on activities

3. Practical applications

   - Bridge theory and practice with numerical examples and real-world scenarios

   - Understand how mechanical engineering principles are applied in real-life situations

4. Integration with ChatGPT

   - Use ChatGPT to verify solutions and clarify concepts

   - Receive immediate, personalized feedback to enhance your learning experience

5. Inclusion of GPT apps

   - Access interactive simulations and problem-solving tools within the book

   - Experience virtual labs and design tools related to statics, dynamics, and vibrations

6. Consistent and structured format

   - Benefit from a clear and organized learning path

   - Each prompt includes an objective, activity, numerical example, and ChatGPT integration.

7. Flexible learning resource

   - Use the book as a primary guide or a supplementary tool for revision.

   - Accommodates different learning styles and educational needs

8. Enhanced interactivity

   - Stay engaged with interactive elements that make learning enjoyable

   - Hands-on approach deepens understanding and retention of complex concepts

9. Personalized learning

   - Receive customized guidance based on your inputs through GPT-powered applications

   - Tailor your learning experience to suit your individual needs

10. Up-to-date content

    - Stay current with the latest developments in mechanical engineering.

    - Content reflects modern practices and technologies in the field.

11. Empowering tool for students and professionals

    - Ideal for students aiming to strengthen their understanding

    - Valuable for professionals seeking to refresh or advance their expertise

12. Accessible advanced computational tools

    - Utilize advanced tools without the need for additional software.

    - Lower barriers to experimentation and exploration in mechanical engineering concepts.

ABOUT THE AUTHOR

Dr. Mehrzad Tabatabaian is currently a senior faculty member at the Mechanical Engineering Department, School of Energy at British Columbia Institute of Technology (BCIT) with several years of teaching, research, and industry experience. He has authored textbooks and published papers in scientific journals and conferences. He holds several registered patents in the clean energy field.

Mehrzad Tabatabaian got his BEng from Sharif University of Technology (formerly AUT), graduated from McGill University (MEng and PhD). He has been an active academic, professor, and engineer in leading traditional and alternative energy industries. Mehrzad has also a Leadership Certificate from the University of Alberta and holds an EGBC P.Eng. license.

PART 1: STATICS

INTRODUCTION TO STATICS

Statics is a fundamental branch of mechanics that deals with the analysis of forces and their effects on bodies in equilibrium. It serves as the cornerstone of engineering disciplines, providing the tools necessary to design structures, machines, and systems that remain stable under various loading conditions. Engineers rely on the principles of statics to ensure the safety, functionality, and reliability of everything from bridges and buildings to mechanical components and aerospace systems.

This part of the book introduces the foundational concepts of statics in a structured manner, covering key principles such as equilibrium, distributed forces, friction, centroids, moments of inertia, and more. Each chapter is designed to progressively build your understanding, with interactive prompts, real-world examples, and numerical problems to enhance your learning experience.

Part 1 is divided into ten chapters, each addressing a specific topic in statics:

Chapter 1. Basic Concepts
Chapter 2. Forces and Equilibrium
Chapter 3. Trusses and Frames
Chapter 4. Distributed Forces
Chapter 5. Center of Gravity and Centroid
Chapter 6. Moments of Inertia
Chapter 7. Analysis of Structures

Chapter 8. Virtual Work and Applications
Chapter 9. Internal Forces
Chapter 10. Real-World Applications in Statics

Each chapter includes approximately ten interconnected prompts, designed to guide readers in exploring the subject matter in greater depth, incorporating numerical examples and additional exploratory questions.

CHAPTER 1

BASIC CONCEPTS

Prompt 1 Explain the difference between scalar and vector quantities. Provide examples in the context of statics.

Prompt 2 Describe the concept of force and how it is represented as a vector.

Prompt 3 Calculate the resultant force of multiple vectors acting on a point.

Prompt 4 Discuss the principle of transmissibility of forces and its applications.

Prompt 5 Explain the concept of a position vector and how it is used in statics problems.

Prompt 6 Describe the dot product of two vectors and its significance in statics.

Prompt 7 Describe the cross product of two vectors and its significance in statics.

Prompt 8 Explain the concept of a unit vector and how it is used to represent direction.

Prompt 9 Calculate the angle between two vectors using the dot product.

Prompt 10 Discuss the concept of a moment arm and its role in calculating moments.

PROMPT 1

Explain the difference between scalar and vector quantities. Provide examples in the context of statics.

Objective

Understand the fundamental difference between scalar and vector quantities and their significance in statics.

Activity

- Define scalar and vector quantities, emphasizing their unique characteristics.
- Provide examples of scalar quantities (e.g., mass, temperature) and vector quantities (e.g., force and displacement) within the context of statics.
- Discuss how vectors are represented and manipulated in engineering applications.

Numerical Example

Scalar Example: a mass of 10 kg, temperature of an object, and volume of a box.

Consider a system where two forces act on a point:

- Force 1: $F_1 = 10\,N$ directed along the $x - axis$.
- Force 2: $F_2 = 15\,N$ at an angle of $60°$ to the $x - axis$.

1. Calculate the resultant force R.
2. Identify whether the magnitude of the resultant force is a scalar or vector quantity.

Solution

1. Break down the forces into components:

$$F_{2x} = 15\cos\left(60°\right) = 7.5\,N, \quad F_{2y} = 15\sin\left(60°\right) = 12.99\,N.$$

The total components are as follows:

$$R_x = F_{1x} + F_{2x} = 10 + 7.5 = 17.5\,N,$$

$$R_y = F_{2y} = 12.99\,N.$$

2. Calculate the magnitude of the resultant force:

$$R = |R| = \sqrt{R_x^2 + R_y^2} = \sqrt{17.5^2 + 12.99^2} = 21.8\,\text{N}$$

3. Conclusion: The magnitude of R is a scalar quantity, but $\vec{R}$ itself is a vector represented by its magnitude and direction.

ChatGPT Integration

▪ Ask ChatGPT to verify the definitions of scalar and vector quantities.

▪ Request examples of scalar and vector quantities in real-world statics problems.

PROMPT 2

Describe the concept of force and how it is represented as a vector.

Objective

Understand the representation of force as a vector, its components, and its significance in statics problems.

Activity

▪ Define force and explain its vector nature, including magnitude, direction, and point of application.

▪ Discuss the representation of force vectors using Cartesian components.

▪ Illustrate how to calculate the resultant force when multiple forces act on a point.

Numerical Example

A block is subjected to two forces:

▪ $F_1 = 30\,\text{N}$ acting at an angle of $45°$ to the horizontal.

▪ $F_2 = 40\,\text{N}$ acting horizontally.

1. Represent F_1 and F_2 as vectors in Cartesian form.

2. Calculate the resultant force R.

3. Determine the direction of R relative to the horizontal.

Solution

1. Represent F_1 and F_2 as vectors:

$$F_1 = 30\cos\left(45°\right)\hat{i} + 30\sin\left(45°\right)\hat{j}$$

$$F_2 = 40\hat{i}$$

2. Add the components to find R:

$$R_x = F_{1x} + F_{2x} = 30\cos\left(45°\right) + 40$$

$$R_y = F_{1y} = 30\sin\left(45°\right)$$

Through substituting, we obtain

$$R_x = 21.21 + 40 = 61.21\,\text{N}, \ R_y = 21.21\,\text{N}$$

3. Calculate the magnitude of R:

$$R = \sqrt{R_x^2 + R_y^2} = \sqrt{61.21^2 + 21.21^2} \approx 64.78\,\text{N}$$

4. Determine the direction of R:

$$\theta = \tan^{-1}\left(\frac{R_y}{R_x}\right) = \tan^{-1}\left(\frac{21.21}{61.21}\right) \approx 19.1°$$

ChatGPT Integration

- Ask ChatGPT for alternative methods to represent and solve force vector problems.
- Experiment with different angles or force magnitudes to observe changes in the resultant vector.

PROMPT 3

Calculate the resultant force of multiple vectors acting on a point.

Objective

Learn to calculate the resultant force by resolving force vectors into components and combining them to find the magnitude and direction of the resultant.

Activity

Resolve three given force vectors into x- and y-components, sum the components, and calculate the magnitude and direction of the resultant force.

Numerical Example

Three forces act on a point:
- F_1=50 N at $30°$ from the positive x-axis.
- F_2=40 N along the negative y-axis.
- F_3=30 N at $120°$ from the positive x-axis.

Calculate the resultant force.

Solution

1. Resolve the forces into components:

$$F_{1x} = F_1 \cos(30°), \quad F_{1y} = F_1 \sin(30°)$$

$$F_{2x} = 0, \quad F_{2y} = -F_2$$

$$F_{3x} = F_3 \cos(120°), \quad F_{3y} = F_3 \sin(120°)$$

2. Substitute the given values:

$$F_{1x} = 50\cos(30°) \approx 43.3\,\text{N}, \quad F_{1y} = 50\sin(30°) = 25\,\text{N}$$

$$F_{2x} = 0, \quad F_{2y} = -40\,\text{N}$$

$$F_{3x} = 30\cos(120°) = -15\,\text{N}, \quad F_{3y} = 30\sin(120°) \approx 25.98\,\text{N}$$

3. Calculate the resultant magnitude:

$$R = \sqrt{R_x^2 + R_y^2} = \sqrt{28.3^2 + 10.98^2} \approx \sqrt{800.89 + 120.56} \approx \sqrt{921.45} \approx 30.36\,\text{N}$$

4. Determine the resultant direction:

$$\theta = \tan^{-1}\left(\frac{R_y}{R_x}\right) = \tan^{-1}\left(\frac{10.98}{28.3}\right) \approx \tan^{-1}(0.388) \approx 21.21°$$

ChatGPT Integration

- Use ChatGPT to verify the vector resolution and calculation of the resultant.
- Ask for further examples, such as the role of combined forces in structural engineering or robotics.

PROMPT 4

Discuss the principle of transmissibility of forces and its applications.

Objective

Understand the principle of transmissibility of forces and its significance in analyzing mechanical systems, particularly in statics and dynamics.

Activity

Define the principle of transmissibility of forces. The principle of transmissibility states that the external effect of a force on a rigid body (such as causing translation or rotation) remains unchanged if the force is shifted along its line of action. This is useful in simplifying problems where determining the effects of forces at specific points is challenging. For example, in calculating torque, the force can be moved to a more convenient point along its line of action without affecting the result. This principle is particularly relevant in mechanical systems like beams, levers, and gears.

Numerical Example

A force of F = 100 N is applied at a distance r = 2 m from a point, at an angle $\theta = 30°$ from the position vector, assuming the force is shifted along its line of action to simplify the calculation.

Solution

The principle of transmissibility states that a force can be moved along its line of action without altering the external effect on the body. In this example, the torque about the point is determined by the following:

$$\tau = r \times F \times \sin(\theta)$$

Substituting the given values results in the following:

$$\tau = 2 \times 100 \times \sin(30°) = 100\,N.m$$

ChatGPT Integration

- Use ChatGPT to verify the application of the principle of transmissibility.
- Ask it to explain why the force's position along its line of action does not affect the resultant torque and provide other real-world examples.

PROMPT 5

Explain the concept of a position vector and how it is used in statics problems.

Objective

Understand the concept of a position vector and its role in representing the location of a point relative to a reference in statics problems.

Activity

Calculate the position vector for a given point in space. A position vector represents the location of a point in space relative to an origin, defined by its components along the coordinate axes. It is a fundamental concept in statics, used to define the line of action of forces and calculate moments. For example, in a 2D system, the position vector can be used to determine the torque produced by a force relative to a reference point.

Numerical Example

A force $F = (50,30)\,N$ acts at a point $P\,(3,4)m$. Calculate the torque produced by the force about the origin using the position vector of point P.

Solution

The position vector r of point P(3,4) relative to the origin is

$$r = (3,4)\,\text{m} = 3\hat{i} + 4\hat{j}$$

The torque τ is given by the cross product of r and F:

$$\tau = r \times F$$

For a 2D system, the magnitude of torque is

$$\tau = r_x F_y - r_y F_x$$

Through substitution, we obtain the following:

$$\tau = (3)(30) - (4)(50) = -110 \, N.m$$

Thus, the torque about the origin is -110 N.m, indicating a clockwise rotation (i.e., negative Z-axis).

ChatGPT Integration

- Use ChatGPT to verify the calculation of torque and explore how position vectors are used in different coordinate systems (e.g., 3D).
- Ask ChatGPT for examples of moments or applications of position vectors in real-world structures.

PROMPT 6:

Describe the dot product of two vectors and its significance in statics.

Objective

Understand the concept of the dot product and how it is used to compute projections of vectors and analyze work or energy in statics problems.

Activity

Calculate the dot product of two vectors. The dot product is a scalar value that represents the projection of one vector onto another. In statics, for example, it is used to calculate work done by a force, orthogonality check of two vectors, projection of a force vector on a specified direction, and determine the angle between two vectors.

Numerical Example

Calculate the following:

1. The dot product of force vectors
   $A = (4,3) = \left(4\hat{i} + 3\hat{j}\right)$ and $B = (2,5) = 2\hat{i} + 5\hat{j}$.

2. The angle θ between the two vectors.

Solution

1. Dot product

   The formula for the dot product is as follows:

   $$A \cdot B = A_x B_x + A_y B_y$$

   We substitute the given values as follows:

   $$A \cdot B = (4)(2) + (3)(5) = 8 + 15 = 23$$

2. Angle between vectors

   The formula for the angle between two vectors is

   $$\cos \theta = \frac{A \cdot B}{|A||B|}$$

   where $|A| = \sqrt{A_x^2 + A_y^2}$, $|B| = \sqrt{B_x^2 + B_y^2}$

   Calculate the magnitudes to obtain the following:

   $$|A| = \sqrt{4^2 + 3^2} = \sqrt{16 + 9} = 5, \quad |B| = \sqrt{2^2 + 5^2} = \sqrt{4 + 25} = \sqrt{29}$$

   We substitute the values to obtain

   $$\cos \theta = \frac{23}{5\sqrt{29}}, \text{ or } \theta = \cos^{-1}\left(\frac{23}{5\sqrt{29}}\right) = 31.33°$$

ChatGPT Integration

- Ask ChatGPT to compute the dot product for other vector pairs or verify the calculations.
- ChatGPT can also explain how the dot product applies to work and energy concepts in engineering applications.

PROMPT 7:

Describe the cross product of two vectors and its significance in statics.

Objective

Understand the concept of the cross product, its mathematical representation, and its application in determining a perpendicular vector in 3D space, such as torque or moment.

Activity

Calculate the cross product of two vectors. The cross product is a vector perpendicular to the plane formed by the two original vectors, with magnitude proportional to the area of the parallelogram defined by the vectors. This concept is commonly used in statics and dynamics to compute torque or rotational effects and angular momentum. For example, the torque generated by a force applied at a distance from a pivot point is a result of the cross product.

Numerical Example

Two vectors are given as $\mathbf{A} = (3,-2,5)$ and $\mathbf{B} = (1,4,-1)$. Calculate the following:

1. The cross-product $\mathbf{A} \times \mathbf{B}$.

2. Verify that the result is perpendicular to both $\mathbf{A}$ and $\mathbf{B}$.

Solution

1. Cross product

    The formula for the cross product is as follows:

    $$A \times B = \begin{vmatrix} i & j & k \\ 3 & -2 & 5 \\ 1 & 4 & -1 \end{vmatrix}$$

    Expand the determinant:

    $$A \times B = i\begin{vmatrix} -2 & 5 \\ 4 & -1 \end{vmatrix} - j\begin{vmatrix} 3 & 5 \\ 1 & -1 \end{vmatrix} + k\begin{vmatrix} 3 & -2 \\ 1 & 4 \end{vmatrix}$$

    Simplify each minor determinant:

    $$A \times B = -18i + 8j + 14k$$

2. Verification of perpendicularity

    A vector $\mathbf{C}$ is perpendicular to $\mathbf{A}$ if $\mathbf{A \cdot C} = 0$ and perpendicular to $\mathbf{B}$ if $\mathbf{B \cdot C} = 0$. In other words, the product $A \times B$ must be aligned with its unit vector.

Check **A·(A×B)**:

$$A \cdot (A \times B) = (3)(-18) + (-2)(8) + (5)(14) = 0$$

Check **B·(A×B)**:

$$B \cdot (A \times B) = (1)(-18) + (4)(8) + (-1)(14) = 0$$

Thus, **A×B** is perpendicular to both **A** and **B**, or equivalently perpendicular to their plane.

ChatGPT Integration

- Use ChatGPT to compute the cross product for other vectors or verify orthogonality.
- Ask ChatGPT to explain applications of cross products in engineering, such as torque and angular momentum.

PROMPT 8

Explain the concept of a unit vector and how it is used to represent direction.

Objective

Understand the concept of a unit vector and its role in defining the direction of a vector in space.

Activity

Define the unit vector as a vector with a magnitude of 1 that points in the direction of a given vector. Explain how to compute a unit vector by dividing a vector by its magnitude. Highlight the importance of unit vectors in Statics, particularly in representing directions independent of magnitude. For example, in mechanics, unit vectors are used to define orientations like axes in Cartesian coordinates, forces, etc.

Numerical Example

A vector A has the components $A = 3i - 4j + 5k$.

Question: Can you find the unit vector in the direction of **A**?

Solution

The magnitude of **A** is calculated as

$$|A| = \sqrt{(3)^2 + (-4)^2 + (5)^2} = \sqrt{50}$$

The unit vector $\hat{\mathbf{A}}$ is given by

$$\hat{\mathbf{A}} = \frac{A}{|A|} = \frac{1}{\sqrt{50}}(3i - 4j + 5k)$$

Simplify to obtain the following:

$$\hat{\mathbf{A}} = \frac{1}{5\sqrt{2}}(3i - 4j + 5k)$$

ChatGPT Integration

- Use ChatGPT to verify the calculation of the magnitude of the vector and the resulting unit vector.
- Ask for real-world examples of the application of unit vectors in engineering and physics.

PROMPT 9

Calculate the angle between two vectors using the dot product.

Objective

Learn how to determine the angle between two vectors using the dot product formula and understand its applications in statics problems.

Activity

Calculate the angle between two given vectors.
The dot product provides a mathematical tool to find the cosine of the angle between two vectors. This is critical in statics to analyze forces and their directions relative to one another. For example, understanding the interaction between two forces acting at an angle is vital in structural analysis.

Numerical Example

Two vectors are given as follows:

$$A = 3i + 4j, B = 2i - j$$

Question: Can you calculate the angle θ between A and B?

Solution

1. Here is the dot product formula:

$$A \cdot B = |A||B|\cos\theta$$

2. Compute A·B:

$$A \cdot B = (3)(2) + (4)(-1) = 2$$

3. Find the magnitudes of A and B:

$$|A| = \sqrt{3^2 + 4^2} = 5$$

$$|B| = \sqrt{2^2 + (-1)^2} = \sqrt{5}$$

4. Substitute this into the dot product formula:

$$\cos\theta = \frac{2}{5\sqrt{5}}$$

5. Solve for θ:

Use the inverse cosine function:

$$\theta = \cos^{-1}\left(\frac{2}{5\sqrt{5}}\right) \approx 79.7°$$

ChatGPT Integration

Ask ChatGPT to verify your calculations for the dot product and magnitudes of the vectors. You can also explore how changing the vectors alters the angle and its implications in real-world scenarios.

PROMPT 10

Discuss the concept of a moment arm and its role in calculating moments.

Objective

Understand the concept of a moment arm and how it is used to calculate moments in statics problems.

Activity

Identify the moment arm for a force applied to a lever. Provide a brief description of the moment arm, explaining its significance in determining the turning effect or torque produced by a force. Include real-world examples, such as a wrench or seesaw, to help readers connect the concept to practical applications.

Numerical Example

A force of $F = 100\,N$ is applied perpendicularly at the end of a 2 m long wrench.
Question: Can you calculate the moment produced by the force about the axis of rotation?

Solution

The formula for the moment is as follows:

$$M = F \cdot d$$

where M is the moment, F is the applied force, and d is the perpendicular distance from the axis of rotation to the line of action of the force.

By substituting the values, we obtain the following:

$$M = 100\,N \cdot 2\,m = 200\,N.m$$

The moment produced by the force about the axis of rotation is 200 Nm.

ChatGPT Integration

- Ask ChatGPT to explain the relationship between the force direction and the moment arm length.
- Use it to verify your calculations and explore how the moment changes with different angles between the force and the lever.

Upon review, these ten prompts cover the essential topics in statics, including the fundamental concepts of vectors, forces, moments, and vector operations. They provide a comprehensive foundation for students beginning their study of statics and include sufficient detail to ensure a thorough understanding of the material. The addition of numerical examples and ChatGPT integration enhances the learning experience by providing practical applications and interactive feedback.

CHAPTER 2

FORCES AND EQUILIBRIUM

Prompt 11 Define equilibrium and explain the conditions for a rigid body to be in equilibrium.

Prompt 12 Describe the concept of a free-body diagram and its importance in solving statics problems.

Prompt 13 State and explain the three equations of equilibrium in statics.

Prompt 14 Explain the difference between equilibrium and non-equilibrium conditions in the context of statics. Provide examples of each type.

Prompt 15 Discuss Newton's First Law of Motion and its application to equilibrium in statics problems

Prompt 16 Discuss the significance of the moment of a force in maintaining equilibrium.

Prompt 17 Explain the concept of concurrent forces and their equilibrium conditions.

Prompt 18 Describe the difference between two-dimensional and three-dimensional equilibrium.

Prompt 19 Solve a problem involving the equilibrium of a rigid body under multiple forces.

Prompt 20 Explain how equilibrium analysis is applied to solve indeterminate beams in engineering.

PROMPT 11

Define equilibrium and explain the conditions for a rigid body to be in equilibrium.

Objective

Understand the concept of equilibrium and the necessary conditions for a rigid body to remain in equilibrium.

Activity

Learn the fundamental principles of equilibrium in statics. A rigid body is in equilibrium when it is at rest or moves with a constant velocity, requiring both translational and rotational forces to balance. Provide a detailed explanation of the conditions for equilibrium: the sum of all forces and the sum of all moments acting on the body must equal zero. Relate this to real-world applications, such as ensuring structural stability in bridges or buildings.

Numerical Example

A uniform beam of length 6 m and weight $W = 300$ N is supported by a hinge at one end and a roller at the other. A vertical load $P = 500$ N is applied at 2 m from the hinge.

Question: Can you determine the reactions at the hinge (R_H) and the roller (R_R)?

Solution

Step 1: Draw the free-body diagram of the beam.

Step 2: Apply the conditions for equilibrium:

- sum of the vertical forces: $R_H + R_R - W - P = 0$
- sum of the moments about the hinge: $6R_R - 2P - 3W = 0$

Step 3: Solve the equations:

- from the moment equation: $R_R = \dfrac{2 \times 500 + 3 \times 300}{6} = 316.67\,N$
- substituting R_R into the vertical force equation:
  $R_H = 300 + 500 - 316.67 = 483.33N \cdot$

ChatGPT Integration

- Use ChatGPT to verify the equations for equilibrium and confirm the calculated reactions.
- Ask for variations, such as changing the position or magnitude of the load, to explore different scenarios.

PROMPT 12

Describe the concept of a free-body diagram and its importance in solving statics problems.

Objective

Understand the purpose of a free-body diagram (FBD) and its role in analyzing forces acting on a body in equilibrium.

Activity

Learn how to construct a free-body diagram by isolating a body and representing all external forces and moments acting on it. Provide a brief explanation of the key elements of an FBD, including forces, reaction forces, and applied loads. Discuss its significance in solving statics problems, such as calculating structural stability or analyzing the mechanics of a machine.

Numerical Example

A block of weight W = 200 N rests on an inclined plane that makes an angle $\theta = 30°$ with the horizontal. A force P is applied parallel to the plane to prevent the block from sliding.

Question: Can you draw the FBD of the block and determine the value of P needed to keep the block stationary, assuming no friction?

Solution

Step 1: Draw the FBD:

- Represent the block with a rectangle.
- Include the weight W = 200 N acting vertically downward.

- Resolve W into components:
  - $W_x = W \sin \theta$ (parallel to the plane)
  - $W_y = W \cos \theta$ (perpendicular to the plane).
- Add the applied force P acting parallel to the plane.

Step 2: Apply equilibrium conditions:

- Sum of forces along the inclined plane: $P - W \sin \theta = 0$
- Solve for P: $P = W \sin \theta$

Step 3: Substitution:

- $P = 200 \sin\left(30^\circ\right) = 100 \text{ N}$

ChatGPT Integration

- Ask ChatGPT to verify the FBD and equilibrium equations.
- Use it to test variations, such as adding friction or changing the incline angle, to explore different scenarios.

PROMPT 13

State and explain the three equations of equilibrium in statics.

Objective

Understand the three equations of equilibrium and their significance in maintaining the stability of a rigid body under various forces and moments.

Activity

Explore the core principles of statics by learning how the three equilibrium equations ensure a rigid body remains stationary.

- Provide the mathematical representation of the equations of equilibrium.
- Discuss their physical meaning and relevance in analyzing engineering structures.
- Use an example of a beam under multiple forces to demonstrate their application.

Numerical Example

A beam of length 6 m is supported at both ends. A load of 100 N is placed 2 m from the left support. Apply the three equations of equilibrium to find the reaction forces at the supports.

Solution

1. Equations of Equilibrium

$$\sum F_x = 0, \quad \sum F_y = 0, \quad \sum M = 0$$

   These represent the conditions for horizontal, vertical, and rotational equilibrium, respectively.

2. Apply the vertical force balance:

   Let reactions at the left and right supports be R_A and R_B:

   $$\sum F_y = 0 \Rightarrow R_A + R_B = 100$$

3. Apply moment balance at A:

   Taking moments about the left support A:

   $$\sum M = 0 \Rightarrow R_B = 100 / 3 \text{ N}$$

4. Solve for R_A:

   Substitute R_B into the vertical force balance equation:

   $$R_A = 100 - R_B = 100 - 33.33 = 66.67 \text{ N}$$

ChatGPT Integration

- Use ChatGPT to explore variations of the problem, such as changing the beam length, load position, or adding additional forces.
- Ask for detailed explanations of how moments are calculated and for additional examples of applying the equilibrium equations.

PROMPT 14

Explain the difference between equilibrium and non-equilibrium conditions in the context of statics. Provide examples of each type.

Objective

Understand the distinction between equilibrium and non-equilibrium conditions and how they relate to the analysis of forces and moments in static systems.

Activity

Define equilibrium and non-equilibrium conditions. Provide a brief description of the important concepts, focusing on the importance of analyzing forces and moments for stability. Include real-world examples, such as a balanced seesaw (equilibrium) and a tipping ladder (non-equilibrium).

In statics, a system is said to be in equilibrium when the sum of all forces and moments acting on it is zero. This implies that the object is either at rest or moving with constant velocity. Conversely, non-equilibrium occurs when the sum of forces or moments is not zero, causing acceleration or instability.

Real-world examples

- Equilibrium: a balanced seesaw where the moments of the children's weights are equal about the fulcrum
- Non-equilibrium: a ladder tipping over due to an uneven distribution of forces

Numerical Example

A uniform beam of length L = 6 m and weight W = 300 N is supported at its ends by two supports. A downward force F = 200 N is applied 1.5 m from the left support. Determine whether the beam is in equilibrium.

Solution

1. Calculate the reactions:

   The beam is in equilibrium if

$$\sum F_y = 0 \quad \text{and} \quad \sum M = 0$$

Reaction forces at the supports are R_A (left) and R_B (right).

2. Equilibrium Equations

   The vertical forces are as follows:

   $$R_A + R_B - W - F = 0$$

   The moments about A are as follows:

   $$L.R_B - 1.5F - W\frac{L}{2} = 0$$

3. Substitute the values:

   From vertical forces, we obtain

   $$R_A + R_B - 300 - 200 = 0 \Rightarrow R_A + R_B = 500 \text{ N}$$

   From moments, we obtain

   $$6R_B - 1.5 \times 200 - 3 \times 300 = 0 \Rightarrow R_B = 200 \text{ N}$$

   We then substitute R_B into the first equation:

   $$R_A + R_B = 500 \text{ N} \Rightarrow R_A = 300 \text{ N}$$

4. Conclusion

   Since all forces and moments balance, the beam is in equilibrium.

ChatGPT Integration

- Use ChatGPT to solve variations of the example, such as changing the location of the applied force or the weight of the beam.
- Ask ChatGPT to explain how these conditions apply to complex systems, like truss bridges or cranes.

PROMPT 15

Discuss Newton's First Law of Motion and its application to equilibrium in statics problems

Objective

Understand Newton's First Law of Motion and how it governs the conditions of equilibrium in statics. Learn to apply the concept to solve statics problems involving particles and rigid bodies.

Activity

Explain Newton's First Law of Motion, stating that a body remains at rest or in uniform motion unless acted upon by an external force. Discuss how this principle leads to the concept of equilibrium in statics, where the sum of forces and moments acting on a system is zero. Use examples such as a book resting on a table or a bridge structure under distributed loads to illustrate its relevance in practical engineering scenarios.

Numerical Example

A 10-kg block is suspended by two cables.

▪ The first cable makes an angle of 30° with the positive x-axis (measured counterclockwise).

▪ The second cable makes an angle of 45° with the negative x-axis (measured clockwise).

Determine the tension in each cable, assuming the system is in equilibrium.

Solution

Step 1: Define the parameters and draw the free-body diagram.

▪ weight of the block: $W = mg = 10 \text{ kg} \times 9.81 \text{ m/s}^2 = 98.1 \text{ N}$.

▪ tensions in cables: T_1 and T_2

▪ angles: $\theta_1 = 30°$, $\theta_2 = 45°$

Step 2: Write the equations of equilibrium.

1. Horizontal equilibrium $\left(\sum Fx = 0\right)$

$$T_1 \cos\theta_1 - T_2 \cos\theta_2 = 0$$

2. Vertical equilibrium $(\Sigma Fy = 0)$

$$T_1 \sin\theta_1 + T_2 \sin\theta_2 - W = 0$$

Step 3: Solve for T_1 and T_2.

From the equilibrium equations, eliminate T_1 to find T_2:

$$T_2 = \frac{W}{\dfrac{\cos\theta_2}{\cos\theta_1}\sin\theta_1 + \sin\theta_2} \Rightarrow T_2 = \frac{98.1}{\dfrac{\cos 45}{\cos 30}\sin 30 + \sin 45} = 87.95\,N$$

Find T_1:

$$T_1 = T_2 \frac{\cos\theta_2}{\cos\theta_1} \Rightarrow T_1 = 87.95 \frac{\cos 45}{\cos 30} = 71.81 \text{ N}$$

ChatGPT Integration

- Ask ChatGPT to verify the calculations for a block suspended by cables at different angles, providing the parameters.
- Request an explanation of real-world applications of equilibrium, such as analyzing forces in trusses or cable-stayed bridges.

PROMPT 16

Discuss the significance of the moment of a force in maintaining equilibrium.

Objective

Understand the concept of the moment of a force and its role in maintaining rotational equilibrium in statics. Learn to calculate moments and apply them to real-world problems involving rigid bodies.

Activity

Explain the concept of the moment of a force as the rotational effect produced by a force about a point or axis. Discuss its mathematical representation ($M = F \cdot d$, where d is the perpendicular distance from the line of action of the force to the axis). Use examples such as a seesaw, a door being pushed open, or a lever system to illustrate how moments contribute to maintaining equilibrium in mechanical systems.

Numerical Example

A uniform 4-meter beam weighing 200 N is supported at its ends. A 300 N load is placed 1 meter from the left support. Find the reaction forces at the supports by applying the conditions of equilibrium, including the moment equation.

Solution

Step 1: Define the parameters.

- weight of the beam (W_b): 200 N, acting at its center (2 meters from each end)
- load (W_l): 300 N, acting 1 meter from the left support
- reaction forces: R_A (at left support) and R_B (at right support)
- length of beam: L = 4 m

Step 2: Apply the equations of equilibrium.

1. Vertical force balance ($\Sigma F_y = 0$):

$$R_A + R_B = W_b + W_l = 0 \Rightarrow R_A + R_B = 200 + 300 = 500\,N$$

2. Moment Balance ($\Sigma M = 0$):
   Take the moments about point A:

$$4R_B - 2W_b - W_l = 0 \Rightarrow R_B = 175\ N$$

3. Solve for R_A.
   Substitute R_B into the vertical force balance equation:

$$R_A + 175 = 500 \Rightarrow R_A = 325\ N$$

ChatGPT Integration

- Explain the significance of the moment of a force in equilibrium analysis with real-world examples.
- Can you help me solve a problem where a force of 300 N acts at 3m from a pivot point at an angle of 45°?

PROMPT 17

Explain the concept of concurrent forces and their equilibrium conditions.

Objective

Understand the concept of concurrent forces and the conditions required for their equilibrium. Learn how to analyze systems where multiple forces act through a common point.

Activity

Explain that concurrent forces are forces whose lines of action pass through a single point. Discuss the equilibrium conditions for such forces, where the vector sum of all forces must equal zero ($\Sigma F = 0$). Provide examples, such as the forces acting on a suspended object or forces at a joint in a truss.

Numerical Example

Three concurrent forces act on a point:

- F_1 = 200 N at an angle of $0°$ (along the positive x-axis)
- F_2 = 150 N at $120°$
- F_3 = 250 N at $270°$

Determine whether the system is in equilibrium.

Solution

1. Resolve the forces into components:

$$F_{1x} = 200, \quad F_{1y} = 0$$

$$F_{2x} = 150 \cos 120° = -75N, \quad F_{2y} = 150 \sin 120° = 129.9N$$

$$F_{3x} = 0, \quad F_{3y} = -250$$

2. Find the summation of the forces:
   - x-direction: $\sum F_x = F_{1x} + F_{2x} + F_{3x} = 200 - 75 + 0 = 125N$
   - y-direction: $\sum F_y = F_{1y} + F_{2y} + F_{3y} = 0 + 129.9 - 250 = -120.1\,N$

3. Check the equilibrium conditions:
   For equilibrium, both $\sum F_x = 0, \quad \sum F_y = 0$
   Since $\sum F_x = 125\,N$ and $\sum F_y = -120.1$, the system is *not in equilibrium*.

ChatGPT Integration

- Analyze a system with three concurrent forces: F_1 = 100 N at $30°$, F_2 = 200 N at $150°$, and F_3 = 150 N at $270°$. Check if the forces are in equilibrium.

- Explain the concept of concurrent forces using an example of a suspended lamp held by multiple cables, and calculate the forces if equilibrium is maintained.

PROMPT 18

Describe the difference between two-dimensional and three-dimensional equilibrium.

Objective

Understand the fundamental differences between two-dimensional (2D) and three-dimensional (3D) equilibrium, focusing on the equations and conditions necessary for maintaining equilibrium in each case.

Activity

Explain that 2D equilibrium involves forces and moments confined to a plane, requiring only three equations of equilibrium. In contrast, 3D equilibrium considers forces and moments in three spatial dimensions, requiring six equations of equilibrium. Use practical examples such as a ladder resting against a wall for 2D and a suspended load stabilized by cables for 3D to illustrate these concepts.

Numerical Example

A 500 N chandelier is suspended by three cables anchored at points A(0,0,0), B(3,0,0), and C(0,4,0) on the ceiling. The cables meet at point P(1,1,–2). Determine the tensions in the cables assuming the system is in equilibrium.

Solution

1. Create the free-body diagram and force representations:
   - The tensions in the cables are T_1, T_2, and T_3 for cables PA, PB, and PC, respectively.
   - The force from the light acts downward: $F = (0,0,-500)$ N.

2. Find the unit vectors for the cable directions.
   - For PA:

   $$\vec{r}_{PA} = P - A = (1,1,-2)$$

   $$\hat{u}_{PA} = \frac{\vec{r}_{PA}}{|\vec{r}_{PA}|} = \frac{(1,1,-2)}{\sqrt{1^2 + 1^2 + (-2)^2}} = \frac{(1,1,-2)}{\sqrt{6}}$$

■ For PB:

$$\vec{r}_{PB} = P - B = (-2, 1, -2)$$

$$\hat{u}_{PB} = \frac{\vec{r}_{PB}}{|\vec{r}_{PB}|} = \frac{(-2, 1, -2)}{\sqrt{9}} = \frac{(-2, 1, -2)}{3}$$

■ For PC:

$$\vec{r}_{PC} = P - C = (1, -3, -2)$$

$$\hat{u}_{PC} = \frac{\vec{r}_{PC}}{|\vec{r}_{PC}|} = \frac{((1, -3, -2))}{\sqrt{14}}$$

3. Solve the equations of equilibrium.

Resolve forces along the x, y, and z axes:

● x-axis: $\dfrac{1}{\sqrt{6}}T_1 - \dfrac{2}{3}T_2 + \dfrac{1}{\sqrt{14}}T_3 = 0$

● y-axis: $\dfrac{1}{\sqrt{6}}T_1 + \dfrac{1}{3}T_2 - \dfrac{3}{\sqrt{14}}T_3 = 0$

● z-axis: $\dfrac{2}{\sqrt{6}}T_1 + \dfrac{2}{3}T_2 + \dfrac{2}{\sqrt{14}}T_3 = 500$

Solve the system of equations:

Using matrix methods or substitution, solve for T_1, T_2, and T_3. The final answers are as follows:

$T_1 = 625 / \sqrt{6}\,\text{N}$, $T_2 = 250\,\text{N}$, and $T_3 = \dfrac{125}{2}\sqrt{14}\,\text{N}$

ChatGPT Integration

■ Solve a 3D equilibrium problem involving three cables supporting a load. Provide the tensions in the cables given their directions and the load magnitude.

■ Explain the differences between 2D and 3D equilibrium with examples. Illustrate how equations of equilibrium differ in complexity.

PROMPT 19

Solve a problem involving the equilibrium of a rigid body under multiple forces.

Objective

Understand how to solve equilibrium problems for a rigid body subjected to multiple forces by applying the conditions of equilibrium.

Activity

Solve a real-world equilibrium problem for a rigid body under the action of multiple forces.

Provide a brief explanation of how to apply the three equations of equilibrium ($\sum Fx = 0$, $\sum Fx = 0$, and $\sum M = 0$) to determine unknown forces and reactions. Use an example from engineering, such as a ladder leaning against a wall or a beam supported at two ends, to demonstrate the solution process.

Numerical Example

A uniform beam of length $L = 6$ m and weight $W = 800$ N is supported by a hinge at point A and a roller at point B, located 4 meters from point A. The beam is subjected to

1. a distributed load of $q = 300$ N/m over its entire length
2. a point load of $P = 600$ N applied at 2 meters from point A

Find the following:

1. the reaction forces at A (R_{Ax} and R_{Ay}) and B (R_B)
2. Verify the equilibrium conditions for the beam.

Solution

The problem involves both translational and rotational equilibrium.

1. **Free-body diagram**

   Draw the beam with all forces:

   - Distributed load q, with a total load of $Q = qL = 6 \times 300 = 1800$ N, acting at the midpoint of the beam (3 meters from A).

   - Point load $P = 600$ N applied at 2 meters from A.

- Weight of the beam $W = 800$ N, acting at its center (3 meters from A).

- Reaction forces R_{Ax}, R_{Ay} at A and R_B at B.

2. Apply equilibrium equations.

- $\Sigma F_x = 0$: $R_{Ax} = 0$ (No horizontal forces act on the beam.)
- $\Sigma F_y = 0$: $R_{Ay} + R_B - Q - W - P = 0$

  Substituting values:

  $$R_{Ay} + R_B - 1800 - 800 - 600 = 0 \Rightarrow R_{Ay} + R_B = 3200 \text{ N}$$

- $\Sigma M_A = 0$:

  Taking moments about A: $3Q + 3W + 2P - 4R_B = 0$

  We substitute values:

  $$3 \times 1800 + 3 \times 800 + 2 \times 600 - 4R_B = 0 \Rightarrow R_B = 2250 \text{ N}$$

  Solving for R_{Ay}: $R_{Ay} = 3200 - 2250 = 950 \text{ N}$

ChatGPT Integration

- Ask ChatGPT to analyze beams with varying load distributions, angles, or additional supports.
- Request step-by-step calculations for equilibrium in more complex scenarios.

PROMPT 20

Explain how equilibrium analysis is applied to solve indeterminate beams in engineering.

Objective

Understand the principles of equilibrium and how they are extended with compatibility conditions to analyze statically indeterminate beams.

Activity

Discuss the limitations of using only equilibrium equations for indeterminate beams. Introduce the concept of compatibility equations and explain how additional conditions (such as deflection or rotation constraints) are used to solve for unknown reactions or internal forces in indeterminate systems. Provide an overview of methods like superposition or the moment distribution method.

Numerical Example

A fixed-fixed beam has a span of 10 m and a uniform load of 2 kN/m. Calculate the reactions at the supports.

Solution

1. Define the problem: For a statically indeterminate beam, the number of unknown reactions exceeds the available equilibrium equations:
   - R_A (vertical reaction at the left support)
   - M_A (moment at the left support)
   - R_B (vertical reaction at the right support)
   - M_B (moment at the right support)

   Total unknowns = 4; equilibrium equations provide only 2:

   $$\sum F_y = 0, \ \sum M_A = 0$$

   Please note that only one independent equation can be written for moment equilibrium.

2. Introduce compatibility condition: Use beam deflection equations to incorporate additional relationships between reactions and moments. For a fixed-fixed beam, deflections and rotations at both supports must be zero:
   - $\delta_A = 0$, deflection
   - $\theta_A = 0$, rotation

3. Simplify using symmetry.
   For a uniform load on a symmetric fixed-fixed beam, reactions and moments are equal at both ends:

   $$R_A = R_B, \ M_A = M_B$$

4. Calculate the reactions and moments using beam formulas.

   Use standard beam formulas for fixed-fixed beams under a uniform load:

   • Reaction:

   $$R_A = R_B = \frac{wL}{2} = \frac{2 \times 10}{2} = 10 \text{ kN}$$

   • Moment:

   $$M_A = M_B = \frac{wL^2}{12} = \frac{2 \times 10^2}{12} = 16.67 \text{ kN.m}$$

   An alternative solution involves using the differential equations for slope and deflection based on the bending moment expression.

ChatGPT Integration

▪ Ask ChatGPT to verify the assumptions and calculations for indeterminate beams using different methods such as moment distribution or finite element analysis.

▪ Request a step-by-step solution for a similar indeterminate beam with varying loads.

CHAPTER 3

TRUSSES AND FRAMES

Prompt 21 Describe the method of joints for analyzing trusses. Solve a basic truss problem.

Prompt 22 Explain the method of sections and apply it to determine the forces in specific truss members.

Prompt 23 Discuss the differences between trusses and frames and solve a basic frame analysis problem.

Prompt 24 Analyze a compound truss using the method of joints and sections.

Prompt 25 Explain how to determine the internal forces in members of a space truss.

Prompt 26 Solve a problem involving the analysis of a planar truss.

Prompt 27 Discuss the stability of trusses and how to ensure their structural integrity.

Prompt 28 Explain the concept of a determinate vs. indeterminate truss.

Prompt 29 Analyze the forces in a truss subjected to moving loads.

Prompt 30 Explain the different types of trusses and their typical structural applications.

PROMPT 21

Describe the method of joints for analyzing trusses. Solve a basic truss problem.

Objective

Understand the method of joints and its application in analyzing forces in truss members. Learn how to solve a simple truss problem step by step.

Activity

The method of joints is a systematic approach used to analyze trusses by isolating each joint and applying equilibrium equations ($\sum F_x = 0$, $\sum F_y = 0$, $\sum M = 0$). This method is critical for calculating the internal forces (tension or compression) in truss members. It is widely used in structural engineering applications such as bridge design and roof trusses.

- Define the method of joints and its underlying assumptions.
- Explain the process of isolating a joint and applying equilibrium equations.
- Highlight the importance of identifying zero-force members and determining member forces (tension or compression).

Numerical Example

Analyze the forces in a simple truss with three members forming a triangle. The truss is supported by a pin at A and a roller at B. A vertical load of 500 N and a horizontal load of +200 N are applied at joint C. The dimensions are as follows:
AB = 4 m, BC = 3 m, AC = 5 m

Solution

GIVEN

AB = 4 m, BC = 3 m, AC = 5 m

Supports: pin at A (A_x, A_y), roller at B (B_y)

Loads at C: +200 N ($\rightarrow$), 500 N ($\downarrow$)

1. Support Reactions

   $\Sigma F_x : A_x + 200 = 0 \Rightarrow A_x = -200$ N (left)

   $\Sigma F_y : A_y + B_y\ 500 = 0$

   $\Sigma M_A : 4B_y - 500(4) - 200(3) = 0 \Rightarrow B_y = 650$ N (up)

   $A_y = 500 - 650 = -150$ N (down)

2. Joint A

   $\Sigma F_y : -150 + \dfrac{3}{5}F_{AC} = 0\ F_{AC} \Rightarrow 250$ N (T)

   $\Sigma F_x : -200 + F_{AB} + \dfrac{4}{5}(250) = 0 \Rightarrow F_{AB} = 0$ N

3. Joint C

   $\Sigma F_x : 200 - 200 = 0$

   $\Sigma F_y : -500 - 150 + F_{BC} = 0 \Rightarrow F_{BC} = 650$ N (C)

ChatGPT Integration

- Use ChatGPT to verify the calculations and discuss how zero-force members could simplify the analysis in more complex trusses.
- Explore variations of the truss problem, such as different load positions or angles.

PROMPT 22

Explain the method of sections and apply it to determine the forces in specific truss members.

Objective

Understand the method of sections and its application in efficiently determining forces in selected truss members without analyzing the entire structure.

Activity

The method of sections is a technique used to analyze trusses by cutting through the structure and isolating a section to determine forces in specific members at the cut section. This approach is particularly useful when you need forces in only a few members, as it avoids the need to analyze every joint. Applications include determining forces in important members of bridge trusses or roof trusses under specific loading conditions.

Numerical Example

A simply supported truss has a span of 12 m, with supports at A (pin) and B (roller). The truss is loaded with a vertical force of 600 N at joint C. Determine the force in members CD, CE, and DE using the method of sections. Set a 2D Cartesian coordinate system (x-y) with the origin at joint A (the left support). All dimensions and positions are relative to this coordinate system.

Joint coordinates

- A: (0, 0) (pin support at the left end of the base)
- B: (12, 0) (roller support at the right end of the base)
- C: (6, 0) (midpoint of the base where the load is applied)
- D: (6, 3) (apex of the triangle, 3 m vertically above C)
- E: (12, 3) (vertical member directly above B)

Solution

Geometry (coordinates in metres)

A (0, 0) C (6, 0) B (12, 0)

D (6, 3) E (12, 3)

External data

- 600 N downward at C
- Pin support A, Roller support B

1. Support reactions (whole truss)

$$\Sigma M_A : B_y(12) - 600(6) = 0 \Rightarrow B_y = 300 \text{ N} \quad \uparrow$$
$$\Sigma F_y : A_y + 300 - 600 = 0 \Rightarrow A_y = 300 \text{ N} \quad \uparrow$$
$$\Sigma F_x : A_x = 0$$

2. Section through CD, CE, DE (retain A–C–D).

Assume tension in each cut member.

Moment about C (CD and CE eliminated)
$$\Sigma M_C : 300(6) - F_{DE} \, 3 = 0 \Rightarrow F_{DE} = 600 \text{ N tension}$$

Moment about D (CD and DE eliminated)

$$d = \frac{9}{\sqrt{45}} \text{ m (perpendicular distance C-E to D)}$$

$$\Sigma M_D : -300(6) + F_{CE} d = 0 \Rightarrow F_{CE} = 300\sqrt{5} \text{ N compression}$$

Vertical equilibrium of left portion

$$\Sigma F_y : 300 - 600 + F_{CD} = 0 \Rightarrow F_{CD} = -300\text{N compression}$$

Therefore:

$$F_{DE} = 600 \text{ N tension}$$

$$F_{CE} = 300\sqrt{5} \text{ N } 671 \text{ N compression}$$

$$F_{CD} = 300 \text{ N compression}$$

ChatGPT Integration

- Use ChatGPT to verify the solution and explore variations, such as different load positions or geometries.
- Request visual explanations or diagrams for a better understanding of the cutting process and equilibrium conditions.

PROMPT 23

Discuss the differences between trusses and frames and solve a basic frame analysis problem.

Objective

Understand the structural and functional differences between trusses and frames. Learn to analyze a simple frame to determine internal forces in its members.

Activity

Trusses are structures composed of two-force members connected at joints, designed to carry loads through axial forces (tension or compression). Frames are rigid structures designed to carry loads through bending, shear, and axial forces. Real-world applications include trusses used in bridges and roofs, and frames used in building structures and machinery.

Numerical Example

Analyze a simple rectangular frame with a fixed support at A and a roller support at B, spanning 6 m horizontally. The frame has a vertical member at B, 3 m tall, and is subjected to a horizontal load of 400 N applied at the midpoint of the vertical member. Determine the reaction forces at the supports and the internal forces in the vertical member.

Solution

1. Define the coordinate system. Set the origin at A $(0, 0)$. The coordinates of the main points are as follows:
   - A$(0, 0)$: fixed support
   - B$(6, 0)$: roller support
   - C$(6, 3)$: top of the vertical member
   - D$(6, 1.5)$: midpoint of the vertical member, where the load is applied

2. Find the horizontal equilibrium
$$\sum F_x = 0 \quad \Rightarrow \quad R_{Ax} - 400 = 0 \quad \Rightarrow \quad R_{Ax} = 400$$

3. Find the vertical equilibrium:
$$\sum F_y = 0 \quad \Rightarrow \quad R_{Ay} + R_{By} = 0$$

4. Find the moment equilibrium about A:
$$\sum M_A = 0 \quad \Rightarrow \quad -400 \times 1.5 + R_{By} \times 6 = 0 \quad \Rightarrow \quad R_{By} = 100$$

5. Solve for R_{Ay} using the vertical equilibrium:
$$R_{Ay} + 100 = 0 \quad \Rightarrow \quad R_{Ay} = -100$$

6. Find the internal forces in the vertical member:
   - At point D (midpoint of BC), the horizontal load of 400 N causes bending of $400 \times 1.5 = 600\,N.m$ in the vertical member, BC and shear 400 N from the midpoint to point B.

ChatGPT Integration

- Ask ChatGPT to confirm the solution and explore variations, such as different load positions or frame geometries.
- Use ChatGPT to visualize shear force and bending moment diagrams for the frame.

PROMPT 24

Analyze a compound truss using the method of joints and sections.

Objective:

Learn to analyze a compound truss by combining the method of joints and the method of sections to efficiently determine internal forces in its members.

Activity

Compound trusses are formed by connecting two or more simple trusses together. Analyzing such trusses requires the systematic application of both the method of joints (to resolve forces at specific joints) and the method of sections (to focus on specific members). This approach is critical in practical applications, such as bridge trusses or roof trusses in large structures.

Numerical Example

Analyze the forces in the members of a compound truss that consists of two simple triangular trusses connected at a common joint.

- The structure is supported at A (pin) and B (roller).
- The total span of the truss is 12 m.
- A vertical load of 800 N is applied at joint C, which is the center of the compound truss.

Truss geometry

- The left triangular truss spans from A to C, with an apex at D.
- The right triangular truss spans from C to B, with an apex at E.
- The heights of both triangles are 3 m.
- Members:
  - Left truss: AD, AC, CD.
  - Right truss: CB, CE, EB

Solution

1. Define the coordinate system.
   - Place the origin at A(0, 0).
   - Main coordinates:
     - A(0, 0): pin support
     - B(12, 0): roller support
     - C(6, 0): common joint between the two trusses, where the load is applied
     - D(3, 3): apex of the left truss
     - E(9, 3): apex of the right truss

2. Determine the support reactions.
   - Vertical equilibrium: $\sum F_Y = 0 \implies R_{Ay} + R_{By} - 800 = 0$
   - Moment about A:
     $$\sum M_A = 0 \implies -6 \times 800 + 12R_{By} = 0 \implies R_{By} = 400N$$
   - Solve for R_{Ay}: $R_{Ay} = 800 - R_{By} = 400N$

3. Analyze the left truss using the Method of Joints:
   Begin at joint A:
   $$\sum F_y = 0 \implies R_{Ay} = F_{AD} \cdot \frac{3}{4.24} \implies F_{AD} = \frac{400}{0.71} = 565.33.$$

   Using symmetry, member BE carries the same amount of force.

   Use the Method of Sections to analyze members in the right truss. Cut through CE, and CB to determine forces in these members.

ChatGPT Integration:

- Use ChatGPT to verify support reactions and confirm the step-by-step calculations.
- Explore variations of the problem, such as changing the load position or modifying the truss geometry.

PROMPT 25

Explain how to determine the internal forces in members of a space truss.

Objective

Understand the process of analyzing internal forces in members of a space truss, including the application of 3D equilibrium equations and principles of structural analysis.

Activity

A space truss is a 3D structure composed of members that carry axial forces only. Determining internal forces involves using equilibrium equations for forces ($\Sigma F_x = 0$, $\Sigma F_y = 0$, $\Sigma F_z = 0$) and moments ($\Sigma Mx = 0$, $\Sigma M_y = 0$, $\Sigma M_z = 0$). Space trusses are used in applications like aircraft hangars, roofs, and communication towers.

Numerical Example

A space truss consists of the following joints and members:
- supports: A (fixed support) and B (roller support)
- load: 600 N applied downward at joint D
- geometry: A(0,0,0), B(6,0,0), C(3,4,0), D(3,2,5)

Determine the forces in members AD, BD, and CD.

Solution

1. Define the coordinate system and geometry.
   - Use the coordinates given

     A(0,0,0), B(6,0,0), C(3,4,0), D(3,2,5)

   - Members connect the joints: AD, BD, CD.

2. Determine the support reactions.
   - Apply 3D equilibrium equations to the entire structure:
     - $\sum F_x = 0 \implies F_{ADx} + F_{BDx} + F_{CDx} = 0$
     - $\sum F_y = 0 \implies F_{ADy} + F_{BDy} + F_{CDy} = 0$
     - $\sum F_z = 0 \implies F_{ADz} + F_{BDz} + F_{CDz} - 600 = 0$

3. Resolve the forces using member geometry.

Find direction cosines for each member:

General form:

$$l_{AD} = \frac{\Delta x_{AD}}{L_{AD}}, m_{AD} = \frac{\Delta y_{AD}}{L_{AD}}, n_{AD} = \frac{\Delta z_{AD}}{L_{AD}}$$

Example for AD:

$$\Delta x_{AD} = 3 - 0 = 3, \Delta y_{AD} = 2 - 0 = 2, \Delta z_{AD} = 5 - 0 = 5$$

$$L_{AD} = \sqrt{3^2 + 2^2 + 5^2} = \sqrt{38} = 6.164$$

$$l_{AD} = \frac{3}{6.164} = 0.487, m_{AD} = \frac{2}{6.164} = 0.324, n_{AD} = \frac{5}{6.164} = 0.811$$

Example for CD:

$$\Delta x_{CD} = 0, \Delta y_{CD} = -2, \Delta z_{CD} = 5$$

$$L_{CD} = \sqrt{4 + 25} = \sqrt{29} = 5.385$$

$$l_{CD} = 0, m_{CD} = \frac{-2}{5.385} = -0.371, n_{CD} = \frac{5}{5.385} = 0.928$$

4. Substitute the direction cosines and solve. Substitute values into the equilibrium equations to solve for F_{AD}, F_{BD}, and F_{CD}.

$$\sum F_x = 0 \quad \Rightarrow \quad F_{ADx} + F_{BDx} + F_{CDx} = 0 \Rightarrow F_{AD}(0.487) + F_{BD}(-0.487)$$
$$+F_{CD}(0) = 0$$

$$\sum F_y = 0 \quad \Rightarrow \quad F_{ADy} + F_{BDy} + F_{CDy} = 0 \Rightarrow F_{AD}(0.324) + F_{BD}(0.324)$$
$$+F_{CD}(-0.371) = 0$$

$$\sum F_z = 0 \quad \Rightarrow \quad F_{ADz} + F_{BDz} + F_{CDz} = 0 \Rightarrow F_{AD}(0.811) + F_{BD}(0.811)$$
$$+F_{CD}(0.928) = 0$$

Solve for the internal forces:

$$F_{AD} = F_{BD} = 185\,N, F_{CD} = 323.2\,N$$

ChatGPT Integration

- Use ChatGPT to verify calculations and help with solving equilibrium equations.
- Request assistance in finding direction cosines for members.
- Explore the effect of changing the load position or support types on member forces.

PROMPT 26

Solve a problem involving the analysis of a planar truss.

Objective

Learn to analyze a planar truss by determining the forces in its members using equilibrium equations and identifying tension and compression members.

Activity

Planar trusses are 2D structures composed of straight members connected at joints. Analyzing such trusses involves determining the internal forces in their members to ensure structural stability. These methods are used in applications such as bridge design and roof trusses in buildings.

Numerical Example

A simple planar truss consists of three joints and three members forming a triangular shape. The supports are

- a pin support at joint A
- a roller support at joint B

Joint C is at the top of the triangle, directly above the midpoint of the base. The truss is subjected to a vertical load of 500 N applied at joint C. The dimensions are as follows:

- The base of the truss (AB) is 6 m.
- The height of the truss (C above AB) is 3 m.

Determine the forces in the members AC, BC, and AB.

Solution

Define the coordinate system: Set the origin at A, with the following coordinates:

- A(0, 0): pin support
- B(6, 0): roller support
- C(3, 3): load application point

Determine the support reactions:

$$\sum F_x = 0 \quad \Rightarrow \quad R_{Ax} = 0$$

$$\sum M_A = 0 \quad \Rightarrow \quad -500 \times 3 + R_{By} \times 6 = 0 \quad \Rightarrow \quad R_{By} = 250\,N$$

Analyze joint C: At joint C, the forces in AC and BC must balance the load. Using geometry, the slope of AC and BC is 3:3 or 1:1.

$$\sum F_y = 0 \quad \Rightarrow \quad F_{AC}\cos 45° + F_{BC}\cos 45° - 500 = 0$$

Solve for F_{AC} and F_{BC} :

$$F_{AC} = F_{BC} = \frac{500}{2\cos 45°} = \frac{500}{\sqrt{2}} = 353.55\,N$$

Analyze joint A: At joint A, the horizontal force in AB must balance the horizontal components of F_{AC}:

$$F_{AB} = F_{AC}\cos 45° = \frac{353.55}{\sqrt{2}} = 250\ N$$

ChatGPT Integration

- Use ChatGPT to verify calculations and explore variations, such as changing the load position or truss dimensions.
- Ask ChatGPT to visualize force distribution in the truss or confirm whether members are in tension or compression.

PROMPT 27

Discuss the stability of trusses and how to ensure their structural integrity.

Objective

Understand the factors affecting the stability of trusses and the design considerations to ensure structural integrity under various loading conditions.

Activity

A stable truss resists collapse by maintaining geometric rigidity under applied loads. Stability depends on the arrangement of members, joint connections, and the relationship between the number of members (m) and joints (j). Real-world examples include designing stable bridge trusses or roof structures to withstand dynamic loads such as wind or earthquakes.

Real-world example: Evaluate the stability of a roof truss exposed to wind loads and propose reinforcements to enhance its rigidity.

Numerical Example

A 2D truss has 7 joints and 11 members. Determine if the truss is stable and explain how to improve its design if it is unstable. Use the following equation for a statically determinate and stable truss:

$$m = 2j - 3$$

Solution

1. Determine the stability using the formula.

   Substitute j = 7 and m = 11 into the equation:

   $$m = 2j - 3 \quad \Rightarrow \quad 11 = 2(7) - 3$$

2. Check the stability.

   Since m = 11 matches the formula, the truss is statically determinate and stable.

3. Explain the design considerations for stability.

   - Geometric rigidity: Ensure the truss forms triangular units, as triangles are inherently rigid shapes.
   - Support placement: Properly position supports to prevent movement or instability.

- Load path: Distribute loads evenly across members to avoid overloading specific areas.

4. Propose reinforcements for stability under dynamic loads.

   - Add bracing or cross-members to reduce deformation.

   - Use materials with high strength-to-weight ratios to improve load-bearing capacity.

   - Consider redundancy by including extra members to prevent collapse in case of member failure.

The stability of a 3D truss is determined using the following formula:

$$m = 3j - 6$$

This equation ensures that a 3D truss is statically determinate and stable.

ChatGPT Integration

- Use ChatGPT to simulate various loading conditions on the truss and evaluate its stability.

- Explore how adding or removing members affects the stability equation and structural integrity.

- Ask ChatGPT for design tips to optimize truss stability for specific applications, such as wind-resistant bridges or earthquake-resistant roofs.

PROMPT 28

Explain the concept of a determinate vs. indeterminate truss.

Objective

Understand the distinction between determinate and indeterminate trusses, including how to identify and analyze them in structural systems.

Activity

A determinate truss is one in which all member forces and reactions can be calculated using only the equilibrium equations. In contrast, an indeterminate truss has more members than necessary for stability, requiring additional equations, such as compatibility or deformation relationships, for analysis. Real-world applications include ensuring the efficiency and stability of trusses in bridges and high-rise buildings.

Numerical Example

A truss has 8 joints and 15 members. Determine whether the truss is determinate or indeterminate using the formula for 2D trusses:

$$m = 2j - 3$$

Solution

1. Apply the formula for 2D trusses.

    Substitute j = 8 and m = 15 into the equation:

    $$m = 2j - 3 \quad \Rightarrow \quad 15 \neq 2(8) - 3$$

2. Interpret the result.
    - The truss has m = 15 members but requires m=13 members to be determinate.

    - The truss is statically indeterminate with 2 additional members.

3. Explain the implications.
    - Determinate trusses: Easier to analyze, as member forces are found using equilibrium equations alone

    - Indeterminate trusses: Require additional methods, such as compatibility equations or matrix methods, to solve for member forces

4. Discuss the practical considerations.
    - Indeterminate trusses are often more rigid and capable of redistributing loads, providing a safety margin in case of member failure.

    - However, they require more complex analysis and precision during construction.

For 3D Trusses

The distinction between determinate and indeterminate 3D trusses is based on the formula:

$$m = 3j - 6$$

If $m > 3j - 6$, the truss is indeterminate; if $m = 3j - 6$, it is determinate.

ChatGPT Integration

- Use ChatGPT to calculate degrees of indeterminacy for various truss configurations.
- Ask ChatGPT to explain the advantages and disadvantages of determinate and indeterminate trusses in specific scenarios, such as earthquake-prone regions or long-span bridges.
- Explore how advanced methods like matrix analysis are applied to indeterminate trusses.

PROMPT 29

Analyze the forces in a truss subjected to moving loads.

Objective

Understand how to analyze truss forces under moving loads, including determining maximum internal forces in specific members.

Activity

A moving load on a truss, such as a vehicle or train, creates varying forces in its members depending on the load's position. This analysis helps identify the most critical load positions for each member. Engineers use this method to design trusses for maximum strength in applications like bridges and overhead cranes.

Numerical Example

A simple truss bridge spans 12 m with joints at A, B, and C.

- A (pin support) and B (roller support) are located at the ends of the truss.
- C is the midpoint of the truss, forming two 6 m segments.
- A moving load of 1000 N travels across the span.
- Analyze the forces in member AC when the load is at joint A, at the midpoint (C), and at joint B.

Solution

Define the coordinate system:
Set the origin at A(0, 0).
- A(0, 0): pin support

- B(12, 0): roller support
- C(6, 0): midpoint

1. **Case 1: Load at Joint A**

$$\sum M_A = 0 \quad \Rightarrow \quad -1000 \times 0 + R_{By} \times 12 = 0 \quad \Rightarrow \quad R_{By} = 0$$

$$R_{Ay} = 1000$$

Member AC carries no vertical load since $R_{By} = 0$.

2. **Case 2: Load at Midpoint (C)**

$$\sum M_A = 0 \quad \Rightarrow \quad -1000 \times 6 + R_{By} \times 12 = 0 \quad \Rightarrow \quad R_{By} = 500N$$

$$R_{Ay} = 500N$$

The load is shared equally by supports, and member AC carries a vertical force proportional to its geometry.

3. **Case 3: Load at Joint B**

$$\sum M_A = 0 \quad \Rightarrow \quad -1000 \times 12 + R_{By} \times 12 = 0 \quad \Rightarrow \quad R_{By} = 1000N$$

ChatGPT Integration

- Use ChatGPT to automate the analysis of forces for multiple load positions.
- Explore how load placement impacts the forces in specific members and discuss strategies for optimizing truss designs for moving loads.

PROMPT 30

Explain the different types of trusses and their typical structural applications.

Objective

Understand the characteristics of various truss types and their suitability for specific engineering applications.

Activity

Explore common truss types such as Pratt, Warren, Howe, and K-trusses. Discuss their design features, load-carrying behavior, and preferred uses in different structural applications. Highlight how each truss type addresses unique challenges, such as load distribution, span length, or environmental conditions.

Numerical Example

Compare the characteristics of two truss types (Pratt and Warren) for a bridge design with the following parameters:

- span length: 30 m
- load condition: uniformly distributed load of 50 kN/m
- material: structural steel with a yield strength of 250 MPa

Analyze the following:

1. Which truss design minimizes material usage?
2. Which truss distributes loads more evenly across its members?

Solution

1. Pratt truss
   - Diagonal members handle tension, while vertical members handle compression.
   - Suitable for longer spans where varying load conditions are common.
   - Material usage is moderate, but tension members can be optimized for efficiency.

2. Warren truss
   - Diagonal members alternate between tension and compression.
   - Provides more uniform load distribution across members.
   - Material usage is typically lower due to the absence of vertical members.

The final choice depends on balancing material cost, ease of construction, and load distribution. Warren trusses are more efficient for uniformly distributed loads, requiring less material.

ChatGPT Integration

- Use ChatGPT to compare other truss types, such as Howe or K-trusses, and their applications. Ask for visual representations of each truss type or examples of real-world structures that use them.
- Explore how specific load conditions affect the choice of truss design for applications like bridges, roofs, or towers.

CHAPTER 4

DISTRIBUTED FORCES

Prompt 31 Define distributed forces and their effects on structures.

Prompt 32 Calculate the centroid of a given distributed load.

Prompt 33 Explain how to find the resultant of a distributed force over a surface.

Prompt 34 Discuss the application of distributed forces in engineering design.

Prompt 35 Solve a problem involving a beam with a linearly varying load and a single concentrated load.

Prompt 36 Calculate the reactions for a beam with a linearly varying distributed load.

Prompt 37 Explain the concept of load intensity and its units.

Prompt 38 Determine the equivalent point load for a given distributed load.

Prompt 39 Solve a problem involving a non-uniform distributed load on a beam.

Prompt 40 Discuss the significance of shear force and bending moment diagrams.

PROMPT 31

Define distributed forces and their effects on structures.

Objective

Understand the concept of distributed forces and their impact on structural elements.

Activity

Define distributed forces and differentiate them from concentrated forces. Explain how distributed forces are represented mathematically and graphically. Discuss their effects on beams and other structures, focusing on deflection, bending, and stress.

Numerical Example

Consider a beam of length $L = 10$ m subjected to a uniformly distributed load of $\omega = 5kN/m$. Determine the equivalent point load and its location on the beam.

Solution

1. Find the representation of the distributed load.

    A uniformly distributed load (UDL) of $\omega = 5kN/m$ acts along the entire length of the beam.

2. Perform the equivalent point load calculation.

    The magnitude of the equivalent point load, F, is given by
    $$F = w \cdot L$$

    By substituting the values, we obtain the following:
    $$F = 5 \times 10 = 50\,kN$$

3. Find the location of the equivalent point load.

    For a UDL, the equivalent point load acts at the centroid of the distributed load. Since the load is uniform, the centroid is at the midpoint of the beam, i.e., at $x = \dfrac{L}{2} = \dfrac{10}{2} = 5m$.

ChatGPT Integration

Ask ChatGPT to

- verify the calculation for the equivalent point load
- provide additional examples of distributed loads (e.g., triangular loads)
- explain how the effects of distributed loads differ from concentrated forces

PROMPT 32

Calculate the centroid of a given distributed load.

Objective

Learn how to determine the centroid of distributed loads on structural elements.

Activity

Explain the concept of centroids in the context of distributed forces. Discuss its importance in analyzing the effects of loads on beams and other structures. Use a real-world example, such as calculating the center of mass of a load on a bridge or beam.

Numerical Example

A beam of length L=8 m is subjected to a trapezoidal distributed load varying from $\omega_1 = 4kN / m$ at the left end to $\omega_2 = 10kN / m$ at the right end. Determine

1. the centroid of the trapezoidal load
2. the equivalent point load and its location

Solution

1. Find the representation of the trapezoidal load.

   A trapezoidal distributed load can be decomposed into a rectangular part and a triangular part.

   - The rectangular load is constant, with intensity $\omega_1 = 4kN / m$.

   - The triangular load has a base length L=8 m and a height $h = \omega_2 - \omega_1 = 10 - 4 = 6kN / m$.

2. Find the area of each component.
   - Rectangular load: $Arect = \omega_1 \times L = 4 \times 8 = 32 \ kN$
   - Triangular load: $A_{tri} = \dfrac{1}{2}(L \times h) = \dfrac{1}{2}(8 \times 6) = 24 \ kN$

3. Find the centroid of each component.
   - Rectangular load: Located at the midpoint of its length, $x_{rect} = \dfrac{L}{2} = \dfrac{8}{2} = 4 \ m$.
   - Triangular load: The centroid of a triangular load is located at ⅔ of its base length from the smaller end, $x_{tri} = \dfrac{2}{3} \times 8 = 5.33 \ m$.

4. Perform the resultant centroid calculation.

   The total centroid is calculated using the principle of moments:

   $$x_c = \frac{A_{rect} \cdot x_{rect} + A_{tri} \cdot x_{tri}}{A_{rect} + A_{tri}}$$

   By substituting the values, we obtain the following:

   $$x_c = \frac{(32 \times 4) + (24 \times 5.33)}{32 + 24} = 4.57 \ m$$

   Equivalent Point Load
   The total equivalent point load is

   $$F = A_{rect} + A_{tri} = 32 + 24 = 56 \ kN$$

   The equivalent point load is located at $x_c = 4.57$ m from the left end.

ChatGPT Integration

Ask ChatGPT to

- verify centroid calculations for composite distributed loads
- explore how varying distributions (e.g., parabolic loads) affect centroid placement

PROMPT 33

Explain how to find the resultant of a distributed force over a surface.

Objective

Learn how to calculate the resultant of distributed forces acting over a surface and its application in engineering analysis.

Activity

Describe the process of integrating distributed forces over a surface to determine the resultant force. Provide a real-world application, such as calculating the total wind load acting on the surface of a building or the pressure distribution on a submerged surface.

Numerical Example

A rectangular plate of dimensions 4 m × 6 m is subjected to a distributed load that varies linearly along its length. The load intensity at the left edge is $q_1 = 2$ kN/m and at the right edge is $q_2 = 8$ kN/m. Determine

1. the resultant force acting on the plate
2. the location of the resultant force along the length of the plate

Solution

1. Find the representation of the distributed force.

    The load distribution is triangular across the width of the plate. The intensity q(x) varies linearly from q_1 to q_2 along the length of the plate.
2. Perform the resultant force calculation.

    The resultant force is the total load acting on the plate, which is the integral of the load distribution over the surface:

    $$F = \int_0^L q(x) \cdot W \, dx$$

    where
    * L = 6 m (length of the plate),
    * W = 4 m (width of the plate)

    $$q(x) = q_1 + (q_2 - q_1)/L \cdot x$$

    By substituting for q(x), we obtain

    $$q(x) = 2 + (8 - 2)/6 \cdot x = 2 + x$$

    Integrating:

    $$F = \int_0^6 4(2 + x) \, dx = 4 \left[2x + \frac{x^2}{2} \right]_0^6 = 120 \text{ kN}$$

3. Find the location of the resultant force.

The location x_c of the resultant force is determined using the moment equation:

$$x_c = \frac{\int_0^L x \cdot q(x) \cdot W \, dx}{\int_0^L q(x) \cdot W \, dx}$$

By substituting $q(x) = 2 + x$ into the equation, we obtain

$$x_c = \frac{\int_0^L x \cdot q(x) \cdot W \, dx}{\int_0^L q(x) \cdot W \, dx} = \frac{4 \int_0^6 x(2 + x) \, dx}{4 \int_0^6 (2 + x) \, dx} = \frac{4}{120} \left[x^2 + \frac{x^3}{3} \right]_0^6 = 3.6 \text{ m}$$

4. Obtain the result.

The resultant force is $F = 120$ kN, and it acts at a distance $x_c = 3.6$ m from the left edge of the plate.

ChatGPT Integration

Ask ChatGPT to

- verify the integral for more complex load distributions
- explore how this method applies to different surfaces, like curved or irregular shapes

PROMPT 34

Discuss the application of distributed forces in engineering design.

Objective

Understand how distributed forces are applied and analyzed in engineering design scenarios.

Activity

Discuss the significance of distributed forces in the design of structures and mechanical components. Provide examples such as bridge design, wind pressure analysis on skyscrapers, and the load distribution on submerged surfaces like dams or gates.

Numerical Example

A semicircular gate of radius $R = 2$ m is vertically (i.e., $\theta = \pi / 2$) located at a depth of $h = 5$ m below the water surface. The water density is $\rho = 1000$ kg/m³, and gravitational acceleration is $g = 9.81$ m/s². Determine

1. the total resultant hydrostatic force acting on the gate
2. the location of the resultant force along the vertical axis of the gate

Solution

1. Find the hydrostatic pressure distribution.

    The pressure at any depth y below the water surface is given by
    $$P = \rho g y$$
    where y is measured from the free surface of the fluid, downward.

2. Find the total resultant hydrostatic force.

    The hydrostatic force is calculated by integrating the pressure over the area of the semicircular gate, or

    $$F = \left(\rho g \bar{y} \right) A$$

    where $\bar{y}$ is the distance to the centroid of the gate from the free surface, and A is the area of the gate. The centroid of the gate is located at $\dfrac{4R}{3\pi}$ from the bottom.

    Through substituting, we obtain

    $$\bar{y} = 5 + 2 - \frac{4 \times 2}{3 \times \pi} = 6.151\,\text{m}$$

    $$F = \left[(1000)(9.81)(6.151) \right] \left(\frac{\pi}{2} \times 4 \right) = 379.135\ \text{kN}$$

3. Find the location of the resultant force.

    The vertical position of the force, or center of pressure, is located at a distance y_p below the centroid of the gate:

    $$y_p = \frac{\rho g \sin \theta}{F} I_c$$

    where, I_c is the moment of inertia of the gate about the axis at its centroid, parallel to the ground.

$$I_c = \left(\frac{\pi}{8} - \frac{8}{9\pi} \right) R^4 = 1.756$$

By substituting, we obtain

$$y_p = \frac{1000 \times 9.81 \times \sin 90°}{379135}(1.756) = 4.54 \text{ cm}$$

Therefore, the distance of the location of the acting hydrostatic force is at $\left(\frac{4 \times 2}{3 \times \pi} - 0.0454 \right) \approx 0.8$ m from the bottom of the gate.

ChatGPT Integration

Ask ChatGPT to

- verify the integral for hydrostatic forces on curved surfaces
- provide examples for gates of different shapes (e.g., rectangular or trapezoidal)
- explore how gate positioning affects the force distribution and location of the resultant force

PROMPT 35

Solve a problem involving a beam with a linearly varying load and a single concentrated load.

Objective

Analyze a beam subjected to a combination of linearly varying distributed load and a single concentrated load to determine reactions, shear force, and bending moment.

Activity

Explain how to solve for reactions and internal forces in a beam under combined loading conditions. Use real-world applications, such as designing beams in bridges or supports for industrial machinery.

Numerical Example

A simply supported beam of length $L = 12$ m is subjected to

1. a linearly varying load from $\omega_1 = 0$ kN at the left end to $\omega_2 = 6$ kN/m at the right end
2. a concentrated load of $P = 10$ kN located 4 m from the left support

Determine

1. the reactions at the supports
2. the expressions for the shear force and bending moment along the beam
3. the maximum bending moment

Solution

1. Determine the reactions at the supports.
    - Resultant of the linearly varying load:

    The linearly varying load is triangular, with a total load:

    $$F_{tri} = \frac{1}{2} \cdot \text{Base} \cdot \text{Height}$$

    Substituting the values of the base (12 m) and height (6 kN/m) gives the following:

    $F_{tri} = \frac{1}{2} 12 \times 6 = 36$ kN. The location of this load is $2/3 \times 12 = 8$ m from the left end.
    - Total load and reactions:

    The total load on the beam is

    $$Ftotal = Ftri + P = 36 + 10 = 46 \text{ kN}$$

    Using moments about the left support (A) to find the reaction at the right support (R_B), we find that the moment about A is as follows: $R_B \times 12 = Ftri \times 8 + P \times 4$

    By substituting values, we obtain

    $12R_B = 36 \times 8 + 10 \times 4 = 288 + 40 = 328$ kN $\Rightarrow R_B = 328/12 = 27.33$ kN

    The reaction at the left support (R_A) is

    $$R_A = Ftotal - R_B = 46 - 27.33 = 18.67 \text{ kN}.$$

2. Find the shear force (V) and bending moment (M).

- Divide the beam into segments to derive expressions for shear force and bending moment.

- Segment 1 ($0 \le x \le 4$):

The distributed load is

$$w(x) = 6/12 \cdot x = 0.5x$$

Shear force:

$$V(x) = R_A - \int_0^x w(x')dx' = 18.67 - 0.5\int_0^x x'\, dx' = 18.67 - 0.25x^2$$

Bending moment:

$$M(x) = \int_0^x V(x')dx' = \int_0^x \left(18.67 - 0.25x'^2\right)dx' = 18.67x - \frac{0.25}{3}x^3$$

- Segment 2 ($4 \le x \le 12$):

Add the effect of the concentrated load at $x = 4$:

Shear force:

$$V = R_A - P - \frac{x^2}{4} = 8.64 - \frac{x^2}{4}$$

Solve similarly for $M(x)$.

$$M(x) = -\frac{x^3}{12} + 8.67x + 40$$

3. Find the maximum bending moment.

The maximum bending moment occurs where $V(x) = 0$. Solve $V(x)$ for each segment and substitute into $M(x)$.

ChatGPT Integration

Ask ChatGPT to

- verify the derived expressions for shear force and bending moment
- discuss how combined loads affect design considerations
- provide examples with non-linear or multi-load scenarios

PROMPT 36

Calculate the reactions for a beam with a linearly varying distributed load.

Objective

Learn how to calculate support reactions for a beam subjected to a linearly varying distributed load.

Activity

Describe the process of determining reactions in a beam with a triangular load distribution. Explain its relevance in engineering, such as analyzing cantilevered beams under wind pressure or tapered loads in structural systems.

Numerical Example

A simply supported beam of length L = 10 m is subjected to a linearly varying load that starts at $\omega_1 = 0$ kN/m at the left end and increases to $\omega_2 = 8$ kN/m at the right end. Determine the reactions at the supports.

Solution

1. Find the resultant of the distributed load.

    The linearly varying load forms a triangular distribution. The total resultant load is the area of the triangle:

    $$F = \frac{1}{2} \cdot \text{Base} \cdot \text{Height}$$

    By substituting the values for the base (L = 10 m) and height ($\omega_2 = 8$ kN/m), we obtain

    $$F = \frac{1}{2} \times 10 \times 8 = 40 \text{ kN}$$

2. Find the location of the resultant force.

    The centroid of a triangular load is located at 2/3 of the base length from the smaller end (left side):

    $$x_c = \frac{2}{3} \times 10 = 6.67 \text{ m}$$

3. Equilibrium Equations

Using the equations of equilibrium to determine the reactions at the supports:

- sum of the vertical forces:

$$R_A + R_B = F = 40 \text{ kN}$$

- sum of the moments about A:

$$R_B \cdot L = F \cdot xc$$

By substituting, we obtain the following:

$$R_B = \frac{40}{10} \times 6.67 = 26.68 \text{ kN}$$

Reaction at A:

$$R_A = F - R_B = 40 - 26.68 = 13.32 \text{ kN}$$

ChatGPT Integration

Ask ChatGPT to

- verify the reaction calculations for varying load scenarios
- provide examples of beams with non-linear distributed loads (e.g., parabolic loads)
- discuss how reaction forces influence beam design and material selection

PROMPT 37

Explain the concept of load intensity and its units.

Objective

Understand the concept of load intensity and its significance in analyzing distributed loads on structures.

Activity

Describe what load intensity means and how it is represented. Discuss the units of load intensity and its role in calculating reactions, shear forces, and bending moments. Provide real-world applications, such as distributed loads on bridges or roofs due to wind, snow, or live loads.

Numerical Example

A cantilever beam of length $L = 8$ m is subjected to a parabolic load that is highest at the fixed support with $w_0 = 6$ kN/m and decreases to zero at the free end. Determine

1. the total load acting on the beam
2. the location of the resultant force
3. the moment at the fixed support
4. the shear force at the fixed support

Solution

1. Find the total load acting on the beam.
   The parabolic load is described by

   $$w(x) = w_0 \left(1 - \frac{x^2}{L^2} \right)$$

   The total load is

   $$F = \int_0^L w(x)\, dx$$

   By substituting $w(x)$, we obtain

   $$F = \int_0^8 6 \left(1 - \frac{x^2}{64} \right) dx$$

   Expanding the integrand and integrating term by term, we obtain

   $$F = 6 \left[x - \frac{1}{64 \times 3} x^3 \right]_0^8 = 32 \text{ kN}$$

2. Find the location of the resultant force.
   The location of the resultant force is given by

   $$x_c = \frac{\int_0^L x \cdot w(x)\, dx}{F}$$

   By substituting $w(x) = 6 \left(1 - \frac{x^2}{64} \right)$, after performing the integration, we obtain

$$x_c = \frac{6\int_0^8 x \cdot \left(1 - \frac{x^2}{64}\right) dx}{32} = \frac{3}{16}\left[\frac{x^2}{2} - \frac{x^4}{4 \times 64}\right]_0^8 = 3 \text{ m}$$

from the fixed support.

3. Find the moment at the fixed support.

The moment at the fixed support is

$$M = F \cdot x_c$$

Substituting $F = 32$ kN and $x_c = 3$ m, we obtain

$$M = 32 \times 3 = 96 \text{ kN} \cdot \text{m}$$

4. Find the shear force at the fixed support.

The shear force at the fixed support is equal to the total load:

$$V = F = 32 \text{ kN}$$

ChatGPT Integration:

Ask ChatGPT to

- verify the integration for parabolic load distributions
- explore different load distributions (e.g., varying non-linearly or asymmetrically)
- discuss practical implications for designing cantilevered beams under such loads

PROMPT 38

Determine the equivalent point load for a given distributed load.

Objective

Learn how to calculate the equivalent point load and its location for a given distributed load to simplify structural analysis.

Activity

Explain the concept of converting a distributed load into an equivalent point load. Discuss its importance in simplifying calculations for reactions, shear forces, and bending moments. Provide real-world applications, such as analyzing bridges or beams under varying load intensities.

Numerical Example

A simply supported beam of length L = 10 m is subjected to a combined distributed load:

1. a triangular load varying from $w_1 = 0$ kN/m at the left end to $w_2 = 5$ kN/m at the midpoint
2. a uniform load of $w_u = 2$ kN/m over the second half of the beam

Determine

1. the total equivalent point load
2. the location of the equivalent point load from the left end

Solution

1. Determine the triangular load on the first half.

   The area of the triangular load gives the total load:

$$F_{tri} = \frac{1}{2} \cdot \text{Base} \cdot \text{Height}$$

   We substitute base = 5 m and height = 5 kN/m into the equation to obtain

$$F_{tri} = \frac{1}{2} \times 5 \times 5 = 12.5 \text{ kN}$$

   The centroid of a triangular load is located at 2/3 Base from the smaller end:

$$x_{tri} = \frac{2}{3} \times 5 = 3.33 \text{ m}$$

2. Determine the uniform load on the second half.

   The total load from the uniform load is

$$F_u = w_u \cdot \text{Length}$$

   We substitute $w_u = 2$ kN/m and length = 5 m to obtain

$$F_u = 2 \times 5 = 10 \text{ kN}$$

The centroid of a uniform load is at the midpoint of the loaded section:

$$x_u = 5 + \frac{5}{2} = 7.5 \text{ m}$$

3. Determine the total equivalent point load.

   The total equivalent point load is

   $$F_{total} = F_{tri} + F_u = 12.5 + 10 = 22.5 \text{ kN}$$

4. Determine the location of the equivalent point load.

   Using the principle of moments, we obtain

   $$x_{total} = \frac{F_{tri} \cdot x_{tri} + F_u \cdot x_u}{F_{total}}$$

   By substituting, we obtain

   $$x_{total} = \frac{12.5 \times 3.33 + 10 \times 7.5}{22.5} = 5.18 \text{ m}$$

ChatGPT Integration

Ask ChatGPT to

- verify the calculations for combined distributed loads
- provide examples for more complex distributed loads (e.g., parabolic or asymmetrical)
- discuss the impact of equivalent point loads on beam design and analysis

PROMPT 39

Solve a problem involving a non-uniform distributed load on a beam.

Objective

Analyze a beam subjected to a non-uniform distributed load to determine the reactions, shear force, and bending moment.

Activity

Describe how to solve for reactions, shear forces, and bending moments in a beam with a non-uniform distributed load. Explain its relevance in real-world engineering, such as modeling pressure distributions on beams or irregularly loaded structures.

Numerical Example

A simply supported beam of length $L = 12$ m is subjected to a load $w(x)$ that varies linearly from $\omega_1 = 2$ kN/m at the left end to $\omega_1 = 8$ kN/m at the right end. Determine

1. the reactions at the supports
2. the expressions for the shear force and bending moment along the beam
3. the location and magnitude of the maximum bending moment

Solution

1. Find the total load acting on the beam.

   The total load F is the area under the load distribution curve:

   $$F = \int_0^L w(x)\,dx$$

   The load function $w(x)$ is

   $$w(x) = \omega_1 + \left(\frac{\omega_2 - \omega_1}{L}\right)x = 2 + \left(\frac{8-2}{12}\right)x = 2 + 0.5x$$

   By substituting $w(x)$, we obtain

   $$F = \int_0^{12}(2 + 0.5x)\,dx = \left[2x + 0.25x^2\right]_0^{12}$$

   $$F = 2(12) + 0.25(144) = 24 + 36 = 60\,\text{kN}$$

2. Find the location of the resultant force.

   The location of the resultant force x_c is given by:

   $$x_c = \frac{\int_0^L x \cdot w(x)\,dx}{F}$$

   By substituting $w(x) = 2 + 0.5x$, we obtain

   $$x_c = \frac{1}{60}\int_0^{12} x(2 + 0.5x)\,dx = \frac{1}{60}\left[x^2 + \frac{0.5}{3}x^3\right]_0^{12} = \frac{1}{60}(144 + 288) = 7.2\,\text{m}$$

3. Find the reactions at the supports.

Using the equations of equilibrium:

- sum of the vertical forces: $R_A + R_B = F = 60$ kN

- moment about A: $12 R_B = F \cdot x_c$

  By substituting, we obtain the following: $R_B = \dfrac{60 \times 7.2}{12} = 36$ kN

  Reaction at A: $R_A = F - R_B = 60 - 36 = 24$ kN

4. Find the shear force and bending moment.

- shear force $V(x)$: $V(x) = R_A - \int_0^x w(x')dx'$

  By substituting $w(x) = 2 + 0.5x$, we obtain

  $$V(x) = 24 - 2x - 0.25x^2$$

- bending moment $M(x)$:

  $$M(x) = \int V(x')dx'$$

By substituting $V(x) = 24 - 2x - 0.25x^2$, we obtain

$$M(x) = \int \left(24 - 2x - 0.25x^2\right)dx$$

$$M(x) = 24x - x^2 - \frac{0.25}{3}x^3$$

5. Find the maximum bending moment.

The maximum bending moment occurs where $V(x) = 0$:

$$24 - 2x - 0.25x^2 = 0 \Rightarrow x = 6.583 \text{ m}$$

Solve this quadratic equation for x. Substituting x into $M(x)$ gives the maximum moment.

$$M(6.583) = 90.883 \text{ kN.m}$$

PROMPT 40

Discuss the significance of shear force and bending moment diagrams.

Objective

Understand the purpose and importance of shear force and bending moment diagrams in structural analysis.

Activity

Explain the role of shear force and bending moment diagrams in understanding the internal forces within a beam. Discuss their significance in designing structural elements such as bridges, floor beams, or cantilevered components. Use real-world examples to highlight their application in ensuring structural integrity.

Numerical Example

A simply supported beam of length L = 10 m is subjected to a uniformly distributed load of w = 4 kN/m over its entire length. Determine the following:

1. the reactions at the supports
2. the expressions for the shear force and bending moment along the beam
3. Plot the shear force and bending moment diagrams.

Solution

1. Find the reactions at the supports.

   The total load on the beam is

   $$F = w \cdot L$$

   Substituting w=4 kN//m and L=10 m, we obtain

   $$F = 4 \times 10 = 40 \text{ kN}$$

   The beam is symmetric, so the reactions are equal:

   $$R_A = R_B = \frac{F}{2} = 40/2 = 20 \text{ kN}$$

2. Find the shear force $V(x)$.

   The shear force varies linearly along the beam. At any point x, the shear force is

   $$V(x) = R_A - w \cdot x$$

   By substituting R_A = 20, we obtain

   $$V(x) = 20 - 4x$$

3. Find the bending moment $M(x)$.

   The bending moment is the integral of the shear force:

   $$M(x) = \int V(x)\,dx$$

   By substituting $V(x) = 20 - 4x$, we obtain

   $$M(x) = \int (20 - 4x)\,dx = 20x - 2x^2 + C$$

   At $x = 0$, $M(0) = 0$, so $C = 0$.

   Therefore,

   $$M(x) = 20x - 2x^2$$

4. Important points on the diagrams:

   - At $x = 0$: $V(0) = 20$ kN, $M(0) = 0$ kN·m

   - At $x = 10$: $V(10) = 20-4·10 = -20$ kN, $M(10) = 20 \times 10-2 \times 102 = 0$ kN·m

   - Maximum moment occurs where $V(x) = 0$: $20-4x = 0 \Rightarrow x = 5$ m

     Substituting into $M(x)$: $M(5) = 20 \times 5 - 2 \times 5^2 = 50$ kN·m

5. Create the shear force diagram.

   The shear force starts at 20 kN at $x = 0$, decreases linearly to 0 kN at $x = 5$, and continues to -20 kN at $x = 10$.

6. Create the bending moment diagram.

   The bending moment starts at 0 kN·m at $x = 0$, increases to a maximum of 50 kN·m at $x = 5$, and returns to 0 kN·m at $x = 10$.

CHAPTER 5

CENTER OF GRAVITY AND CENTROID

Prompt 41 Define the concept of center of gravity and centroid.

Prompt 42 Calculate the centroid of a composite area using integration.

Prompt 43 Explain the centroid of a plane figure and its importance in structural design.

Prompt 44 Solve a problem involving the centroid of a symmetric object.

Prompt 45 Discuss the role of the center of gravity in analyzing stability and balance.

Prompt 46 Calculate the centroid of a solid object with varying density.

Prompt 47 Explain how the centroid of a volume is determined and its applications in design.

Prompt 48 Solve a problem involving the center of gravity of a multi-material system.

Prompt 49 Discuss the concept of centroid for irregular shapes and its practical importance.

Prompt 50 Explain how the center of gravity and centroid influence the design of vehicles and machinery.

PROMPT 41

Define the concept of center of gravity and centroid.

Objective

Understand the concepts of center of gravity and centroid and their differences, along with their practical applications.

Activity

Describe the center of gravity as the point where the weight of an object is concentrated and the centroid as the geometric center of an area. Highlight their differences, such as the center of gravity depending on weight distribution and the centroid being purely geometric. Provide examples such as the center of gravity in a vehicle and the centroid in structural design.

Numerical Example

A uniform rectangular plate with dimensions b=4 and h=2 m is suspended horizontally. Determine the location of its centroid. If an additional weight of 50 kg is placed 1 m from the right edge, calculate the new center of gravity.

Solution

1. Find the centroid of the plate.

   The centroid of a uniform rectangular plate is located at its geometric center:

$$x_c = \frac{b}{2}, \quad y_c = \frac{h}{2}$$

   By substituting $b = 4$ and $h = 2$ m, we obtain

$$x_c = 4/2 = 2\,\text{m}, \quad y_c = 2/2 = 1\,\text{m}$$

2. Find the new center of gravity.

   When an additional weight is placed, the center of gravity shifts. Assuming the plate weighs W = 80 kg, we find

$$x_{cg} = \frac{W \cdot x_{plate} + W_{extra} \cdot x_{extra}}{W + W_{extra}}$$

   By substituting W = 80 kg, W_{extra} = 50 kg, x_{plate} = 2 m, and x_{extra} = 3 m, we obtain

$$x_{cg} = \frac{80 \times 2 + 50 \times 3}{80 + 50} = 2.38 \text{ m}$$

The new center of gravity is located at $x_{cg} = 2.38$ m, $y_{cg} = 1$ m (assuming the weight is added symmetrically along the y-axis).

ChatGPT Integration

Ask ChatGPT to

- explore how non-uniform density affects the centroid
- provide additional examples comparing center of gravity and centroid
- verify calculations for complex shapes and weight distributions

PROMPT 42

Calculate the centroid of a composite area using integration.

Objective

Learn to determine the centroid of a composite area using mathematical integration techniques.

Activity

Explain the process of calculating the centroid of a composite area by dividing it into simpler shapes. Use integration to find the centroid for cases where the area cannot be divided into standard geometric shapes. Provide applications in engineering, such as locating the centroid of complex cross-sections in beams.

Numerical Example

Determine the centroid of a region bounded by the x-axis, the curve $y = x^2$, and the vertical line $x = 2$.

Solution

1. The setup of the problem is as follows.

   The region is defined as a parabola between $x = 0$ and $x = 2$.

   The area A is calculated using the following formula:

$$A = \int_{x_1}^{x_2} y \, dx = \int_{0}^{2} x^2 \, dx = \left[\frac{x^3}{3} \right]_{0}^{2} = \frac{8}{3}$$

2. Find the centroid coordinates.

   x-coordinate of the centroid (x_c):

   $$x_c = \frac{1}{A}\int_{x_1}^{x_2} x \cdot y \, dx = \frac{3}{8}\int_0^2 x^3 \, dx = \frac{3}{8}\left[\frac{x^4}{4}\right]_0^2 = 1.5$$

   y-coordinate of the centroid (y_c):

   $$y_c = \frac{1}{A}\int_{x_1}^{x_2} \frac{y^2}{2} \, dx = \frac{3}{16}\int_0^2 x^4 = \frac{3}{16}\left[\frac{x^5}{5}\right]_0^2 = 1.2$$

ChatGPT Integration

Ask ChatGPT to

- verify the integrals and calculations for non-standard areas
- explore the centroid of regions with different bounding curves
- discuss the application of centroids in beam design and analysis

PROMPT 43

Explain the centroid of a plane figure and its importance in structural design.

Objective

Understand the concept of the centroid of a plane figure and how it plays a role in structural analysis and design.

Activity

Describe the centroid of a plane figure as the geometric center of its area, which acts as a balance point. Explain its relevance in structural design, such as determining the neutral axis in beams or ensuring uniform load distribution. Use practical examples, including cross-sectional shapes of beams and plates.

Numerical Example

Find the centroid of a T-shaped cross-section composed of two rectangles:

- a vertical rectangle with a width of b_v = 2 m and height h_v = 6 m
- a horizontal rectangle with a width of b_h = 8 m and height h_h = 2 m, positioned at the top of the vertical rectangle

Solution

Using the symmetry, the x-coordinate of the centroid x_c, is equal to zero. We assume a coordinate system with its origin at the symmetry axis and located at the bottom fiber.

1. Divide the T-section into two rectangles.
   - Rectangle 1 (vertical)

     Area and centroid coordinates with reference to the bottom fiber. Centroid coordinates:

     $$A_1 = b_v \cdot h_v = 12 \text{ m}^2, \quad y_1 = \frac{h_v}{2} = 3 \text{ m}$$

   - Rectangle 2 (horizontal):

     The centroid coordinates are as follows, with reference to the bottom fiber:

     $$A_2 = b_h \cdot h_h = 16 \text{ m}^2, \quad y_2 = h_v + \frac{h_h}{2} = 7 \text{ m}$$

2. Calculate the centroid of the composite shape.

   Use the weighted average formula for the centroid:

   $$y_c = \frac{\sum (A_i \cdot y_i)}{\sum A_i}$$

3. Substitute values for the areas and centroids of the two rectangles:

   $$y_c = \frac{(12 \times 3) + (16 \times 7)}{12 + 16} = \frac{148}{28} = 5.29 \text{ m}$$

ChatGPT Integration

Ask ChatGPT to

- verify centroid calculations for composite or irregular shapes
- explore how centroids influence the placement of neutral axes in beams
- provide additional examples of composite sections like I-beams or L-sections

PROMPT 44

Solve a problem involving the centroid of a symmetric object.

Objective

Learn how to determine the centroid of a symmetric object using symmetry and mathematical calculations.

Activity

Explain the role of symmetry in determining the centroid of an object. Highlight that for symmetric objects, the centroid lies along the axis or plane of symmetry, simplifying calculations. Use examples such as I-beams, circular plates, or rectangular sections.

Numerical Example

Determine the centroid of a semicircular plate of radius R = 4 m lying in the xy-plane with its flat edge along the x-axis.

Solution

1. Understand the symmetry.
   - A semicircular plate is symmetric about the y-axis, so the x-coordinate of the centroid is $x_c = 0$.
   - The y-coordinate of the centroid (y_c) must be calculated using integration.
2. Perform the centroid calculation for y_c.
   The formula for y_c is

$$y_c = \frac{\int y \, dA}{A}$$

The differential area is a horizontal strip of thickness dy at height y. Its width is

$$dA = 2\sqrt{R^2 - y^2}\, dy$$

Therefore,

$$y_c = \frac{2\int_0^R y\sqrt{R^2 - y^2}\, dy}{A}$$

But area A is

$$A = \frac{\pi R^2}{2} = \frac{16\pi}{2}, \quad A = 25.13 \text{ m}^2$$

After writing $y = R\sin\theta$ and $dy = R\cos\theta\, d\theta$, we can calculate the integral as

$$2\int_0^R y\sqrt{R^2 - y^2}\, dy = 2R^3 \int_0^{\pi/2} \sin\theta \cos^2\theta\, d\theta = 2R^3 \left[-\frac{1}{3}\cos^3\theta \right]_0^{\pi/2} = \frac{2}{3}R^3.$$

By substituting in values, we obtain

$$y_c = \frac{\dfrac{2}{3}R^3}{\dfrac{\pi R^2}{2}} = \frac{4R}{3\pi} = 1.698 \text{ m}$$

ChatGPT Integration

Ask ChatGPT to

- verify the integration process for the numerator
- discuss the significance of the centroid in designing curved structures
- provide alternative methods or formulas for other shapes, such as quarter circles or elliptical plates

PROMPT 45

Discuss the role of the center of gravity in analyzing stability and balance.

Objective

Understand the significance of the center of gravity in determining the stability and balance of structures and systems.

Activity

Explain the concept of the center of gravity (CG) as the point where the total weight of an object or system is concentrated. Discuss how the CG influences stability, balance, and tipping behavior. Provide examples from engineering, such as vehicles, buildings, and machinery, highlighting the role of CG in design and performance.

Numerical Example

An inverted L-shaped block is made of two uniform rectangular sections:

1. Vertical section: Width b_1 = 2m, height h_1 = 4 m, and mass m_1 = 300 kg.
2. Horizontal section: Width b_2 = 6m, height h_2 = 1 m, and mass m_2 = 200 kg. The horizontal section is welded to the top right corner of the vertical section.

The block is placed on a rough inclined plane. Determine

1. the center of gravity of the L-shaped block
2. the angle at which the block tips over

Solution

1. Find the center of gravity (CG):

   The CG is calculated using the weighted average of the individual sections.

   • Centroid of the vertical section:

   $$x_1 = \frac{b_1}{2} = 1, \quad y_1 = \frac{h_1}{2} = 2$$

   • Centroid of the horizontal section:

   The horizontal section's centroid is measured from the base of the vertical section:

   $$x_2 = \frac{b_2}{2} + b_1 = 5, \quad y_2 = h_1 + \frac{h_2}{2} = 4.5$$

2. Find the composite center of gravity (x_c, y_c).
   The CG is

$$x_c = \frac{m_1 x_1 + m_2 x_2}{m_1 + m_2}, \quad y_c = \frac{m_1 y_1 + m_2 y_2}{m_1 + m_2}$$

By substituting in values, we obtain

$$x_c = \frac{(300 \times 1) + (200 \times 5)}{300 + 200} = \frac{300 + 1000}{500} = 2.6 \text{ m}$$

$$y_c = \frac{(300 \times 2) + (200 \times 4.5)}{300 + 200} = \frac{600 + 900}{500} = 3.0 \text{ m}$$

3. Find the critical angle for tipping.
   The block tips over when the line of action of the weight passes outside the base.

   - The tipping point is the edge of the vertical section $(x_t = b_1 = 2 \text{ m})$.
   - Tipping occurs when

$$\tan\theta = \frac{x_c - x_t}{y_c} = \frac{2.6 - 2}{3} = 0.2, \quad \theta = \arctan(0.2) \approx 11.31°$$

ChatGPT Integration

Ask ChatGPT to

- verify the center of gravity for other L-shaped configurations
- explore how stability changes with varying section dimensions or mass
- provide insights on real-world applications, such as stability in heavy machinery or stacked objects

PROMPT 46

Calculate the centroid of a solid object with varying density.

Objective

Learn how to calculate the centroid of a solid object with non-uniform density distribution.

Activity

Explain the process of determining the centroid for a solid object where density varies throughout the volume. Highlight its importance in engineering applications, such as analyzing non-uniform materials or structures. Use an example of a varying density system to demonstrate the concept.

Numerical Example

A solid triangular prism has a base width of b = 6 m, height h = 3 m, and length L = 10 m. The density of the prism varies linearly with height as $\rho(y) = \rho_0(1 + y/h)$, where $\rho_0 = 500$ kg/m^3. Determine the centroid of the prism.

Solution

1. Find the total volume of the prism, having $dV = Lb\left[\left(1 - \dfrac{y}{h}\right)\right]dy$

$$V = Lb\int_0^h\left[\left(1 - \frac{y}{h}\right)\right]dy = \frac{1}{2}bhL = \frac{1}{2}\times 6\times 3\times 10 = 90 \text{ m}^3$$

2. Find the mass of the prism, which is calculated using the varying density.

   Differential volume of a horizontal strip (width decreases linearly to the apex):

$$M = \int_0^h \rho(y)dV$$

By substituting $\rho(y) = \rho_0\left(1 + \dfrac{y}{h}\right)$ and $dV = Lb\left[\left(1 - \dfrac{y}{h}\right)\right]dy$, we obtain the following:

$$M = \rho_0 bL\int_0^h\left(1 - \left(\frac{y}{h}\right)^2\right)dy = \frac{2}{3}\rho_0 bhL$$

3. Evaluate the integral, by plugin the values:

$$M = \frac{2}{3}\rho_0\, bhL = \frac{2}{3}(500)(6)(3)(10) = 6.0 \times 10^4\,\text{kg}$$

4. Determine the centroid height (y_c), which is given by

$$y_c = \frac{1}{M}\int_0^h y\rho(y)\,dV = \frac{\rho_0\,Lb}{M}\int_0^h y\left[1-\left(\frac{y}{h}\right)^2\right]dy$$

$$y_c = \frac{3h}{8} = 1.125\ \text{m above the base}$$

ChatGPT Integration

Ask ChatGPT to

- verify the centroid for other density variations
- discuss how non-uniform density affects material properties in design
- explore similar examples for varying density distributions

PROMPT 47

Explain how the centroid of a volume is determined and its applications in design.

Objective

Understand the concept of the centroid of a volume and how it is determined using integration techniques. Explore its significance in design and structural analysis.

Activity

Explain the centroid of a volume as the geometric center, accounting for the distribution of the volume in three dimensions. Discuss the mathematical method of determining the centroid using triple integrals and provide examples of applications in design, such as balancing irregular 3D components or optimizing material use.

Numerical Example

Determine the centroid of a solid hemisphere of radius $R = 5$ m, with its flat circular face lying on the xy-plane.

Solution

1. Understand the symmetry.
   - The hemisphere is symmetric about the z-axis, so $x_c = 0$ and $y_c = 0$.
   - The z-coordinate of the centroid (z_c) must be calculated.
2. Use the centroid formula.
   The formula for z_c is

$$z_c = \frac{\int_V z\, dV}{\int_V dV}, \quad \int_V dV = V = \frac{2}{3}\pi R^3$$

3. Determine the differential volume element (dV).
   Use spherical coordinates:

$$dV = r^2 \sin\theta\, dr\, d\theta\, d\phi$$

$$z = r\cos\theta$$

   The limits of integration are
   - $r: [0, R]$
   - $\theta: [0, \pi/2]$ (upper hemisphere)
   - $\phi: [0, 2\pi]$ (full rotation around z-axis).
4. Integral for z_c:
   Substitute z and dV into the formula for z_c:

$$z_c = \frac{\int_0^{2\pi} \int_0^{\pi/2} \int_0^R r^3 \cos\theta \sin\theta\, dr\, d\theta\, d\phi}{\frac{2}{3}\pi R^3}$$

5. Solve each integral.
   - Integrate with respect to r: $\int_0^R r^3\, dr = \left.\frac{r^4}{4}\right|_0^R = \frac{R^4}{4}$
   - Integrate with respect to θ: $\int_0^{\pi/2} \cos\theta \sin\theta\, d\theta = \left.\frac{1}{2}\sin^2\theta\right|_0^{\pi/2} = \frac{1}{2}$
   - Integrate with respect to ϕ: $\int_0^{\pi/2} d\phi = 2\pi$

6. Combine the results.

By substituting, we obtain

$$z_c = \frac{2\pi \times \dfrac{1}{2} \times \dfrac{R^4}{4}}{\dfrac{2}{3}\pi R^3} = \frac{3R}{8}, \quad z_c = \frac{3 \times 5}{8} = 1.875\,\text{m}$$

ChatGPT Integration

Ask ChatGPT to

- verify calculations for other 3D shapes, such as cones or paraboloids
- discuss how centroid placement affects structural design, such as in domes or storage tanks
- explore applications of 3D centroids in balance and stability

PROMPT 48

Solve a problem involving the center of gravity of a multi-material system.

Objective

Learn to determine the center of gravity (CG) of a composite system composed of multiple materials, including a void (hole).

Activity

Explain how to calculate the CG of a multi-material system by treating the hole as an absence of material. Discuss applications in engineering, such as designing lightweight structures with voids for reduced material usage and cost.

Mathematical Formulation: Weighted Summation Method

The center of gravity (CG) coordinates are calculated as follows:

$$x_c = \frac{\sum M_i x_i}{\sum M_i}, \quad y_c = \frac{\sum M_i y_i}{\sum M}$$

Numerical Example

A rectangular block is composed of two materials with a circular hole at the border of the two materials:

1. Material 1 (left side): Width b_1 = 3 m, height h = 4 m, and density ρ_1 = 800 kg/m³.

2. Material 2 (right side): Width b_2 = 5 m, height h = 4 m, and density ρ_2 = 600 kg/m³.

3. Circular hole: Radius R = 1 m, centered at the border between the two materials (x = 3 m, y = 2 m).

Determine the center of gravity (x_c, y_c) of the composite block. Assume 1 m for the thickness.

Solution

1. Calculate the areas and volumes.

   Each section has the same height (h = 4 m), and the hole subtracts area and volume but does not contribute mass.

   • Material 1:

   $$A_1 = b_1 \times h = 3 \times 4 = 12 m^2, \quad V_1 = A_1 \times 1 = 12 \text{ m}^3$$

Mass:

$$M_1 = V_1 \cdot \rho_1 = 12 \times 800 = 9600 \text{ kg}$$

   • Material 2:

   $$A_2 = b_2 \cdot h = 5 \times 4 = 20, \quad V_2 = A_2 \times 1 = 20 \text{ m}^3$$

Mass:

$$M_2 = V_2 \cdot \rho_2 = 20 \times 600 = 12000 \text{ kg}$$

Mass of the Circular Hole (Distributed Between Two Materials)

The hole is split equally between Material 1 and Material 2, so we calculate the mass of each part of the hole:

$$M_{\text{hole\_1}} = \pi R^2 \times (1/2)(800) = \pi\left(1^2\right) \times 400 = 400\pi \approx 1256.64 \text{ kg}$$

$$M_{\text{hole\_2}} = \pi R^2 \times (1/2)(600) = \pi\left(1^2\right) \times 300 = 300\pi \approx 942.48 \text{ kg}$$

2. Determine the centroid coordinates of each section.

- Material 1:

$$x_1 = \frac{b_1}{2} = 1.5 \text{ m}, \quad y_1 = \frac{h}{2} = 2 \text{ m}$$

- Material 2:

$$x_2 = b_1 + \frac{b_2}{2} = 5.5 \text{ m}, \quad y_2 = \frac{h}{2} = 2 \text{ m}$$

- Circular Hole:

Each half of the hole's CG is located at $\dfrac{4R}{3\pi}$ from the border line at $x = 3$ m. Therefore:

$$x_{\text{hole\_1}} = \left(3 - \frac{4}{3\pi}\right) = 2.576 \text{ m}, \quad x_{\text{hole\_2}} = \left(3 + \frac{4}{3\pi}\right) = 3.424 \text{ m}$$

3. Calculate the composite CG.

The center of gravity is calculated using the weighted average of the coordinates.

- x-coordinate of CG:

$$x_c = \frac{(M_1 \cdot x_1) + (M_2 \cdot x_2) - (M_{\text{hole\_1}} \cdot x_{\text{hole\_1}}) - (M_{\text{hole\_2}} \cdot x_{\text{hole\_2}})}{\sum M} = \frac{73936}{19400.9} = 3.81 \text{ m}$$

- y-coordinate of CG: using symmetry, we obtain

$$y_c = 2 \text{ m}$$

ChatGPT Integration

Use ChatGPT to

- verify the calculations by providing the dimensions, densities, and hole details for similar multi-material systems
- discuss how the placement of the hole impacts the stability and functionality of the composite block
- explore real-world examples, such as the design of lightweight beams, aerospace components, or vehicle parts that integrate voids for weight reduction without compromising structural integrity

PROMPT 49

Discuss the concept of centroid for irregular shapes and its practical importance.

Objective

Understand how to calculate the centroid for irregular shapes and why it is critical in engineering applications.

Activity

Explain the centroid as the geometric center of an object and discuss its determination using integration for irregular shapes. Highlight its importance in structural analysis, stability evaluation, and design optimization in engineering.

Numerical Example

Determine the centroid of an irregular pentagonal plate with vertices at $(0, 0)$, $(10, 0)$, $(8, 6)$, $(4, 8)$, and $(1, 4)$. Units are meters.

Solution

1. Use the definition of the centroid.

   The centroid (x_c, y_c) of a 2D object is calculated as

   $$x_c = \frac{\int_A x\, dA}{\int_A dA}, \quad y_c = \frac{\int_A y\, dA}{\int_A dA}$$

2. Divide the pentagon into triangles.

   To simplify, divide the pentagon into three triangles:

   - Triangle 1: $(0, 0)$, $(10, 0)$, $(8, 6)$
   - Triangle 2: $(0, 0)$, $(8, 6)$, $(4, 8)$
   - Triangle 3: $(0, 0)$, $(4, 8)$, $(1, 4)$

3. Find the area of each triangle.

   - Triangle 1: Substituting
     $$(x_1, y_1) = (0,0), (x_2, y_2) = (10,0), (x_3, y_3) = (8,6).$$
     $$A_1 = \frac{1}{2}\left| x_1 (y_2 - y_3) + x_2 (y_3 - y_1) + x_3 (y_1 - y_2) \right| = 30 \text{ m}^2$$

- Triangle 2: Substituting
  $$(x_1, y_1) = (0,0), (x_2, y_2) = (8,6), (x_3, y_3) = (4,8)$$

  $$A_2 = \frac{1}{2} \left| x_1 (y_2 - y_3) + x_2 (y_3 - y_1) + x_3 (y_1 - y_2) \right| = 20 \text{ m}^2$$

- Triangle 3: Substituting
  $$(x_1, y_1) = (0,0), (x_2, y_2) = (4,8), (x_3, y_3) = (1,4)$$

  $$A_3 = \frac{1}{2} \left| x_1 (y_2 - y_3) + x_2 (y_3 - y_1) + x_3 (y_1 - y_2) \right| = 4 \text{ m}^2$$

1. Find the centroid of each triangle.

$$x_{c1} = \frac{x_1 + x_2 + x_3}{3}, \quad y_{c1} = \frac{y_1 + y_2 + y_3}{3}$$

$$x_{c1} = 6, \quad y_{c1} = 2$$

$$x_{c2} = 4, \quad y_{c2} = 4.67$$

$$x_{c3} = 1.67, \quad y_{c3} = 4$$

2. Find the composite centroid.

   Combine the centroids using the areas as weights:

$$x_c = \frac{A_1 x_{c1} + A_2 x_{c2} + A_3 x_{c3}}{A_1 + A_2 + A_3}, \quad y_c = \frac{A_1 y_{c1} + A_2 y_{c2} + A_3 y_{c3}}{A_1 + A_2 + A_3}$$

   By substituting, we obtain

$$x_c = \frac{266.67}{54} = 4.94, \quad y_c = \frac{169.33}{54} = 3.14$$

ChatGPT Integration

Ask ChatGPT to

- verify centroids for irregular shapes using different methods
- explore applications where centroid accuracy impacts design, such as in bridges or cantilevered structures
- discuss techniques for finding centroids in 3D objects or parametric shapes

PROMPT 50

Explain how the center of gravity and centroid influence the design of vehicles and machinery.

Objective

Understand the role of the center of gravity (CG) and centroid in the stability, performance, and design optimization of vehicles and machinery.

Activity

Discuss the concepts of CG and centroid, emphasizing their significance in engineering design. Explain how their location affects stability, maneuverability, and load distribution in practical applications like automobiles, aircraft, and industrial machines.

Numerical Example

A truck with a rectangular flatbed (10 m×2.5 m) is loaded with two containers:

1. Container 1: Mass 5000 kg, center located at (2 m, 1.25 m).
2. Container 2: Mass 8000 kg, center located at (7 m, 1.25 m).

Determine the center of gravity of the loaded truck.

Solution

1. Use the definition of center of gravity (CG).

   The CG is the point where the total weight of a system acts. For multiple discrete objects, the coordinates of the CG (x_c, y_c) are calculated as

   $$x_c = \frac{\sum(m_i \cdot x_i)}{\sum m_i}, \quad y_c = \frac{\sum(m_i \cdot y_i)}{\sum m_i}$$

2. Calculate x_c.

   $$x_c = \frac{(5000 \times 2) + (8000 \times 7)}{5000 + 8000} = \frac{66000}{13000} = 5.08 \text{ m}$$

3. Calculate y_c.

   $$y_c = \frac{(5000 \times 1.25) + (8000 \times 1.25)}{5000 + 8000} = \frac{16250}{13000} = 1.25 \text{ m}$$

Importance in Design

- Stability: A lower CG improves vehicle stability, reducing the risk of rollover. This is crucial for trucks, SUVs, and machinery operating on uneven terrain.
- Maneuverability: CG placement affects handling. For example, sports cars have a lower CG for better cornering.
- Load distribution: Proper CG location ensures even load distribution, reducing wear on tires and suspension. In machines, it prevents tipping during operation.

ChatGPT Integration

Use ChatGPT to

- verify CG calculations for various load configurations on vehicles or machinery.
- discuss how improper CG placement can lead to failure or instability in systems.
- provide examples of CG design considerations in airplanes, cranes, and robotics.

CHAPTER 6

MOMENTS OF INERTIA

Prompt 51 Define the concept of the moment of inertia and its significance in engineering.

Prompt 52 Derive the formula for the moment of inertia of a rectangular area about its centroidal axis.

Prompt 53 Explain the parallel axis theorem and its application.

Prompt 54 Derive the moment of inertia of a circular area about its centroidal axis.

Prompt 55 Calculate the polar moment of inertia for a solid shaft.

Prompt 56 Explain the concept of the radius of gyration and its significance.

Prompt 57 Solve a problem involving the moment of inertia of a composite area.

Prompt 58 Discuss the importance of moments of inertia in structural stability and dynamics.

Prompt 59 Derive the moment of inertia for a hollow cylindrical section.

Prompt 60 Explain the relationship between moments of inertia and torsional rigidity.

PROMPT 51

Define the concept of the moment of inertia and its significance in engineering.

Objective

Understand the concept of the moment of inertia, its mathematical representation, and its importance in engineering applications.

Activity

The moment of inertia quantifies an object's resistance to deformation (bending or torsion) or its resistance to rotational motion about an axis. It is widely applied in structural and mechanical engineering to analyze the deflection and bending stress of beams, the stiffness and dynamics of shafts, and the stability of rotating machinery like flywheels and turbines. Engineers also use it in the automotive and aerospace industries for optimizing material use and improving performance. Users can explore how the moment of inertia varies with different shapes, such as circles, rectangles, and ellipses, and analyze how increasing or decreasing a beam's cross-sectional height affects its bending resistance.

Numerical Example

Calculate the moment of inertia of a hollow circular section with an outer diameter $do = 500\,mm$ and an inner diameter $di = 300\,mm$ about its centroidal axis.

Solution

1. Use the definition of the moment of inertia.

The moment of inertia (I) quantifies an object's resistance to deformation (bending or torsion) or its resistance to rotational motion. For a 2D area, it is mathematically defined as

$$I = \int_A y^2 \, dA$$

where

- A is the cross-sectional area.
- y is the perpendicular distance from the axis of rotation to the differential area element dA.

2. Use the moment of inertia of a hollow circular section.

   For a hollow circular section with outer radius $R_o = \dfrac{d_o}{2}$ and inner radius $R_i = \dfrac{d_i}{2}$, the moment of inertia about its centroidal axis is

   $$I = \frac{\pi}{4}\left(R_o^4 - R_i^4\right)$$

3. Perform numerical calculations.

   Given $d_o = 500$ mm and $d_i = 300$ mm, convert to meters:

   $$I = \frac{\pi}{4}\left((0.25)^4 - (0.15)^4\right) = 0.00267 \text{ m}^4$$

4. Significance in Engineering
   - Bending resistance: The moment of inertia governs how beams resist bending under loads. Larger moments of inertia reduce deflection.
   - Torsional stiffness: It determines the resistance of shafts and rotating machinery to twisting.
   - Structural stability: Engineers use it to ensure the stability of bridges, buildings, and other load-bearing structures.

ChatGPT Integration

Ask ChatGPT to

- derive the moments of inertia for irregular shapes like L-sections or trapezoidal sections
- explore how altering dimensions impacts structural behavior, like deflection or stability
- provide exercises for calculating the polar moment of inertia for different rotational applications

PROMPT 52

Derive the formula for the moment of inertia of a rectangular area about its centroidal axis.

Objective

Understand the derivation of the moment of inertia for a rectangular area about its centroidal axis and its applications in engineering. In general, for any shape an area differential element is defined and then used for calculating the moment of inertia using integration.

Activity

The moment of inertia for a rectangle is critical for structural and mechanical design. It quantifies how the rectangle resists bending or torsion when subjected to loads. Engineers use this parameter extensively in beam analysis and design optimization. Explore its derivation step by step, considering the symmetry of the shape. Practice applying this formula to rectangular beams of varying dimensions in structural analysis.

Numerical Example

Derive the formula for the moment of inertia of a rectangular area with a base b=200 mm and height h=300 mm about its horizontal centroidal axis.

Solution

1. Definition of the moment of inertia:

   The moment of inertia about the horizontal axis is given by

   $$I = \int_A y^2 \, dA$$

   where y is the perpendicular distance from the axis, and dA is the differential area.

2. Use the geometry and symmetry of the rectangle.

   The rectangle is symmetric about its centroidal horizontal axis. Let the base b lie along the x-axis and height h extend along the y-axis. Divide the rectangle into differential horizontal strips of width dx and height dy.

3. Use the differential area element.

   A differential strip of the rectangle has

   $$dA = b \, dy$$

   where dy is the strip's height and b is the rectangle's width.
4. Perform the moment of inertia calculation.

   Substitute dA into the formula:

   $$I = \int_{-h/2}^{h/2} y^2 \, b \, dy, \quad I = b \int_{-h/2}^{h/2} y^2 \, dy$$

   Integrate:

   $$\int y^2 \, dy = \frac{y^3}{3}, \quad I = b \left[\frac{y^3}{3} \right]_{-h/2}^{h/2} = \frac{bh^3}{12}$$

5. Numerical Example

   Given $b = 200$ mm $= 0.2$ m and $h = 300$ mm $= 0.3$ m:

   $$I = \frac{0.2 \times (0.3)^3}{12} = 0.00045 \text{ m}^4$$

ChatGPT Integration

Ask ChatGPT to

- derive moments of inertia for other shapes, such as circles, triangles or ellipses
- explore how the moment of inertia affects bending and deflection in beams with varying cross-sections
- solve problems involving the moments of inertia of composite shapes like I-beams

PROMPT 53

Explain the parallel axis theorem and its application.

Objective

Understand the parallel axis theorem, its derivation, and its importance in calculating the moment of inertia about axes that are not centroidal.

Activity

The parallel axis theorem is a fundamental concept in mechanics that allows engineers to calculate the moment of inertia about any axis parallel to the centroidal axis. This is particularly useful for beams, rotating machinery components, and other structures where the axis of rotation or bending is offset. To better understand, derive the theorem step-by-step, and apply it to practical examples. Consider exploring its use in composite sections where the axis of interest does not pass through the centroid.

Numerical Example

Calculate the moment of inertia of a rectangular section with $b = 200$ mm and $h = 300$ mm about an axis located 100 mm above the centroidal axis.

Solution

1. Consider the statement of the parallel axis theorem.

    The parallel axis theorem relates the moment of inertia about an axis parallel to the centroidal axis (I_x) to the moment of inertia about the centroidal axis (I_c) as

    $$I_x = I_c + Ad^2$$

    where
    * I_x: moment of inertia about the parallel axis
    * I_c: moment of inertia about the centroidal axis
    * A: area of the section
    * d: perpendicular distance between the centroidal axis and the parallel axis

2. Determine the moment of inertia about the centroidal axis (I_c):

   For a rectangle,

   $$I_c = \frac{bh^3}{12}, \quad A = bh$$

   By substituting in values, we obtain

   $$I_c = \frac{0.2 \times (0.3)^3}{12} = 0.00045\,\mathrm{m^4}, \quad A = 0.2 \times 0.3 = 0.06\,\mathrm{m^2}$$

   Apply the parallel axis theorem:

   $$I_x = 0.00045 + (0.06 \times 0.1^2) = 0.00105\,\mathrm{m^4}$$

ChatGPT Integration

Ask ChatGPT to

- derive moments of inertia for irregular cross-sections using the parallel axis theorem
- explore real-world applications, such as stability analysis of wind turbine blades or rotating machinery
- help with practice using the theorem for composite sections with multiple materials or varying geometries

PROMPT 54

Derive the moment of inertia of a circular area about its centroidal axis.

Objective

Understand the derivation of the moment of inertia for a circular area about its centroidal axis and its applications in engineering.

Activity

The moment of inertia for a circular area is fundamental in analyzing rotational systems, such as wheels, gears, and flywheels, and in evaluating the bending and torsional stiffness of cylindrical structures. By deriving it from first principles, users can understand its dependence on radius and appreciate its role in beam deflection, rotational dynamics, and torsional rigidity. Practice deriving the moment of inertia for solid and hollow circular sections and apply these principles to engineering problems.

Numerical Example

Derive the formula for the moment of inertia of a solid circular section with a radius $R = 0.25$ m.

Solution

1. Use the definition of the moment of inertia.

   The moment of inertia (I) of a 2D object about its centroidal axis is defined as

   $$I_x = \int_A y^2 \, dA$$

   In polar coordinates (r, ϕ), $y = r\sin(\theta)$, and $dA = r \, dr \, d\theta$. Therefore,

   $$I_x = \int_0^{2\pi}\int_0^R (r\sin\theta)^2 \cdot r \, dr \, d\theta$$

2. Evaluate the integral.
   Simplify:

   $$\int_0^R r^3 \, dr = \frac{r^4}{4}\Big|_0^R = \frac{R^4}{4}, \quad I_x = \frac{R^4}{4}\int_0^{2\pi}\sin^2\theta \, d\theta$$

   But $\sin^2\theta = \frac{1}{2}(1 - \cos(2\theta))$. Hence,

   $$I_x = \frac{R^4}{8}\underbrace{\int_0^{2\pi}(1 - \cos(2\theta)) \, d\theta}_{=2\pi} = \frac{\pi R^4}{4}.$$

   Note that the integral of $\cos(2\theta)$ over 0 to 2π is zero due to periodicity.

3. Determine the final formula.

   The moment of inertia of a solid circular section about its centroidal axis is

   $$I_x = \frac{\pi R^4}{4}$$

   Due to symmetry, the moment of inertia is identical about any axis passing through the center of the circle.

ChatGPT Integration

Ask ChatGPT to

▪ derive moments of inertia for hollow circular sections and compare them with solid sections

▪ explore applications of circular sections in pipelines, shafts, and wheels

▪ provide exercises for calculating moments of inertia for elliptical and irregular curved shapes

PROMPT 55

Calculate the polar moment of inertia for a solid shaft.

Objective

Understand the concept of the polar moment of inertia, its derivation for a solid shaft, and its application in analyzing torsional stresses and deformations.

Activity

The polar moment of inertia quantifies an object's resistance to torsional deformation about its axis. It is critical for designing shafts, axles, and other components subjected to twisting loads. The derivation demonstrates its dependency on the radius of the cross-section. Users can practice deriving the polar moment of inertia for solid and hollow sections and apply the results to calculate torsional stresses in engineering applications.

Numerical Example

Calculate the polar moment of inertia of a solid shaft with a diameter $d = 300$ mm.

Solution

1. Use the definition of the polar moment of inertia.

   The polar moment of inertia (J) for a solid shaft about its centroidal axis is defined as

   $$J = \int_A r^2 \, dA$$

where
- r is the radial distance from the axis.
- dA is the differential area, usually defined in a polar coordinate system.

2. Express dA in polar coordinates.

In polar coordinates, the differential area is

$$dA = r\,dr\,d\theta$$

Substitute dA into the formula:

$$J = \int\limits_{0}^{2\pi R}\int\limits_{0}^{R} r^2 \cdot r\,dr\,d\theta$$

3. Evaluate the integrals.

$$\int\limits_{0}^{2\pi} d\theta \int\limits_{0}^{R} r^3\,dr = 2\pi\left[\frac{r^4}{4}\Big|_0^R\right] = 2\pi\frac{R^4}{4}$$

$$J = \frac{\pi R^4}{2}$$

4. Consider a numerical example.

Given $d = 300$ mm, the radius is

$$R = \frac{d}{2} = \frac{300 \text{ mm}}{2} = 0.15 \text{ m}$$

$$J = \frac{\pi(0.15)^4}{2} = 0.000795 \text{ m}^4$$

ChatGPT Integration

Ask ChatGPT to

- derive the polar moment of inertia for hollow shafts and compare them to solid shafts
- explore its applications in designing axles, propeller shafts, and machine tools
- provide exercises for calculating torsional stress and angle of twist in shafts

PROMPT 56

Explain the concept of the radius of gyration and its significance.

Objective

Understand the concept of the radius of gyration, its mathematical definition, and its role in engineering design.

Activity

The radius of gyration represents the distribution of an area (or mass) about an axis, simplifying complex shapes into an equivalent distance from the axis. It is widely used in structural and mechanical engineering to evaluate stability, buckling resistance, and dynamic response. Explore its relationship with the moment of inertia and how it is applied in practical engineering problems. Practice calculating the radius of gyration for standard and composite sections.

Numerical Example

Calculate the radius of gyration for a solid circular cross-section with a diameter d = 300 mm.

Solution

1. Use the definition of the radius of gyration.

    The radius of gyration (k) is defined as

    $$k = \sqrt{\frac{I}{A}}$$

    where

    - I: moment of inertia about the specified axis
    - A: area of the cross-section

2. Use the moment of inertia (I) for a circular section.

    The moment of inertia about the centroidal axis is

    $$I = \frac{\pi R^4}{4}, \quad R = \frac{d}{2}$$

    The area is

    $$A = \pi R^2$$

3. Use the radius of gyration formula for the circular section.
   Substitute I and A into the formula for k:

$$k = \sqrt{\frac{\frac{\pi R^4}{4}}{\pi R^2}} = \sqrt{\frac{R^2}{4}} = \frac{R}{2}$$

4. Consider a numerical example.
   Given $d = 300$ mm, the radius is

$$k = \frac{R}{2} = \frac{0.15}{2} = 0.075\,\text{m}$$

Significance in Engineering

- Buckling analysis: The radius of gyration helps in evaluating the slenderness ratio of columns, critical for determining their buckling load.
- Structural stability: It provides insight into how material is distributed relative to an axis, influencing stability under bending or torsion.
- Dynamic response: In rotating systems, the radius of gyration is crucial for analyzing angular momentum and energy distribution.

ChatGPT Integration

Ask ChatGPT to

- calculate the radius of gyration for composite sections like I-beams or hollow cylinders
- explore how changes in the radius of gyration affect column buckling
- provide practical examples where the radius of gyration determines structural design efficiency

PROMPT 57

Solve a problem involving the moment of inertia of a composite area.

Objective

Learn how to calculate the moment of inertia for a composite area by summing the contributions of individual components about a common axis.

Activity

The moment of inertia of composite areas is essential for analyzing structures made of multiple sections, such as I-beams, T-beams, or other composite shapes. Engineers often use the parallel axis theorem to adjust the individual moments of inertia when the centroidal axes do not align. Explore how to decompose complex sections into simpler shapes, calculate their contributions, and combine them for the total moment of inertia. Apply this method to practical engineering designs involving built-up structural members.

Numerical Example

A composite area consists of two rectangles:

1. Rectangle 1: Width b_1 = 300 mm, height h_1 = 100 mm, with its centroid located y_1 = 50 mm from the reference axis.
2. Rectangle 2: Width b_2 = 200 mm, height h_2 = 150 mm, with its centroid located y_2 = 175 mm from the reference axis.

Calculate the total moment of inertia of the composite area about the reference axis.

Solution

1. Use the formula for composite areas.

    The total moment of inertia (I) is the sum of the moments of inertia of each individual shape:

    $$I = \sum \left(I_c + Ad^2 \right)$$

    where

    - I_c: moment of inertia about the centroidal axis of the shape
    - A: area of the shape
    - d: distance between the centroidal axis of the shape and the reference axis

2. Perform the calculations. First, calculate the areas (A_1, A_2):

$$A_1 = b_1 \cdot h_1 = 300 \times 100 = 30000 \text{ mm}^2, \quad A_2 = b_2 \cdot h_2 = 200 \times 150 = 30000 \text{ mm}^2$$

Second, calculate the centroidal moments of inertia (I_{c1}, I_{c2}):

$$I_{c1} = \frac{b_1 h_1^3}{12} = 25 \times 10^6 \text{ mm}^4, \quad I_{c2} = \frac{b_2 h_2^3}{12} = 56.25 \times 10^6 \text{ mm}^4$$

Third, calculate the total moment of inertia (I):

$$I = I_{c1} + A_1 d_1^2 + I_{c2} + A_2 d_2^2 = 1075 \times 10^6 \text{ mm}^4$$

Finally, substitute in the following values: ($d_1 = 50$ mm, $d_2 = 175$ mm).

$$I = 25 \times 10^6 + 75 \times 10^6 + 56.25 \times 10^6 + 918.75 \times 10^6 = 1075 \times 10^6 \, mm^4 = 1.075 \, m^4$$

ChatGPT Integration

Ask ChatGPT to

- derive moments of inertia for irregular composite sections like L-shapes or T-beams
- solve problems involving asymmetric load distributions
- discuss how composite area analysis influences real-world designs, such as aircraft frames or crane arms

PROMPT 58

Discuss the importance of moments of inertia in structural stability and dynamics.

Objective

Understand the critical role of moments of inertia in determining structural stability and dynamic performance and explore their applications in engineering design.

Activity

The moment of inertia is fundamental for analyzing bending, torsion, and rotational stability in structures and mechanical systems. It defines how a cross-section resists deformation and influences the distribution of stresses under applied loads. Engineers use moments of inertia to design beams, shafts, and rotating machinery with optimal strength and stiffness. Explore its use in buckling analysis, vibration studies, and load distribution in real-world applications like bridges, aircraft, and turbines.

Numerical Example

A cantilever beam with a rectangular cross-section (b = 200 mm, h = 400 mm) is subjected to a point load of P = 10 kN at its free end. Determine the maximum bending stress (σ_{max}) using the moment of inertia.

Solution

1. Use the relationship between the bending stress and moment of inertia.

   The maximum bending stress in a beam is given by

   $$\sigma_{max} = \frac{M_{max}c}{I}$$

   where

   - M_{max}: maximum bending moment
   - c: distance from the neutral axis to the outermost fiber
   - I: moment of inertia about the neutral axis

2. Calculate the maximum bending moment (M_{max}).

   For a cantilever beam subjected to a point load at its free end, we find

   $$M_{max} = P \cdot L = 10 \times 2 = 20 \text{ kN} \cdot \text{m}$$

3. Find the moment of inertia of the rectangular section (I).

   The moment of inertia about the centroidal axis is

   $$I = \frac{bh^3}{12} = \frac{0.2 \times (0.4)^3}{12} = 0.001067 \text{ m}^4$$

4. Calculate the maximum bending stress (σ_{max}).

   Substitute the following into the bending stress formula: (c = h/2).

   $$\sigma_{max} = \frac{M_{max}c}{I} = \frac{20000 \times 0.2}{0.001067} = 3.75 \text{ MPa}$$

ChatGPT Integration

Ask ChatGPT to

- explore how moments of inertia influence beam deflection under different loading conditions

- derive moments of inertia for non-standard sections like L-beams or circular segments
- discuss practical applications in structural design, such as analyzing bridge girders or high-rise columns

PROMPT 59

Derive the moment of inertia for a hollow cylindrical section.

Objective

Understand the derivation of the moment of inertia for a hollow cylindrical section and its applications in engineering design.

Activity

Hollow cylindrical sections are widely used in applications requiring strength with reduced weight, such as in pipes, pressure vessels, and hollow shafts. The moment of inertia helps engineers evaluate their bending resistance and dynamic behavior. Derive the formula for the moment of inertia of a hollow cylinder by subtracting the moment of inertia of the inner hollow region from the outer solid region. Apply this concept to structural and rotational designs to optimize material use and performance.

Numerical Example

Derive the moment of inertia of a hollow cylindrical section with an outer radius R_o = 0.5 m and an inner radius R_i = 0.3 m about its centroidal axis.

Solution

1. Use the definition of the moment of inertia.

   The moment of inertia about the centroidal axis is

   $$I = \int_A y^2 \, dA, \quad I = I_{outer} - I_{inner}$$

   where
   - I_{outer} : moment of inertia of the outer solid cylinder
   - I_{inner} : moment of inertia of the inner hollow region

2. Find the moment of inertia for a solid cylinder.

   The moment of inertia for a solid circular section about its centroidal axis is

   $$I_{solid} = \frac{\pi R^4}{4}, \quad I_{outer} = \frac{\pi R_o^4}{4}, \quad I_{inner} = \frac{\pi R_i^4}{4}$$

3. Calculate the moment of inertia for the hollow cylinder.

   Substitute into the formula to obtain

   $$I = \frac{\pi}{4}\left(R_o^4 - R_i^4\right)$$

4. Consider a numerical example.

   Substitute R_o = 0.5 m and R_i = 0.3 m into the formula to obtain

   $$I = \frac{\pi}{4}\left((0.5)^4 - (0.3)^4\right) = 0.0427 \text{ m}^4$$

ChatGPT Integration

Ask ChatGPT to

- derive moments of inertia for irregular hollow sections, like elliptical or trapezoidal shapes
- explore the advantages of hollow vs. solid sections in practical designs, such as crane booms or bike frames
- practice calculating torsional stress for hollow cylindrical shafts using polar moments of inertia

PROMPT 60

Explain the relationship between moments of inertia and torsional rigidity.

Objective

Understand the relationship between moments of inertia, particularly the polar moment of inertia, and torsional rigidity in engineering applications.

Activity

Torsional rigidity determines a material's resistance to twisting when subjected to torque. The polar moment of inertia plays a crucial role in calculating torsional stress and angular deformation in shafts and other structural components. Explore how torsional rigidity depends on the material's modulus of rigidity (G) and the geometry of the cross-section. Practice applying these concepts to solve problems involving shafts, pipes, and other cylindrical components subjected to twisting loads.

Numerical Example

A solid cylindrical shaft with a radius $R = 0.1$ m and length $L = 2$ m is subjected to a torque $T = 500$ Nm. The material has a modulus of rigidity $G = 80$ GPa. Calculate the angular twist (θ).

Solution

1. Use the torsional rigidity equation.

   The angular twist (θ) of a shaft subjected to torque is given by

$$\theta = \frac{TL}{GJ}$$

   where

   - T: applied torque
   - L: length of the shaft
   - G: modulus of rigidity of the material
   - J: polar moment of inertia of the cross-section

2. Determine the polar moment of inertia (J).

   For a solid cylindrical shaft, the polar moment of inertia is

$$J = \frac{\pi R^4}{2} = \frac{\pi (0.1)^4}{2} = 0.00015708 \text{ m}^4$$

3. Substitute values into the torsional rigidity formula.

$$\theta = \frac{500 \times 2}{\left(80 \times 10^9\right) \times 0.00015708} = 7.96 \times 10^{-5} \text{ radians}$$

$$\theta(\text{degrees}) = \theta(\text{radians}) \cdot \frac{180}{\pi} = 0.00457°$$

ChatGPT Integration

Ask ChatGPT to

- derive the torsional rigidity for hollow cylindrical sections and compare with solid sections
- explore real-world applications, such as designing drive shafts and wind turbine blades
- provide additional exercises for calculating angular deformation in shafts of varying cross-sections

CHAPTER 7

ANALYSIS OF STRUCTURES

Prompt 61 Explain the importance of load paths in structural systems and how to analyze them.

Prompt 62 Describe the different types of connections in structural analysis (pinned, fixed, and roller) and their effects on load distribution.

Prompt 63 Analyze a two-span continuous beam using the method of superposition.

Prompt 64 Analyze the deflection of a cantilever beam subjected to multiple point loads and UDLs.

Prompt 65 Evaluate the impact of support settlements on the internal forces and deflections in structural systems.

Prompt 66 Discuss the application of energy methods in structural analysis for determining deflections and internal forces.

Prompt 67 Solve a problem involving the analysis of a cantilever frame under multiple loads.

Prompt 68 Explain the difference between statically determinate and indeterminate beams with examples.

Prompt 69 Discuss the use of moment distribution methods in structural analysis.

Prompt 70 Solve a problem involving the influence lines for shear and moment in a continuous beam.

PROMPT 61

Explain the importance of load paths in structural systems and how to analyze them.

Objective

Understand the concept of load paths in structural systems, their significance in ensuring stability and safety, and methods for analyzing them.

Activity

Load paths are the routes that forces and loads take through a structure to reach its supports or foundation. Proper design ensures that these paths are efficient, minimizing material usage while maintaining safety. Mismanaged load paths can lead to instability or failure. In this activity, you will explore how to identify load paths in various structures and evaluate their effectiveness. Apply the concept to analyze load distribution in simple beams and trusses and optimize designs for structural integrity.

Numerical Example

A beam of length L = 6 m is simply supported at both ends and subjected to a point load P = 20 kN at its center. Determine the load paths and reactions at the supports.

Solution

1. Define the load path.

   The load path begins at the point where the force P is applied. The force is transferred along the beam to the supports at both ends. These supports provide reactions (R_A and R_B) that balance the applied load.

2. Determine the reactions using equilibrium equations.
   - Sum of the vertical forces:
   $$R_A + R_B = P, \quad R_A + R_B = 20$$
   - Sum of the moments about A: $\sum M_A = 0 \Rightarrow R_B \cdot L - P \cdot \dfrac{L}{2} = 0$
     Substitute L = 6 m: $R_B = 10$ kN

   - Substitute RB into the first equation to obtain
     $R_A + 10 = 20 \Rightarrow R_A = 10$ kN.

3. Identify the load path.

   - The point load P generates a bending moment and shear forces in the beam.
   - These forces are distributed internally along the beam to maintain equilibrium.
   - Reactions at the supports balance the applied load, completing the load path.

4. Perform the shear force and bending moment analysis:

   - Shear force:

   $$V(x) = \begin{cases} R_A = 10 \text{ kN}, & 0 \le x < 3 \text{ m} \\ -10 \text{ kN}, & 3 < x \le 6 \text{ m} \end{cases}$$

   - Bending moment:

   $$M(x) = \begin{cases} R_A.x, & 0 \le x < 3 \text{ m} \\ (L-x)R_B, & 3 < x \le 6 \text{ m} \end{cases}$$

Significance in Engineering:

- Structural stability: Understanding load paths ensures that loads are efficiently transferred to foundations without overstressing components.
- Design optimization: Identifying critical load paths helps minimize material usage while maintaining safety.
- Failure prevention: Poorly designed load paths can lead to uneven load distribution and structural failure.

ChatGPT Integration

Ask ChatGPT to

- analyze load paths for complex systems, such as multi-span beams or space trusses
- discuss how material properties affect load distribution in structural members
- solve problems involving unevenly distributed loads or asymmetric supports

PROMPT 62

Describe the different types of connections in structural analysis (pinned, fixed, and roller) and their effects on load distribution.

Objective

Understand the role of different types of connections in structural systems and their effects on load distribution and stability.

Activity

Connections are critical components in structural analysis as they define how forces and moments are transmitted between members. Each type of connection—pinned, fixed, and roller—offers unique properties that influence load transfer and stability. Explore these connections' roles in various structures, such as beams, trusses, and frames. Analyze how they affect reaction forces and moment distribution under external loads.

- Pinned supports allow rotation but resist translation; they are commonly used in trusses and simply supported beams.
- Roller supports resist vertical forces while allowing horizontal movement, and are ideal for bridge supports to accommodate thermal expansion.
- Fixed supports provide the highest stability, resisting forces and moments in all directions, and are used in cantilever beams and rigid frames.

Numerical Example

Analyze the reactions at supports for two cases of a beam subjected to a triangular distributed load:

- Case (a): A simply supported beam of length L = 8 m with a pinned connection at one end (A) and a roller at the other end (B). The triangular load has a maximum intensity of $w_{max} = 3k$ N / m at A and tapers to zero at B.
- Case (b): A cantilever beam of length L = 8 m, fixed at one end (A) and free at the other end (B), subjected to the same triangular load as in case (a).

Solution

Case (a): Pinned and Roller Supports

1. Find the resultant force (F_R):
    The resultant force of the triangular load is

    $$F_R = \frac{1}{2} \cdot w_{max} \cdot L$$

    Substitute $w_{max} = 3$ kN/m and $L = 8$ m into the formula to obtain

    $$F_R = \frac{1}{2} \times 3 \times 8 = 12 \text{ kN}$$

2. Determine the location of the resultant force (d).
    The resultant force acts at a distance d from the support A:

    $$d = \frac{1}{3} \cdot L = \frac{8}{3} = 2.67 \text{ m}$$

3. Use the equilibrium equations.
    Sum of the vertical forces:

    $$\sum F_z = 0 \Rightarrow R_A + R_B = F_R = 12 \text{ kN}$$

    Sum of the moments about A:

    $$\sum M_A = 0 \Rightarrow R_B \cdot L - F_R \cdot d = 0$$

    $$R_B = \frac{12 \times 2.67}{8} = 4 \text{ kN}$$

Solve for R_A:

$$R_A + 4 = 12 \Rightarrow R_A = 8 \text{ kN}$$

Case (b): Fixed Support
The resultant force and its location remain the same as in Case (a):
$F_R = 12$ kN, $d = 2.67$ m.

1. Use the equilibrium equations at the fixed support (A).
    *   Vertical reaction (R_A):
        The fixed support must resist the entire vertical load.
        $$R_A = F_R = 12 \text{ kN}$$

■ Moment reaction (M_A):

The fixed support resists the bending moment caused by the load.

$$M_A = F_R \cdot d = 12 \times 2.67 = 32.04 \text{ kN} \cdot \text{m}$$

ChatGPT Integration

Ask ChatGPT to

■ analyze the effects of different connections in a multi-span beam

■ compare the structural behavior of pinned and fixed connections under identical loads

■ solve advanced problems involving hybrid connections in complex frames

PROMPT 63

Analyze a two-span continuous beam using the method of superposition.

Objective

To understand the analysis of a two-span continuous beam by applying the method of superposition, focusing on calculating reactions, moments, and deflections under given loading conditions.

Activity

The method of superposition is a powerful tool for analyzing indeterminate beams, especially in cases of multiple spans or varying load conditions. By decomposing the beam into simpler determinate systems, the effects of different load types can be independently calculated and then combined to satisfy equilibrium and compatibility conditions. This approach allows engineers to analyze the system more effectively while maintaining structural continuity at critical points such as supports or joints.

Numerical Example

A two-span continuous beam A–B–C has spans of 6 m (AB) and 4 m (BC).

■ AB: subjected to a uniformly distributed load of 10 kN/m

■ BC: subjected to a point load of 20 kN at its midpoint

Both supports at A and C are pinned, while B is a roller. Analyze the beam using the method of superposition.

Solution

Remove support B and calculate the deflection at point B for the deteminate beam AC. The total defelction reads:

$$\delta_B = \delta_{B\_P} + \delta_{B,P}$$

where, δ_{B_U} is deflection at B due to uniform continuous load and $\delta_{B,P}$ due to the point load.

Or ($a = AB = 6\ m$, $L = AC = 10\ m$):

$$\delta_{B\_U} = \frac{\omega a^3 (4L^2 - 7La + 3a^2)}{24EIL}$$

$$\delta_{B\_P} = \frac{Pab(L^2 - b^2 - a^2)}{6EIL}$$

Where, $b = 2$, distance from the point-load to C. Plugin the numerical values, gives:

$$\delta_{B\_U} = \frac{10 \times 6^3 (4 \times 10^2 - 7 \times 10 \times 6 + 3 \times 6^2)}{24EIL} = \frac{7920}{EIL}$$

$$\delta_{B\_P} = \frac{(20 \times 6 \times 2)(10^2 - 4 - 36)}{6EIL} = \frac{2400}{EIL}$$

Or

$$\delta_B = \frac{7920}{EIL} + \frac{2400}{EIL} = \frac{10320}{EIL}$$

But for satisfying kinematic compatibility, the deflection at B must be zero. Therefore, the deflection to an upward point load of R_B must equal, in magnitude, to the total deflection δ_B.

$$\delta_B = \frac{R_B a^2 (L-a)^2}{3EIL} = \frac{10320}{EIL}$$

Therefore,

$$R_B = 53.8\ kN$$

Now we can write the equilibrium equations for calculating the reactions at A and C.

$$R_A + 53.8 + R_C = 60 + 20 = 80 \text{ and } 10R_C + 6 \times 53.8 = 20 \times 8 + 10 \times 6 \times 3$$

After solving, gives:

$$R_C = 1.72 \text{ kN}$$
$$R_B = 24.5 \text{ kN}$$

ChatGPT Integration

- Verify the calculations for reactions, moments, and deflections for each span.
- Use ChatGPT to derive the compatibility condition between spans and explain its significance.
- Request additional examples of continuous beam analysis under different loading scenarios.
- Ask ChatGPT to provide visual aids, such as shear force and bending moment diagrams, for the given beam.

PROMPT 64

Analyze the deflection of a cantilever beam subjected to multiple point loads and UDLs.

Objective

To calculate the deflection of a cantilever beam subjected to a combination of point loads and uniformly distributed loads (UDLs), understanding the effect of loading patterns on deformation.

Activity

The deflection of cantilever beams is a critical aspect of design in structural and mechanical engineering. The combination of point loads and UDLs introduces varying moments along the length of the beam, affecting its deformation profile. By applying principles of beam theory and superposition, we can determine the deflection at any point on the beam. This analysis ensures the beam meets deflection limits and serviceability criteria.

The activity involves deriving the equations for deflection using standard formulas for beam bending and applying them to combined loading scenarios. Engineers use these calculations to ensure designs comply with serviceability and performance requirements.

Numerical Example

A cantilever beam of length $L = 6$ m is subjected to

1. a UDL of $w = 5 \text{kN} / \text{m}$ over the entire length
2. a point load $P = 10$ kN at $x = 4$ m from the fixed end

The beam has a flexural rigidity $EI = 2 \times 10^4$ kN·m². Calculate the deflection at the free end.

Solution

1. Determine the deflection due to UDL.

   The deflection at the free end due to a UDL over the entire length is

$$\delta_{\text{UDL}} = \frac{wL^4}{8EI}$$

   By substituting $L = 6$ m and $EI = 200$ kN·m², we obtain

$$\delta_{\text{UDL}} = \frac{5 \times 6^4}{8 \times 20000} = 0.041 \text{ m}$$

2. Determine the deflection due to the point load.

   The deflection at the free end due to a point load P at distance a from the fixed end is

$$\delta_P = \frac{Pa^2 (3L - a)}{6EI}$$

   By substituting $P = 10$ kN, a = 4 m, $L = 6$ m, and $EI = 200$ kN·m² into the formula, we obtain

$$\delta_P = \frac{10 \times 4^2 (3 \times 6 - 4)}{6 \times 20000} = 0.0187 \text{ m}$$

3. Calculate the total deflection at the free end.

   Using the principle of superposition, we obtain

$$\delta_{\text{total}} = \delta_{\text{UDL}} + \delta_P$$

   By substituting values, we obtain

$$\delta_{\text{total}} = 0.041 + 0.0187 = 0.0597 \text{ m}$$

ChatGPT Integration

- Ask ChatGPT to verify the calculations for each load type and check for alternative methods of analysis.
- Use ChatGPT to explore how deflection limits influence beam design in real-world applications.
- Request additional examples involving more complex loading scenarios for cantilever beams.

PROMPT 65

Evaluate the impact of support settlements on the internal forces and deflections in structural systems.

Objective

Understand how support settlements influence internal forces and deflections in statically determinate and indeterminate structures. Learn how to analyze and quantify these effects using equilibrium and compatibility equations.

Activity

Support settlement occurs when one or more supports of a structure move vertically, either upward or downward. This displacement introduces additional internal forces and deformations that must be accounted for in structural analysis. Engineers analyze support settlements to prevent excessive stress, deformation, and potential failure in structures such as bridges, buildings, and pipelines.

This activity involves

1. identifying the types of structures affected by support settlements (e.g., beams, trusses, and frames)
2. applying equilibrium and compatibility conditions to solve for internal forces and deflections
3. using numerical methods or analytical solutions for statically indeterminate structures

Numerical Example

A continuous beam ABC of length L = 6 m is supported at points A (fixed hinge), B (roller support at 3 m), and C (roller support at 6 m). The beam carries a uniformly distributed load of 4 kN/m across its entire span.

After construction, support B settles by 10 mm downward, while supporta A and C remain at their original position. Determine

1. the reactions at supports A, B, and C due to the settlement
2. the bending moments at key locations (A, B, and C)
3. the maximum deflection in the beam

Use the following information:

- Flexural Rigidity (EI) = 50×10^3 kN·m²
- Assume small deflections and linear behavior.

Solution

1. Determine the static reactions (ignoring the settlement, initially).

   The reactions at A, B, and C can be calculated assuming no settlement and solved using static equilibrium equations:

   $$R_A + R_B + R_C = \int_0^6 4dx = 24 \text{ kN}$$

   $$3R_B + 6R_C = \int_0^6 4xdx = 72 \text{ kN·m}$$

2. Choose R_B as the redundant and evaluate the flexibility coefficient.

   For the simplysupported beam AC (supports only at A and C):
   - distance to B: a=3 m, b = L−a = 3 m
   - flexural rigidity: EI = 50×10^3 kN/m²

   The displacement at B caused by a unit upward load at B is

   $$f_{BB} = \frac{a^2 b^2}{3EIL} = \frac{3^2 \, 3^2}{3(50 \times 10^3)6} = 9.0 \times 10^{-5} \text{ m per kN}$$

3. Compute the downward deflection of B due to the UDL (with $R_B = 0$)

   For a uniformly loaded simply supported span, the deflection at a point xxx is

   $$\delta(x) = \frac{wx\left(L^3 - 2Lx^2 + x^3\right)}{24EI}$$

   At $x = 3$ m:

   $$\delta_w(B) = \frac{4 \times 3\left(6^3 - 2 \times 6 \times 3^2 + 3^3\right)}{24 \times 50 \times 10^3} = 1.35 \text{ mm downward}$$

4. Compatibility equation including the settlement $s = 10$ mm $= 0.010$ ms = 10\text{ mm} = 0.010\text{ m}$s = 10$ mm $= 0.010$ m (downward)

   Downward positive:

   $$\delta_w(B) - R_B f_{BB} = s$$

   Substitute for $\delta_w(B)$ and f_{BB}:

   $$1.35 \times 10^{-3} - R_B\left(9 \times 10^{-5}\right) = 0.01$$

   Therefore,
   $$R_B = -96 \, kN \, downward.$$

   Because we defined **upward** as positive, a negative value means a **downward** reaction—i.e. the beam would have to be in tension with the support to keep contact. so the physical solution is

   $$R_B = 0 \, kN$$

5. Final reactions (beam lifts off the settled support).

   With $R_B = 0$, we have:

   $$R_A = R_C = 12 \, kN$$

6. Bending moments at key points

   With the beam now simply supported between A and C:

   $$M_A = M_C = 0,$$

   $$M_B = \frac{wL^2}{8} = \frac{4(6)^2}{8} = 18 \, kN$$

7. Maximum beam deflection (ignoring the 10 mm ground movement)

For a uniform load on a 6 m simply supported span,

$$\delta_{max} = \frac{5wL^4}{384EI} = 1.35\,mm\,(at\,mid-span)$$

Because the 10 mm foundation settlement is far larger than the beam's own elastic deflection, the beam loses contact with support B and behaves as a simple span.

ChatGPT Integration

You can use ChatGPT to

* verify calculations for different beam configurations and material properties
* analyze a real-world scenario, such as support settlement in a truss or multi-span bridge
* explore alternative methods, such as finite element analysis (FEA) for more accurate settlement modeling

PROMPT 66

Discuss the application of energy methods in structural analysis for determining deflections and internal forces.

Objective

To understand how energy methods, such as the work-energy principle and Castigliano's theorem, are applied to determine deflections and internal forces in mechanical structures.

Activity

Energy methods are powerful tools for analyzing mechanical systems, especially when traditional equilibrium approaches become cumbersome. These methods use principles of work and strain energy to relate external forces and moments to displacements and rotations in structures. Castigliano's theorem, for example, simplifies the process of calculating deflections and internal forces in beams, frames, and mechanical assemblies by deriving relationships between energy and applied loads.

In mechanical engineering, energy methods are particularly useful for systems with complex geometries or redundant supports, such as robotic arms, machine frames, and automotive suspension components. Engineers use these methods to predict deflections and stresses, ensuring that designs meet safety and performance criteria.

Numerical Example

A cantilever beam of length $L = 1.5$ m is subjected to

1. a vertical point load $P = 500$ N at the free end, and
2. a clockwise moment $M = 200$ N·m applied at the free end

The beam has a rectangular cross-section with a width $b = 50$ mm and height $h = 100$ mm. Determine the vertical deflection and slope at the free end using Castigliano's theorem. Assume modulus of elasticity is $E = 200$ GPa.

Solution

1. Calculate the moment of inertia.

    The moment of inertia I for a rectangular cross-section is

$$I = \frac{bh^3}{12} = \frac{50 \times 100^3}{12} = 4.167 \times 10^{-6}\, m^4$$

2. Write the strain energy equation.

    The strain energy U in the beam is due to bending:

$$U = \int_0^L \frac{M^2(x)}{2EI}\,dx$$

    where $M(x)$ is the bending moment at a distance x from the fixed end.

$$M(x) = P \cdot x + M$$

3. Substitute this value into the strain energy equation and integrate:

$$U = \int_0^L \frac{(P \cdot x + M)^2}{2EI}\,dx$$

$$U = \frac{1}{2EI} \int_0^L \left(P^2 x^2 + 2PMx + M^2\right) dx$$

$$U = \frac{1}{2EI}\left[\frac{P^2 L^3}{3} + PML^2 + M^2 L\right]$$

Substituting for the values, we find

$$U = \frac{1}{2 \times 200 \times 10^9 \times 4.167 \times 10^{-6}} \left(\frac{500^2 \times 1.5^3}{3} + 500 \times 200 \times 1.5^2 + 200^2 \times 1.5 \right) = 0.34 \text{ J}$$

4. Apply Castigliano's theorem.

   The deflection δ at the free end is

$$\delta = \frac{\partial U}{\partial P}$$

We substitute the computed strain energy function:

$$\delta = \frac{1}{EI} \left(\frac{PL^3}{3} + \frac{ML^2}{2} \right)$$

By substituting with the given values, we obtain

$$\delta = \frac{1}{\left(200 \times 10^9 \times 4.167 \times 10^{-6} \right)} \times \left(\frac{500 \times (1.5)^3}{3} + \frac{200 \times (1.5)^2}{2} \right) = 0.945 \text{ mm}$$

5. Compute the slope at the free end.

   The slope θ at the free end is given by

$$\theta = \frac{\partial U}{\partial M}$$

From the strain energy equation, we obtain

$$\theta = \frac{1}{EI} \left(\frac{PL^2}{2} + ML \right)$$

By substituting given values, we obtain

$$\theta = \frac{1}{\left(200 \times 10^9 \times 4.167 \times 10^{-6} \right)} \times \left(\frac{500 \times (1.5)^2}{2} + 200 \times 1.5 \right) = 0.001035 \text{ rad}$$

Please note that Castigliano's method yields the deflection along in the same direction as applied load.

ChatGPT Integration

You can use ChatGPT to

- verify calculations for strain energy, deflections, and internal forces in beams, trusses, and frames
- analyze real-world applications, such as structural deformation in aircraft wings or bridges under variable loads
- explore alternative methods, such as symbolic computation for energy-based solutions or numerical integration for complex geometries

PROMPT 67

Solve a problem involving the analysis of a cantilever frame under multiple loads.

Objective

To understand the analysis of internal forces (shear, axial, and bending moment) in a cantilever frame subjected to multiple loads using equilibrium equations.

Activity

Cantilever frames are commonly used in structural engineering applications such as industrial buildings, bridges, and tower structures. These frames resist bending, shear, and axial forces due to applied loads. This activity focuses on

1. identifying the equilibrium conditions for a rectangular cantilever frame
2. determining support reactions and internal forces at key sections
3. using moment equations to find bending moments in members
4. understanding axial and shear force distributions in frames

Numerical Example

A rectangular cantilever frame consists of a vertical column (AB) and a horizontal beam (BC). The frame is fixed at A and is subjected to the following:

- a horizontal compressive force H = 500 N at C (free end of the beam)
- a vertical force P = 1000 N at C
- The frame has dimensions: h=3 m (height of column AB) and w=4 m (length of beam BC).

Determine

1. the support reactions at A
2. the internal forces (shear, axial, and bending moment) at section B

Solution

1. Determine the support reactions at A.

   Since A is a fixed support, it has three unknown reactions:
   - R_{Ax}: horizontal reaction
   - R_{Ay}: vertical reaction
   - M_A: moment reaction

   Using static equilibrium equations, we obtain the following:
   - Sum of the forces in the horizontal direction ($\Sigma F_x = 0$)

$$R_{Ax} - H = 0$$
$$R_{Ax} = 500 \text{ N}$$

   - Sum of the forces in vertical direction ($\Sigma F_y = 0$)

$$R_{Ay} - P = 0$$

$$R_{Ay} = 1000 \text{ N}$$

   - Sum of the moments about A ($\Sigma M_A = 0$)

$$M_A - P \cdot w + H \cdot h = 0$$
$$M_A = (1000 \times 4) - (500 \times 3)$$
$$M_A = 2500 \text{ N·m}$$

2. Determine the internal forces at section B (the base of the beam).

   To analyze the internal forces at B, we make a section cut just above B and analyze the horizontal beam BC.

   - Shear force at B (V_B)

     The shear force is equal to the vertical reaction at A:

     $$V_B = R_{Ay} = 1000 \text{ N}$$

   - Axial force at B (N_B)

     The axial force at B is equal to the horizontal reaction at A:

     $$N_B = R_{Ax} = 500 \text{ N}$$

   - Bending moment at B (M_B)

     The bending moment at B is found using the moment caused by the applied forces at C:

     $$M_B = (P \times w)$$

     $$M_B = (1000 \times 4) = 4000 \text{ N·m}$$

ChatGPT Integration

You can use ChatGPT to

- verify calculations for reaction forces, shear forces, axial forces, and bending moments in cantilever frames
- analyze real-world applications, such as frame structures in buildings, robotic arms, and industrial crane supports
- explore alternative methods, such as finite element analysis (FEA) for stress distribution and automated load analysis for optimized frame design

PROMPT 68

Explain the difference between statically determinate and indeterminate beams with examples.

Objective

To understand the distinction between statically determinate and indeterminate beams, how to identify them, and the implications for structural analysis.

Activity

Beams in engineering structures are classified as statically determinate or statically indeterminate, depending on the number of unknown support reactions compared to the available equilibrium equations.

- Statically determinate beams
  - All support reactions and internal forces can be found using only equilibrium equations.
  - Easier to analyze because they do not require additional compatibility conditions
  - Examples: simply supported beams, cantilever beams, and overhanging beams with sufficient supports
- Statically indeterminate beams
  - The number of unknown reactions exceeds the number of equilibrium equations.
  - Require additional compatibility conditions (such as deflection continuity) to solve
  - Examples: fixed beams, continuous beams over multiple supports, and propped cantilevers

This activity involves

1. identifying whether a beam is determinate or indeterminate using structural equations
2. solving a numerical example to demonstrate the distinction
3. discussing the implications of each type for real-world structural applications

Numerical Example

A beam of length $L = 6$ m is supported in the following ways:

1. Case 1: simply supported at both ends
2. Case 2: fixed at both ends
3. Case 3: fixed at one end and supported by a roller at the other end

Determine

1. whether each case is statically determinate or indeterminate
2. the degree of static indeterminacy for indeterminate cases

Solution

1. Define the static equilibrium equations.

   For 2D beam structures, equilibrium equations provide the following:

   $$\sum F_x = 0, \quad \sum F_y = 0, \quad \sum M_z = 0$$

   This gives a maximum of three independent equations for solving unknowns.

   For 3D beam structures, the equilibrium equations expand to include forces and moments in three dimensions:

   $$\sum F_x = 0, \quad \sum F_y = 0, \quad \sum F_z = 0$$
   $$\sum M_x = 0, \quad \sum M_y = 0, \quad \sum M_z = 0$$

   This gives a maximum of six independent equations for solving unknowns.

2. Analyze each case.

   Case 1: Simply Supported Beam
   - Supports: One pin support (A) + one roller support (B)
   - Unknown reactions:
     - Pin support (A): R_{Ax}, R_{Ay} (2 unknowns)
     - Roller support (B): R_{By} (1 unknown)
     - Total unknowns: 3

   Available equilibrium equations: 3

   Conclusion: The beam is statically determinate.

   Total unknowns = 3, *Available equations* = 3, *Statically Determinate*

   Case 2: Fixed-Fixed Beam
   - Supports: Two fixed supports (A and B)
   - Unknown reactions:
     - Fixed support (A): R_{Ax}, R_{Ay}, M_A (3 unknowns)
     - Fixed support (B): R_{Bx}, R_{By}, M_B (3 unknowns)
     - Total unknowns: 6

   Available equilibrium equations: 3

Degree of Indeterminacy:

$$Total\ unknowns = 6, \quad Available\ equations = 3$$
$$Unknowns - Equations = 6 - 3 = 3$$

Conclusion: The beam is statically indeterminate to the third degree.

Case 3: Propped Cantilever Beam (Fixed-Roller)

▪ Supports: One fixed support (A) and one roller support (B)
▪ Unknown reactions:
   • Fixed support (A): R_{Ax}, R_{Ay}, M_A (3 unknowns)
   • Roller support (B): R_{By} (1 unknown)
   • Total unknowns: 4

Available equilibrium equations: 3.
Degree of Indeterminacy:

$$Indeterminacy = Unknowns - Equations = 4 - 3 = 1$$

Conclusion: The beam is statically indeterminate to the first degree.

Case 4: 3D Beam Structure with Fixed and Roller Supports
Consider a 3D beam supported as follows:

▪ Fixed support at A
▪ Roller support at B, allowing movement along the x-axis only
▪ The beam is subjected to external forces in all three directions.

Determine whether this beam is statically determinate or indeterminate and calculate its degree of indeterminacy.

For a fixed support at A, the reactions include

▪ Forces: R_{Ax}, R_{Ay}, R_{Az} (3 unknowns)
▪ Moments: M_{Ax}, M_{Ay}, M_{Az} (3 unknowns)

For a roller support at B, which allows motion in the xxx-direction but resists forces in y and z:

▪ **Forces:** R_{By}, R_{Bz} (2 unknowns)

Total unknown reactions: $3 + 3 + 2 = 8$
Total available equations: 6

The degree of static indeterminacy is:

$$Indeterminacy = Unknowns - Equations = 8 - 6 = 2$$

ChatGPT Integration

You can use ChatGPT to

- verify calculations for different beam configurations, including varying support conditions
- analyze real-world applications, such as continuous bridge structures or building frames where redundancy improves safety
- explore alternative methods, such as matrix structural analysis or finite element methods (FEM), for solving statically indeterminate beams

PROMPT 69

Discuss the Use of the Three-Moment Method (Clapeyron's theorem) in Structural Analysis

Objective

To understand the moment distribution method and how it is applied in analyzing statically indeterminate beams and frames.

Activity

The three-moment method is a classical analytical approach used to determine bending moments in continuous beams subjected to various loading conditions. Based on applying moment equilibrium at three successive supports, the method leads to a system of linear equations that can be solved directly, even for statically indeterminate structures. This activity focuses on the application of the three-moment equations to two-span and multi-span continuous beams, taking into account uniform and concentrated loads, settlement, and support conditions. The method is especially useful for hand calculations when analyzing structures with constant or piecewise constant flexural rigidity.

This activity involves

Understanding the formulation of the three-moment equation and its underlying assumptions applying the method to analyze beams with two or more spans under different loading scenarios exploring how load combinations and continuity affect the resulting internal moment distribution

Numerical Example

A continuous beam ABC has a span of 6 m between supports A and B and 4 m between supports B and C. The beam is

- fixed at A and simply supported at B and C
- subjected to a uniform distributed load of 4 kN/m on the entire span
- EI is constant throughout the beam.

Determine

1. The bending moment at A and B
2. The reaction forces at the supports.

Solution

1. Geometry and Data
   Span AB: L_1 = 6 m
   Span BC: L_2 = 4 m
   Uniformly distributed load: w = 4 kN/m
   EI = constant
   Support A is fixed, B and C are simple supports.

2. Fixed-End Moments
   For a span with both ends simply supported, FEM = 0.
   For span AB (A fixed, B simple), use the three-moment method with M_C = 0.

3. Right-hand side terms
   The term on the right-hand side of the three-moment equation for a uniformly distributed load w over length L is:
   H = (w * L³) / 24
   For span AB: H_1 = (4 * 6³) / 24 = 36
   For span BC: H_2 = (4 * 4³) / 24 = 10.67

4. Apply Three-Moment Equation
   General form: M_A * L_1 + 2 * M_B * (L_1 + L_2) + M_C * L_2 = − 6 * (H_1 / L_1 + H_2 / L_2)
   Substitute:
   M_A * 6 + 2 * M_B * (6 + 4) + 0 * 4 = -6 * (36 / 6 + 10.67 / 4)
   6M_A + 20M_B = −6 * (6 + 2.667) = −6 * 8.667 = −52.00

5. Apply Boundary Condition at A

   Since A is fixed, the equation involving the imaginary span left of A:
   $2 * M\_A + M\_B = -6 * H_1 / L_1 = -6 * 36 / 6 = -36$

6. Solve the System of Equations

   From above:
   Equation (1): $6M\_A + 20M\_B = -52$
   Equation (2): $2M\_A + M\_B = -36$
   Solving:
   Multiply (2) by 6: $12M\_A + 6M\_B = -216$
   Subtract from (1): $(6M\_A + 20M\_B) - (12M\_A + 6M\_B) =$
   $-52 + 216$
   $-6M\_A + 14M\_B = 164$
   $\Rightarrow M\_B = (164 + 6M\_A) / 14$
   Substitute back to find M_A and M_B:
   $M\_A = -25.14$ kN·m, $M\_B = +14.29$ kN·m

7. Final Moments

   $M\_A = -25.14$ kN·m (hogging)
   M_B (on AB) $= +14.29$ kN·m (sagging)
   M_B (on BC) $= -14.29$ kN·m (hogging)
   $M\_C = 0$ (hinged support)

8. Reaction Forces at Supports

   Using equilibrium and the known moments, we compute vertical reactions at supports A, B, and C.

   Span AB:
   Length = 6 m, w = 4 kN/m, $M\_A = -25.14$ kN·m (hogging), M_B $= +14.29$ kN·m (sagging)

   Reaction at A (R_A):
   $R\_A = wL/2 - (M\_B - M\_A)/L = (4×6)/2 - (14.29 - (-25.14))/$ $6 = 12 - (39.43)/6 = 5.43$ kN

   Reaction at B from span AB (R_B_left):
   $R\_B\_left = wL - R\_A = 24 - 5.43 = 18.57$ kN

   Span BC:
   Length = 4 m, w = 4 kN/m, $M\_B = -14.29$ kN·m (hogging), $M\_C = 0$
   Reaction at B from span BC (R_B_right):
   $R\_B\_right = wL/2 + (M\_B - M\_C)/L = 8 + (-14.29)/4 = 8 - 3.57$ $= 4.43$ kN

Total Reaction at B:
R_B = R_B_left + R_B_right = 18.57 + 4.43 = 23.00 kN

Reaction at C:
R_C = total load – (R_A + R_B) = 40 – (5.43 + 23.00) = 11.57 kN

ChatGPT Integration

You can use ChatGPT to

- verify moment distribution calculations for different beam configurations and loading conditions
- analyze real-world applications, such as multi-span bridges, building frames, and machine supports
- explore alternative methods, such as matrix stiffness method or direct stiffness method for structural analysis

PROMPT 70

Solve a problem involving the influence lines for shear and moment in a continuous beam.

Objective

To understand influence lines for shear and moment in continuous beams and how they help determine critical locations for maximum shear force and bending moments under moving loads.

Activity

Influence lines represent the variation of reaction, shear force, or bending moment at a specific point in a structure due to a moving unit load. They are essential in analyzing

1. bridge and highway design, where loads are dynamic
2. structural elements subjected to moving loads, such as cranes and railways
3. determining maximum responses in indeterminate beams

This activity involves

1. constructing influence lines for shear and moment at a specific location in a continuous beam

2. determining the critical points where the structure experiences the highest internal forces

3. using influence lines to find the maximum shear and moment values under a given loading scenario

Numerical Example

A continuous beam ABC has

- Span AB = 6 m and Span BC = 6 m.
- fixed at A and simply supported at B and C
- A unit moving load travels from A to C.

Determine

1. the influence line for shear at B
2. the influence line for moment at B

Solution

1. Define the structural model.
   - The beam is continuous, meaning it is statically indeterminate.
   - Influence lines for indeterminate structures are obtained using Müller-Breslau's Principle, which states the following:
   - To find the influence line for shear, release the shear force at B by virtually inserting a roller at B and applying a unit displacement.
   - To find the influence line for moment, release the moment at B by virtually inserting a hinge at B and applying a unit rotation.
2. Construct the influence line for the shear at B.

To determine the influence line for shear at B, we

1. release shear at B by introducing a roller
2. apply a unit force at different points on the beam and compute shear reactions

   For a unit force moving from A to C, the influence values at B are

$$
V_B(x) = \begin{cases} \dfrac{x^2(6-x)}{54}, & 0 \le x \le 6, \\[3mm] -\dfrac{(x-6)^2(12-x)}{54}, & 6 \le x \le 12, \end{cases}
$$

- The shear influence line is positive in Span AB and negative in Span BC.
- The maximum shear occurs when the load is just before B.

3. Construct the influence line for the moment at B.

To determine the influence line for moment at B, we

1. introduce a hinge at B to allow rotation

2. apply a unit moment at B and compute the moments in spans AB and BC

The influence line equation for the moment at B is

$$M_B(x) = -\frac{V_B(x)L}{4} \begin{cases} -\dfrac{x^2(6-x)^2}{216}, & 0 \le x \le 6, \\[2ex] -\dfrac{(x-6)^2(12-x)^2}{216}, & 6 \le x \le 12, \end{cases}$$

- The moment influence line is parabolic.
- The maximum moment occurs when the load is at B.

ChatGPT Integration

You can use ChatGPT to

- verify influence line calculations for different beam spans and support conditions
- analyze real-world applications, such as bridges under dynamic truck loads
- explore alternative methods, such as finite element modeling for influence line validation

CHAPTER 8

VIRTUAL WORK AND APPLICATIONS

Prompt 71 Explain the principle of virtual work and its application in analyzing mechanical systems.

Prompt 72 Solve a problem involving the calculation of deflections in a beam using the virtual work method.

Prompt 73 Discuss the role of virtual work in determining reactions in statically indeterminate structures.

Prompt 74 Apply the principle of virtual work to calculate forces in a truss structure.

Prompt 75 Use virtual work to determine the displacement of a spring-mass system.

Prompt 76 Explain the use of virtual work in the design of robotic and machine elements.

Prompt 77 Discuss the advantages of virtual work compared to traditional force equilibrium methods.

Prompt 78 Explain the concept of complementary virtual work and its application in structural optimization.

Prompt 79 Solve a real-world problem using virtual work, such as deflections in a cantilever beam under combined loads.

Prompt 80 Discuss how virtual work is integrated into software tools for structural and mechanical analysis.

PROMPT 71

Explain the principle of virtual work and its application in analyzing mechanical systems.

Objective

To understand the principle of virtual work and explore how it is applied to analyze forces and displacements in mechanical systems, including both statically determinate and indeterminate systems.

Activity

The principle of virtual work states that if a system is in equilibrium, the total work done by all forces during a virtual (infinitesimally small..., consistent, and hypothetical) displacement is zero. This method is especially useful in determining unknown forces, moments, and displacements in mechanical systems. It offers a powerful alternative to traditional equilibrium equations, simplifying the analysis of complex systems such as trusses, beams, and machines.

In this task, we explore the derivation and fundamental ideas behind virtual work, including its application to structural and mechanical systems. Examples of its use include analyzing trusses for internal forces or calculating beam deflections under load.

Numerical Example

A simple truss consists of three members forming a right triangle. Joint A is pinned, and joint B is on a roller. Joint C is loaded with a vertical force of $F = 1000$ N.

- The length of $AB = 3$ m, and $BC = 4$ m.
- Use the principle of virtual work to determine the vertical deflection at C.

Solution

1. Real axial forces in the truss (due to $F = 1000$N at C).

   Using the method of joints (right triangle AB = 3 m, BC = 4 m, AC = 5 m):

   $$N_{AC} = 0, \quad N_{AB} = 0, \quad N_{BC} = 1000 \ N$$

2. Virtual axial forces (unit downward load at C).
   Apply $F^* = 1\,N$ downward at C:

   $$N^*_{AC} = 0, \; N^*_{AB} = 0, \qquad N^*_{BC} = 1\;N(\text{tension})$$

3. Vertical deflection at C by the unitload theorem

   $$\delta_C = \sum_i \frac{N_i N^*_i L_i}{AE} = \frac{1000 \times 1 \times 4}{AE} = \frac{4000}{AE}\;\text{m}$$

4. Numerical value

With $A = 200\,|\,\text{mm}^2 = 2.00 \times 10^{-4}\,|\,\text{m}^2$, $|\,E = 200\,|\,\text{GPa} = 2.00 \times 10^{11}\,|\,\text{N/m}^2$:

$$\delta_C = \frac{4000}{\left(2.00 \times 10^{-4}\right)\left(2.00 \times 10^{11}\right)} = 0.10\,|\,\text{mm (downward)}$$

ChatGPT Integration

* Use ChatGPT to verify the virtual work equations for internal and external forces.
* Request additional examples where virtual work simplifies the analysis of trusses or beams.
* Ask ChatGPT for a step-by-step solution to similar problems to strengthen understanding.

PROMPT 72

Solve a problem involving the calculation of deflections in a beam using the virtual work method.

Objective

To apply the virtual work method to calculate deflections in beams under various loading conditions and understand its advantages in structural analysis.

Activity

The virtual work method is particularly useful for determining deflections in beams and other structural elements. By applying an imaginary virtual force or moment, we can calculate the resulting displacements or rotations due to actual loads. This approach simplifies the calculation process and avoids solving complex differential equations.

In this task, we calculate the vertical deflection at the free end of a cantilever beam subjected to a uniformly distributed load (UDL) using the principle of virtual work.

Numerical Example

A cantilever beam of length L = 4 m is subjected to a uniformly distributed load w = 2 kN/m over its entire length. The beam has a rectangular cross-section with b = 50 mm and h = 200 mm. Determine the vertical deflection at the free end using the virtual work method. Assume modulus of elasticity, E = 210 GPa.

Solution

1. Calculate the moment of inertia.

   The moment of inertia I for a rectangular cross-section is

   $$I = \frac{bh^3}{12}$$

   By substituting b = 50 mm and h = 200 mm, we obtain

   $$I = \frac{50 \cdot (200)^3}{12} = 3.33 \times 10^{-5} \, m^4$$

2. Determine the virtual load application.

   Apply a unit vertical force $F_v = 1N$ at the free end to compute the virtual moment:

   $$M_{virtual}(x) = F_v \cdot (L - x)$$

3. Determine the actual moment due to UDL.

   The actual bending moment $M_{actual}(x)$ at a section x from the fixed end is

   $$M_{actual}(x) = \frac{w}{2} \cdot (L - x)^2$$

   By substituting w = 2 kN/m and L = 4 m, we obtain

   $$M_{actual}(x) = \frac{2}{2}(4 - x)^2 = (4 - x)^2$$

4. Find the strain energy using virtual work.

Using the principle of virtual work, we obtain

$$\delta_v = \int_0^L \frac{M_{\text{virtual}}(x) \cdot M_{\text{actual}}(x)}{EI} dx = \int_0^L \frac{(L-x)^3}{EI} dx$$

By substituting, we obtain

$$\delta_v = \int_0^4 \frac{(4-x)^3}{210 \times 10^9 \times 3.33 \times 10^{-5}} dx =$$

Solve the integrals, and simplify:

$$\delta_v = 9.14 \text{ mm}$$

ChatGPT Integration

▪ Use ChatGPT to verify the virtual work calculations for the beam deflection.

▪ Explore alternative methods to compute deflections, such as moment-area theorems.

▪ Ask ChatGPT to solve similar problems involving different beam geometries or loadings.

PROMPT 73

Discuss the role of virtual work in determining reactions in statically indeterminate structures.

Objective

To explore how the principle of virtual work can be applied to solve for unknown reactions in statically indeterminate structures and compare it to other methods of analysis.

Activity

Statically indeterminate structures have more unknown reactions than the available equilibrium equations. The principle of virtual work provides an efficient way to analyze such systems by introducing compatibility conditions that relate internal deformations to external reactions. This method is particularly useful in mechanical and structural systems, where displacements or rotations need to be determined alongside reactions.

In this task, we analyze a statically indeterminate beam subjected to combined loads, using virtual work to solve for the unknown reactions and deflections.

Numerical Example

A propped cantilever beam of length $L = 6$ m is fixed at A and supported by a roller at B (end of the beam). The beam is subjected to a uniformly distributed load $w = 4$ kN/m over its entire length. Using the principle of virtual work, determine the reaction at the roller support R_B.

Solution

1. Apply a virtual load at B.

   Introduce a unit virtual force $F_v = 1$ N in the vertical direction at B. This force induces a virtual moment along the beam.

2. Write the internal virtual work equation.

   The internal virtual work is

   $$U_{int} = \int_0^L \frac{M_{actual}(x) \cdot M_{virtual}(x)}{EI}$$

3. Calculate the actual bending moment.

   The actual bending moment $M_{actual}(x)$ is:

   $$M_{actual}(x) = R_B \cdot x - \frac{w}{2} \cdot x^2$$

4. Calculate the virtual bending moment.

   The virtual bending moment $M_{virtual}(x)$ is

   $$M_{virtual}(x) = x$$

5. Substitute the values into the virtual work equation.

   We substitute $M_{actual}(x)$ and $M_{virtual}(x)$ into the equation:

   $$\int_0^L \frac{\left(R_B \cdot x - \frac{w}{2} \cdot x^2\right) \cdot x}{EI} dx = 0$$

6. Solve for R_B.

Solve the integral and calculate R_B:

$$\frac{R_B \cdot L^3}{3EI} - \frac{w \cdot L^4}{8EI} = 0$$

$$R_B = \frac{3w \cdot L}{8} = \frac{3 \times 4 \times 6}{8} = 9kN$$

ChatGPT Integration

- Use ChatGPT to verify the virtual work calculations and integrals for accuracy.

- Explore alternative methods for analyzing statically indeterminate structures, such as the force method or displacement method.

- Request examples of mechanical systems that rely on virtual work for design optimization.

PROMPT 74

Apply the principle of virtual work to calculate forces in a truss structure.

Objective

To understand how the principle of virtual work can be used to calculate internal forces in truss members efficiently, particularly in statically determinate trusses.

Activity

The principle of virtual work is an effective tool for analyzing forces in truss members. By applying a virtual force at the point of interest and ensuring the compatibility of displacements, we can calculate internal forces without solving a large system of equilibrium equations. This method is particularly useful for complex trusses where direct analysis can be cumbersome.

In this task, we analyze a simple truss using virtual work to determine the axial force in one of its members.

Numerical Example

A right-triangle truss ABC is supported at A (pin) and B (roller).

- Geometry
  - $AB = 4$ m, $AC = 3$ m, $BC = 5$ m
  - All members have the same $A = 200$ mm² and $E = 200$ GPa.
- Loading
  - A vertical load $P = 600$ N acts *downward* at node C.

Determine the axial force in member AC using the principle of virtual work.

Solution

Step 1. Support reactions (static equilibrium)

$$\sum M_A = B_y (4\,\text{m}) - 600(0) = 0 \mid \Rightarrow \mid B_y = 0$$

$$\sum F_y = A_y + B_y - 600 = 0 \mid \Rightarrow \mid A_y = 600\,\text{N}$$

$$\sum F_x = A_x = 0$$

Step 2. Real member forces (method of joints)

Joint C

Let N_{AC}, N_{BC} be tensile forces in members AC, BC (positive in tension). Horizontal equilibrium:

$$\frac{4}{5} N_{BC} = 0 \mid \Rightarrow \mid N_{BC} = 0$$

Vertical equilibrium:

$$-N_{AC} - 600 + \frac{-3}{5} N_{BC} = 0$$

So member AC is in compression with magnitude 600 N.

Joint A

$$A_x + N_{AB} = 0 \mid \Rightarrow \mid N_{AB} = 0$$

Step 3. Virtual system (unit axial load in member AC)

To obtain the virtual internal forces N_i^v:

- Remove the 600 N external load.
- Apply a pair of equal and opposite unit forces +1N along the line of member AC (tension).
  - At joint C the unit force acts downward along AC.
  - At joint A it acts upward along AC.

Solving the (very similar) joint equations gives:

$$N_{AC}^v = -1\,\text{N}, \qquad N_{AB}^v = 0, \qquad N_{BC}^v = 0$$

Step 4. Principle of virtual work check

For a determinate truss, the external virtual work W_{ext}^v plus the internal virtual work W_{int}^v must vanish:

$$W_{ext}^v + W_{int}^v = 0.$$

$$W_{int}^v = \sum_{i=1}^{n} \frac{N_i\, N_i^v\, L_i}{A\,E}$$

Here only member AC contributes:

$$W_{int}^v = \frac{(-600)(-1)(3)}{(2\times10^{-4})(200\times10^9)} = 4.5\times10^{-5}\ \text{m}$$

The external virtual work is the unit virtual load at C multiplied by the (unknown) real shortening of AC; its magnitude is $-\delta_{AC}$. The equality

$$-\delta_{AC} + 4.5\times10^{-5} \mid = \mid 0 \mid \delta_{AC} \mid = \mid 4.5\times10^{-5}\ \text{m}$$

ChatGPT Integration

- Verify equilibrium equations and virtual work steps using ChatGPT.
- Ask for similar examples involving truss structures.
- Explore how virtual work simplifies force and deflection analysis in trusses.

PROMPT 75

Use virtual work to determine the displacement of a spring-mass system.

Objective

To apply the principle of virtual work to analyze the displacement of a spring-mass system under applied forces, illustrating its relevance in mechanical systems.

Activity

The principle of virtual work is a powerful tool for solving problems in mechanics, especially for systems involving elastic components like springs. By applying a virtual displacement and ensuring compatibility between the internal and external work, we can calculate the displacement of a system under load efficiently.

In this task, we calculate the displacement of a mass connected to a series of springs subjected to a vertical force using the virtual work principle.

Numerical Example

A mass $m = 50$ kg is connected to two linear springs in parallel.

- Spring 1 has a stiffness $k_1 = 1000$ N/mk, and Spring 2 has a stiffness $k_2 = 1500$ N/m.
- The system is subjected to a vertical force $F = 2000$ N.

Determine the displacement of the mass using the virtual work method.

Solution

1. Assume the virtual displacement.

   Assume a small virtual displacement δ in the direction of the applied force.

2. Determine the external virtual work.

   The external virtual work done by the applied force is

   $$W_{ext} = F.\delta$$

3. Determine the internal virtual work.

   The internal virtual work is the energy stored in the springs:

   $$W_{int} = \frac{1}{2} \cdot k_1 \cdot \delta^2 + \frac{1}{2} \cdot k_2 \cdot \delta^2$$

4. Use the virtual work equation.

Equating the external and internal virtual work gives us

$$F \cdot \delta = \frac{1}{2} \cdot (k_1 + k_2) \cdot \delta^2$$

5. Solve for δ.

Rearrange to find δ:

$$\delta = \frac{2F}{k_1 + k_2} = \frac{2 \times 2000}{1000 + 1500} = 1.6 \, \text{m}$$

ChatGPT Integration

▪ Use ChatGPT to validate the calculations and explore alternative methods to find displacements in spring-mass systems.

▪ Request similar examples with different configurations, such as springs in series or a non-linear spring.

▪ Ask ChatGPT to simulate the behavior of the system under varying loads.

PROMPT 76

Explain the use of virtual work in the design of robotic and machine elements.

Objective

To understand how the principle of virtual work is applied in the design and analysis of robotic systems and machine elements, enabling the calculation of displacements, forces, and moments in complex mechanical systems.

Activity

Virtual work plays a vital role in the design and optimization of robotic and machine elements. In robotics, it helps analyze linkages and actuators to ensure precise motion and minimal energy consumption. For machine components, virtual work aids in calculating loads and deflections to verify that the elements perform as intended under operational forces. This method is particularly useful in the analysis of parallel manipulators, robotic arms, and compliant mechanisms, where multiple forces interact to achieve desired motions.

This task explores the application of virtual work in calculating the actuator force required to lift a robotic arm under load.

Numerical Example

A robotic arm is modeled as a simple 2D linkage consisting of two rigid bars AB and BC, each of length L = 0.5 m.

- Joint A is fixed.
- A vertical force P = 100 N acts at point C.
- The arm is held in equilibrium by an actuator providing a horizontal force F_x at joint B.
- The angle between AB and the horizontal is θ = 45°.

Determine the required actuator force F_x using the principle of virtual work.

Solution

1. Use the virtual displacement.

   Apply a small virtual displacement δ_x at joint B in the horizontal direction due to F_x. The corresponding virtual displacement δ_y at C in the vertical direction is related by the geometry of the linkage. The relation between virtual displacements is as follows:

$$\delta_y = \delta_x \cdot \tan(\theta)$$

2. Determine the external virtual work.

   The external virtual work is

$$W_{ext} = P \cdot \left(\delta_x \cdot \tan(\theta)\right)$$

3. Determine the internal virtual work.

   The internal virtual work done by the actuator force F_x is:

$$W_{int} = F_x \cdot \delta_x$$

4. Use the virtual work equation.

   Equating internal and external virtual work, we obtain

$$F_x \cdot \delta_x = P \cdot \left(\delta_x \cdot \tan(\theta)\right)$$

Simplify to solve for Fx:

$$F_x = P \cdot \tan(\theta) = 100 \tan\left(45°\right) = 100N$$

ChatGPT Integration

- Use ChatGPT to explore how virtual work applies to multi-link robotic arms and compliant mechanisms.
- Ask ChatGPT for examples of robotic designs optimized using energy and work principles.
- Verify the calculations and extend the analysis to dynamic conditions with accelerations.

PROMPT 77

Discuss the advantages of virtual work compared to traditional force equilibrium methods.

Objective

To understand the unique benefits of the principle of virtual work in mechanical system analysis and compare it with traditional equilibrium-based approaches.

Activity

Virtual work provides a versatile and elegant framework for solving mechanical problems. Unlike traditional force equilibrium methods, which rely solely on balancing forces and moments, virtual work incorporates the relationship between forces, displacements, and energy. This enables the analysis of complex systems such as statically indeterminate structures, mechanisms, and elastic deformations. Virtual work is particularly advantageous in cases where displacements or deflections are the primary quantities of interest, as it eliminates the need for lengthy differential equations or iterative solutions.

Examples of its applications include analyzing robotic linkages, calculating deflections in beams, and optimizing mechanical systems for efficiency.

Comparison of Virtual Work and Equilibrium Methods

1. Virtual work:
   - relates forces and displacements directly
   - efficient for calculating deflections and elastic deformations
   - handles statically indeterminate systems naturally via compatibility conditions
   - simplifies the analysis of systems with distributed loads or complex geometries

2. Force equilibrium:
   - focuses on balancing forces and moments
   - best suited for determining reaction forces in statically determinate systems
   - requires additional equations for deflections or compatibility in indeterminate systems
   - can be cumbersome for systems with multiple interconnected components

Numerical Example

Compare the use of virtual work and traditional equilibrium methods in analyzing the deflection of a simply supported beam of length L = 4 m subjected to a concentrated load P = 500 N at midspan. The beam has a rectangular cross-section with b = 100 mm, h = 200 mm, and E = 200 GPa.

Solution

1. Use virtual work:
   - Apply a virtual unit force F_v = 1 N at the location of interest.
   - The bending moment due to the actual load is

$$M_{actual}(x) = \begin{cases} P.x/2 & for\, 0 \leq x \leq L/2 \\ \dfrac{P.(L-x)}{2} & for\, L/2 \leq x \leq L \end{cases}$$

   - The bending moment due to the virtual load is

$$M_{virtual}(x) = \begin{cases} x/2 & for\, 0 \leq x \leq L/2 \\ \dfrac{(L-x)}{2} & for\, L/2 \leq x \leq L \end{cases}$$

- The strain energy using virtual work is

$$\delta = \int_0^{L/2} \frac{M_{\text{actual}}(x) \cdot M_{\text{virtual}}(x)}{EI} dx + \int_{L/2}^{L} \frac{M_{\text{actual}}(x) \cdot M_{\text{virtual}}(x)}{EI} dx$$

$$\delta = \frac{(PL^3)}{(48EI)}$$

2. Use the force equilibrium:
   - Solve for reactions R_A and R_B using
   $$\sum F_y = 0, \quad \sum M_A = 0$$

   - Use the moment-curvature relation to derive the deflection equation:

   $$EI \frac{d^2y}{dx^2} = M(x)$$

Integrate twice and solve for constants using boundary conditions.

ChatGPT Integration

- Use ChatGPT to compare results from both methods for accuracy and efficiency.
- Request alternative examples where virtual work outperforms equilibrium-based methods.
- Explore dynamic systems to see how virtual work extends beyond static cases.

PROMPT 78

Explain the concept of complementary virtual work and its application in structural optimization.

Objective

To understand the concept of complementary virtual work and explore how it is applied in structural optimization to improve performance and minimize material use in mechanical systems.

Activity

The principle of complementary virtual work expands on the standard virtual work principle by focusing on stress-strain relationships. It considers both internal stresses and external forces, ensuring compatibility between deformations and applied loads. This principle is widely used in structural optimization, allowing engineers to design lightweight yet strong structures by distributing material efficiently.

Applications include the optimization of truss structures, beams, and machine components where load paths and stress distributions are critical to minimizing material use while maintaining safety and performance.

Numerical Example

A truss structure consists of two members, AB and BC, forming a right triangle.

- $AB = 3$ m, $BC = 4$ m, $AC = 5$ m.
- Joint C is pinned, and joint B is a roller support.
- The truss is subjected to a vertical load $P = 2000$ N at joint A.

Using the complementary virtual work principle, determine how the stresses in the truss members can be redistributed to minimize deflection.

Solution

1. Define complementary virtual work.

   Complementary virtual work relates the virtual internal stresses to virtual strains:

   $$W_{comp} = \int_V \sigma \cdot \delta\epsilon \, dV$$

2. Identify the virtual strain-displacement relationship.

   For a truss member under axial load, we have

   $$\epsilon = \frac{\delta L}{L}$$

3. Find the complementary work in a truss.

   The complementary virtual work for a truss member becomes

   $$W_{comp} = \sum \frac{\sigma \cdot \delta L \cdot A}{L}$$

where
- σ: stress in the member
- δL: virtual elongation
- A: cross-sectional area
- L: member length

4. Calculate the stresses in members.

   Using equilibrium, calculate the forces in AB and BC. For example,

   $$F_{AB} = \frac{P \cdot BC}{AC} = \frac{2000 \times 4}{5} = 1600 \text{ N}$$

5. Optimize the stress distribution.

   Adjust member areas A_{AB} and A_{BC} to achieve uniform stress and minimize deflection. The optimal area distribution satisfies the following:

   $$\frac{\sigma_{AB}}{\sigma_{BC}} = \frac{A_{BC}}{A_{AB}}$$

ChatGPT Integration

- Use ChatGPT to validate stress-strain relationships and explore alternative optimization techniques.
- Request examples of complementary virtual work applied in real-world scenarios.
- Simulate stress redistribution for complex truss systems using complementary virtual work.

PROMPT 79

Solve a real-world problem using virtual work, such as deflections in a cantilever beam under combined loads.

Objective

To apply the principle of virtual work in solving practical engineering problems, such as calculating deflections in a cantilever beam subjected to multiple loads.

Activity

The principle of virtual work is widely used in structural and mechanical engineering to predict deformations under various loading scenarios. This method simplifies the calculation of deflections and rotations in beams, particularly when multiple loads are present. This task demonstrates how to use virtual work to calculate the deflection of a cantilever beam under combined point and distributed loads.

Numerical Example

A cantilever beam of length $L = 6$ m is subjected to

1. a point load P = 400 N at the free end
2. a uniformly distributed load $w = 2$ kN/m over the entire length

The beam has a rectangular cross-section with width $b = 100$ mm, height $h = 200$ mm, and $E = 210$ GPa. Determine the vertical deflection at the free end.

Solution

1. Calculate the moment of inertia.

   The moment of inertia III for a rectangular cross-section is

   $$I = \frac{bh^3}{12} = \frac{100(200)^3}{12} = 6.67 \times 10^{-6} \text{ m}^4$$

2. Solve the virtual work equation.

   The vertical deflection at the free end is given by

   $$\delta = \int_0^L \frac{M_{actual}(x) \cdot M_{virtual}(x)}{EI} dx$$

3. Find the bending moments.

   - Moment due to the point load:

     $$M_{actual,P}(x) = P \cdot x$$

   - Moment due to the UDL:

     $$M_{actual,w}(x) = \frac{w}{2} \cdot x^2$$

   - Virtual moment:

$$M_{\text{virtual}}(x) = x$$

4. Combine the moments.

The total actual moment is

$$M_{\text{actual}}(x) = M_{\text{actual},P}(x) + M_{\text{actual},w}(x)$$

5. Solve for the deflection.

Substitute into the virtual work equation and integrate:

$$\delta = \int_0^L \frac{\left(P \cdot x + \dfrac{w}{2} \cdot x^2\right) \cdot x}{EI}\, dx$$

ChatGPT Integration

- Use ChatGPT to validate the deflection calculations for the cantilever beam under combined loads.
- Request alternative approaches to solving the problem, such as using moment-area theorems or numerical methods.
- Ask ChatGPT for detailed explanations of how virtual work simplifies deflection analysis compared to traditional methods.

PROMPT 80

Discuss how virtual work is integrated into software tools for structural and mechanical analysis.

Objective

To explore how modern software tools use the principle of virtual work to perform structural and mechanical analyses efficiently.

Activity

Virtual work is embedded into many structural analysis and finite element method (FEM) software tools. FEM is a computational technique that divides a structure or system into smaller elements, solving for unknowns (such as displacements or stresses) by applying energy principles, including virtual work. This method is particularly powerful for analyzing complex systems with irregular geometries, material properties, or boundary conditions.

By automating the application of virtual displacements and computing work-energy relationships, FEM-based tools such as COMSOL, ANSYS, Abaqus, and SAP2000 enable engineers to perform highly detailed analyses. These tools are widely used for both linear and nonlinear problems in fields ranging from aerospace to robotics.

ChatGPT Integration

- Use ChatGPT to explain how virtual work principles are applied in finite element analysis (FEA) for solving deflection and stress problems.

- Explore how ChatGPT can simulate virtual work concepts for trusses, beams, and machine components.

- Discuss how FEM software automates the application of virtual work in both static and dynamic analyses.

- Request examples where FEM software uses virtual work for structural optimization.

Connecting Statics and Strength of Materials in Design

Statics forms the foundation of engineering analysis by determining the forces and moments necessary to maintain equilibrium in mechanical and structural systems. These principles provide essential insights into internal forces, such as shear, axial, and bending forces, which are critical for structural and mechanical design. However, statics alone does not address how materials respond to these forces or assess their ability to safely withstand operational loads. This is where the strength of materials becomes indispensable. By evaluating stresses, strains, and deformations in materials under various loading conditions, it extends the scope of analysis to encompass practical material behavior. Important topics include torsional stress and strain, bending and shear stress in beams, combined stresses, stress and strain transformation, and the stability of compressed elements, such as column buckling. The strength of materials ensures that engineering designs meet rigorous safety and performance criteria by incorporating material properties like yield strength, Young's modulus, and ultimate tensile strength. Together, statics and the strength of materials empower engineers to design systems that are not only in equilibrium but also structurally sound, reliable, and resource-efficient, bridging the gap between theoretical analysis and real-world application.

In the remaining chapters of **Part 1**, we do not aim to cover the entire scope of the strength of materials, as that would require a separate, dedicated exploration. Instead, we will focus on linking the principles of statics and Strength of Materials from a technical perspective, providing a foundational understanding of how these disciplines interact. This connection ensures that the transition to the strength of materials is both logical and grounded in the concepts already established in statics.

CHAPTER 9

INTERNAL FORCES

Prompt 81 Define internal forces in the context of static equilibrium and their role in analyzing mechanical structures.

Prompt 82 Derive the internal force equations for beams under common loads, including point loads and uniformly distributed loads.

Prompt 83 Discuss the relationship between shear force, bending moment, and load intensity, and derive the governing differential equations.

Prompt 84 Explain the conditions for internal force equilibrium in beams and their application in section analysis.

Prompt 85 Solve a problem involving internal forces in a cantilever beam with a varying distributed load.

Prompt 86 Explain the concept of internal force equilibrium at a section cut and its application in truss and frame analysis.

Prompt 87 Discuss the significance of the maximum bending moment and its location in beam design.

Prompt 88 Analyze internal forces in a simple frame structure using equilibrium and section cuts.

Prompt 89 Explain the role of internal forces in ensuring the stability and safety of mechanical systems.

Prompt 90 Solve a real-world problem involving internal force analysis in a multi-span beam.

PROMPT 81

Define internal forces in the context of static equilibrium and their role in analyzing mechanical structures.

Objective

To understand the concept of internal forces in mechanical structures and their significance in maintaining static equilibrium.

Activity

Internal forces are the forces that develop within a structural or mechanical element in response to external loads, moments, or other forces. These forces arise to maintain the structure in equilibrium according to Newton's Third Law of Motion. Internal forces include axial forces, shear forces, and bending moments, each playing a critical role in ensuring the stability and functionality of mechanical systems.

In this prompt, you will define internal forces, explain their classifications, and explore their role in mechanical structures such as beams, trusses, and frames. For instance, shear forces ensure that sections of a beam can resist sliding past each other, while bending moments provide resistance against curvature under loads. Consider examples of real-world structures like bridges or cranes, where internal forces are critical for design.

Numerical Example

A simply supported beam of length $L = 6$ m is subjected to a point load $P = 500$ N at its midpoint. Determine the internal shear force and bending moment at a section located $x = 2$ m from the left support.

Solution

1. Determine the reactions at the supports.

   Using equilibrium equations, we obtain

   $$R_A + R_B = P$$

   $$R_A \times 6 = P \times 3$$

   $$R_A = R_B = \frac{P}{2} = 250\,\text{N}$$

2. Use the shear force at $x = 2$ m.

To the left of the section, only R_A acts as follows:

$$V = R_A = 250 \text{ N}$$

3. Use the bending moment at $x = 2$ m.

The bending moment is

$$M = R_A \cdot x = 250 \times 2 = 500 \text{ N} \cdot \text{m}$$

ChatGPT Integration

- Ask ChatGPT to explain internal force classifications with additional real-world examples.
- Use ChatGPT to solve similar problems for trusses or frames.
- Explore the role of internal forces in different loading conditions, such as moving loads or varying distributed loads.

PROMPT 82

Derive the internal force equations for beams under common loads, including point loads and uniformly distributed loads.

Objective

To derive equations for internal forces in beams subjected to typical loading scenarios, focusing on shear forces and bending moments.

Activity

Internal force equations are fundamental for analyzing how beams respond to applied loads. These equations describe shear forces (V) and bending moments (M) at any point along the beam, based on the loading and support conditions. In this task, we derive these equations for two common scenarios: a simply supported beam under a point load and a beam with a uniformly distributed load (UDL). The derivation illustrates how equilibrium principles are used to determine internal force distributions.

Numerical Example

Derive the shear force and bending moment equations for a simply supported beam of length $L = 6$ m, subjected to

1. a point load $P = 1000$ N at midspan
2. a uniformly distributed load $w = 500$ N/m across the entire length

Solution

1. Determine the reactions at the supports for both cases.

   For equilibrium:
   - $R_A + R_B = P$ (vertical force equilibrium)
   - $R_A \cdot L = P \cdot L/2$ (moment about one support)|

   For the point load:

   $$R_A = R_B = \frac{P}{2} = 500\,\text{N}$$

   For the UDL:

   $$R_A = R_B = \frac{wL}{2} = 1500\,\text{N}$$

2. Use the shear force equations.

   For the point load:
   - Left of the load ($0 \leq x \leq L/2$):

   $$V = R_A = 500\,\text{N}$$

   - Right of the *load* ($L/2 \leq x \leq L$):

   $$V = R_A - P = 500 - 1000 = -500\,\text{N}$$

   - For the UDL: ($0 \leq x \leq L$)

   $$V = R_A - w \cdot x = 1500 - 500x$$

3. Use the bending moment equations.

   For the point load:
   - Left of the load ($0 \leq x \leq L/2$):

   $$M = R_A \cdot x = 500 \cdot x$$

- Right of the load $(L/2 \le x \le L)$:

$$M = R_A \cdot x - P \cdot \left(x - \frac{L}{2} \right)$$

- For the UDL: $(0 \le x \le L)$

$$M = R_A \cdot x - \frac{w \cdot x^2}{2}$$

ChatGPT Integration

- Use ChatGPT to validate the derivations and solve variations of this problem with different load configurations.
- Explore interactive tools to visualize shear force and bending moment diagrams for both loading scenarios.
- Request further examples, such as beams with overhangs or varying cross-sections.

PROMPT 83

Discuss the relationship between shear force, bending moment, and load intensity, and derive the governing differential equations.

Objective

To explore the mathematical relationship between shear force, bending moment, and load intensity, and to derive the governing differential equations that describe these interactions in beams.

Activity

In statics, the shear force (V) and bending moment (M) are interrelated through the load intensity $(w(x)$, a function of $x)$. These relationships are expressed through first-order and second-order differential equations. Understanding these relationships helps engineers predict the internal force distributions within a beam, which is critical for safe and efficient structural design. This prompt focuses on deriving these equations using fundamental principles of equilibrium and provides examples of their application.

Numerical Example

A simply supported beam of length $L = 6$ m is subjected to a uniformly distributed load $w = 500$ N/m. Derive the equations relating load intensity, shear force, and bending moment for this beam, and calculate the shear force and bending moment at $x = 2$ m.

Solution

1. Determine the differential relationship between $w(x)$, $V(x)$, and $M(x)$.

   • The load intensity $w(x)$ is related to the shear force $V(x)$ by

   $$\frac{dV}{dx} = -w(x)$$

   • The shear force $V(x)$ is related to the bending moment $M(x)$ by

   $$\frac{dM}{dx} = V(x)$$

2. Apply the boundary conditions for the beam.

   For a simply supported beam with a UDL

   • Reactions at the supports:

   $$R_A = R_B = \frac{wL}{2} = 1500\,\text{N}$$

3. Solve the shear force equation.

   By integrating $\dfrac{dV}{dx} = -w(x)$ and substituting $R_A = 1500$ N and $w = 500$ N/m, we obtain

   $$V(x) = R_A - w \cdot x = 1500 - 500 \cdot x$$

4. Solve the bending moment equation.

   By integrating $\dfrac{dM}{dx} = V(x)$, we obtain

   $$M(x) = \int V(x)\,dx = \int (1500 - 500 \cdot x)\,dx = 1500 \cdot x - \frac{500 \cdot x^2}{2} + C$$

   We apply the boundary condition $M(0) = 0$ to solve for C, and use $C = 0$:

   $$M(x) = 1500 \cdot x - 250 \cdot x^2$$

5. Calculate shear force and bending moment at $x = 2$ m.

- Shear force:

$$V(2) = 1500 - 500 \cdot 2 = 500 \text{ N}$$

- Bending moment:

$$M(2) = 1500 \cdot 2 - 250 \cdot (2)^2 = 2000 \text{ N} \cdot \text{m}$$

ChatGPT Integration

- Use ChatGPT to derive differential equations for beams with varying load intensities or supports.
- Request visualizations for shear force and bending moment diagrams generated using the derived equations.
- Explore examples with combined loading conditions to see how the relationships evolve.

PROMPT 84

Explain the conditions for internal force equilibrium in beams and their application in section analysis.

Objective

To understand the equilibrium conditions for internal forces in beams and explore how they are applied in section analysis to determine shear force, bending moment, and axial force distributions.

Activity

Section analysis is a fundamental approach in statics used to determine internal forces within beams. By cutting a beam at a specified location and applying the conditions for static equilibrium ($\Sigma F = 0$, $\Sigma M = 0$), one can calculate the shear force, bending moment, and axial force at the section. This method is critical for analyzing beams subjected to multiple loading types, ensuring accurate determination of internal force distributions.

In this task, you will derive the equilibrium equations for a cut section, solve a numerical example, and discuss the importance of section analysis in structural design.

Numerical Example

A simply supported beam of length $L = 10$ m is subjected to

1. a point load $P = 2000$ N at $x = 3$ m from the left support
2. a uniformly distributed load $w = 400$ N/m across the entire length

Determine the internal shear force and bending moment at $x = 6$ m using section analysis.

Solution

1. Find the reactions at the supports.

   Using the equilibrium equations, we find the following:

   - Sum of the vertical forces:

   $$R_A + R_B = P + w \cdot L = 2000 + 400 \times 10 = 6000\,N$$

   - Sum of the moments about A:

   $$R_B \cdot 10 = P \cdot 3 + w \cdot 10 \cdot \frac{10}{2}$$

   By substituting values, we obtain

   $$R_B \times 10 = 2000 \times 3 + 400 \times 10 \times 5 = 6000 + 20000 = 26000$$

   Sove for R_B:

   $$R_B = \frac{26000}{10} = 2600\,N$$

   Calculate R_A:

   $$R_A = 6000 - 2600 = 3400\,N$$

2. Perform the section analysis at $x = 6$ m.

   Cut the beam at $x = 6$ m and consider the left section.

   Vertical force equilibrium ($\Sigma F_y = 0$):

   $$V = R_A - w \cdot x - P$$

   By substituting, we obtain

   $$V = 3400 - 400 \cdot 6 - 2000 = -1000\,N$$

   Moment equilibrium ($\Sigma M = 0$):

   $$M = R_A \cdot x - w \cdot \frac{x^2}{2} - P(x - 3)$$

By substituting, we obtain

$$M = 3400 \times 6 - 400\frac{6^2}{2} - 2000(6-3) = 20400 - 7200 = 13200\,\text{N}\cdot\text{m}$$

ChatGPT Integration

- Use ChatGPT to explore section analysis for beams under different support and loading conditions.
- Request step-by-step explanations for solving problems with combined point and distributed loads.
- Ask for additional examples involving frames or trusses to expand understanding.

PROMPT 85

Solve a problem involving internal forces in a cantilever beam with a varying distributed load.

Objective

To calculate the internal forces (shear force and bending moment) at specified points along a cantilever beam subjected to a linearly varying distributed load.

Activity

A cantilever beam with a varying distributed load presents an essential challenge in structural analysis. The load's intensity changes linearly along the beam, making the calculations for shear force and bending moment non-trivial. By using equilibrium equations and integration techniques, you will analyze the internal forces and moments. This task highlights the importance of understanding varying loads in practical applications like cantilevered balconies or overhanging beams in machinery.

Numerical Example

A cantilever beam of length $L = 6$ m is fixed at one end (point A) and free at the other. It is subjected to a linearly varying distributed load, starting from $w_0 = 0$ N/m at the free end to $w_L = 600$ N/m at the fixed end. Determine

1. the reactions at the fixed end (R_A and M_A)
2. the shear force and bending moment equations
3. the shear force and bending moment at $x = 4$ m. x is measured from the fixed end.

Solution

Step 1: Determine the reactions at the fixed end.

1. Identify the resultant force of the distributed load.
    The total force is the area of the triangular load:

    $$F_{total} = \frac{1}{2} \cdot w_L \cdot L$$

    By substituting, we obtain

    $$F_{total} = \frac{1}{2} \times 600 \times 6 = 1800 \text{ N}$$

2. Identify the location of the resultant force.
    For a triangular load, the centroid is located at $2L/3$ from the smaller end (i.e., free end):

    $$x_{centroid} = \frac{2}{3} \times 6 = 4 \text{ m}$$

3. Solve for the reactions at the fixed end.
    - Vertical reaction (R_A):
    $$R_A = F_{total} = 1800 \text{ N}$$

    - Moment reaction (M_A):
    $$M_A = F_{total} \cdot (L - x_{centroid}) = 1800 \cdot 2 = 3600 \text{ N} \cdot \text{m}$$

Step 2: Use the shear force and bending moment equations.

1. Load intensity at any x.
    The load intensity varies linearly:

    $$w(x) = w\left(\frac{1-x}{L}\right) = 600 - 100x$$

2. Solve the shear force equation ($V(x)$).
    The shear force is obtained by integrating the load intensity:

    $$V(x) = R_A - \int_0^x w(x')dx'$$

Substituting, $w(x') = 600 - 100x'$ and perform the integration:

$$V(x) = 1800 + 50x^2 - 600x$$

3. Solve the bending moment equation $(M(x))$.

The bending moment is obtained by integrating the shear force:

$$M(x) = M_A - \int_0^x V(x')dx'$$

Substituting, $V(x')$ and perform the integration:

$$M = \frac{50}{3}x^3 - 300x^2 + 1800x - 3600$$

Step 3: Solve for the internal forces at $x = 4$ m.

1. For the shear force at x = 4,
$$V(4) = 200\,\text{N}$$

2. For the bending moment at x = 4,
$$M(4) = -133.33\ \text{N.m}$$

ChatGPT Integration

- Use ChatGPT to derive similar shear and moment equations for beams with different varying load intensities.
- Request step-by-step solutions for complex cantilever problems.
- Ask for comparisons of the effects of uniform and linearly varying loads on internal forces.

PROMPT 86

Explain the concept of internal force equilibrium at a section cut and its application in truss and frame analysis.

Objective

To understand the principle of internal force equilibrium at a section cut and how it is applied to analyze internal forces in truss and frame structures.

Activity

In statics, analyzing internal forces involves "cutting" a structure at a specific section and applying equilibrium equations ($\Sigma F = 0$, $\Sigma M = 0$) to solve for unknown forces. This technique is essential for determining axial forces, shear forces, and bending moments in trusses and frames. For trusses, it simplifies the analysis by isolating a few members, while for frames, it identifies bending and axial forces at specific sections.

In this prompt, you will explore the equilibrium conditions at a section cut, derive equations for internal forces, and solve a numerical example involving a truss and a frame.

Numerical Example

1. Analyze the axial forces in a simple truss using the method of sections. The truss consists of three members forming a triangle: a horizontal base of length $L = 6$ m, a vertical height of $h = 3$ m, and a diagonal member connecting the top of the vertical member to the far end of the base. A point load $P = 3000$ N is applied at the top joint where the vertical and diagonal members meet, with the base joints being pinned at one end and on rollers at the other.

2. Determine the bending moment and axial forces in a rectangular frame of height $h = 3$ m and width $w = 6$ m, fixed at the bottom corners, and subjected to a horizontal load $H = 500$ N applied at the top corner.

Solution

Truss Analysis (Method of Sections)

1. Determine the equilibrium at the supports.
   - The supports are pinned and roller, giving vertical reactions R_A and R_B.
   - Using the equilibrium,

$$R_A + R_B = P$$

   By substituting in $P = 3000$ N, we obtain

$$R_A = 1500, \ R_B = 1500$$

2. Determine the section cut and force equilibrium.
   - Cut through members AB, BC, and AC, isolating the left section.

- Apply equilibrium equations:
  - $\Sigma F_x = 0$: Forces along the horizontal direction.
  - $\Sigma F_y = 0$: Forces along the vertical direction.
  - $\Sigma M = 0$: Moments about the section cut.

Solving for member forces, we obtain
  - axial force in AB = $1732\,$N (tension)
  - axial force in AC = $2000\,$N (compression)
  - axial force in BC = $1000\,$(tension)

Frame Analysis

1. Solve for the reaction forces.
   - For a rectangular frame fixed at the bottom with a horizontal force H:
   - Vertical reaction $R_V = 0$.
   - Horizontal reactions balance the applied load:

$$H_A = H = 500 \text{ N}$$

2. Determine the bending moment at the base.
   - Moment at the fixed support:

$$M = H \cdot h = 500 \times 3 = 1500 \text{ N.m}$$

3. Determine the axial forces in the members.
   - Solve using equilibrium equations for horizontal and vertical forces at each joint.

ChatGPT Integration

- Use ChatGPT to apply the method of sections for complex trusses with varying spans and loads.
- Explore examples of frame analysis under combined loading conditions.
- Request derivations of bending moment and axial forces in multi-bay frames.

PROMPT 87

Discuss the significance of the maximum bending moment and its location in beam design.

Objective

To understand the concept of the maximum bending moment in beams, its significance in design, and the factors influencing its location.

Activity

The maximum bending moment in a beam is a critical parameter in structural design as it determines the stress experienced by the material and directly impacts the beam's size, material selection, and safety. The location of the maximum bending moment depends on the type and distribution of loads and the beam's support conditions. For instance, in a simply supported beam with a single point load, the maximum bending moment occurs under the load, while for a uniformly distributed load (UDL), it occurs at midspan.

In this prompt, you will analyze the conditions that lead to the maximum bending moment, derive the equations for its location, and discuss its implications for material strength and safety margins in design.

Numerical Example

A simply supported beam of length $L = 8$ m is subjected to

1. a point load $P = 2000$ N at $x = 3$ m from the left support
2. a uniformly distributed load $w = 500$ N/m across the entire length

Determine

1. the maximum bending moment and its location
2. the design implications of the maximum bending moment

Solution

1. Determine the reactions at the supports.
   Use the equilibrium equations:
   - Sum of the vertical forces:
   $$R_A + R_B = P + w \cdot L$$

Substituting values, we obtain

$$R_A + R_B = 2000 + 500 \times 8 = 6000 \, \text{N}$$

• Sum of the moments about A:

$$R_B \times 8 = P \times 3 + w \times 8 \times \frac{8}{2}$$

Substituting values, we obtain

$$R_B \cdot 8 = 2000 \cdot 3 + 500 \cdot 8 \cdot 4 = 6000 + 16000 = 22000$$

Solving for R_B:

$$R_B = \frac{22000}{8} = 2750 \, \text{N}$$

Solve for the reaction at R_A:

$$R_A = 6000 - 2750 = 3250 \, \text{N}$$

2. Solve the bending moment equations.

• For the segment $0 \le x \le 3$ (left of the point load),

$$M(x) = R_A \cdot x - w \frac{x^2}{2}$$

Substituting $R_A = 3250$, we obtain

$$M(x) = 3250 \cdot x - 250x^2$$

• For the segment $3 \le x \le 8$,

$$M(x) = R_A \cdot x - P \cdot (x - 3) - \frac{wx^2}{2}$$

Substituting values, we obtain

$$M(x) = 3250x - 2000 \cdot (x - 3) - 250x^2$$

3. Find the location of the maximum bending moment.
   Maximum bending moment occurs at $x_{max} = 3$ m:

4. Determine the maximum bending moment.
   At $x = 3$,
   By substituting $R_A = 3250$ and $x = 3$, we obtain

$$M_{max} = 3250(3) - 250(3)^2 = 7500 \, \text{N} \cdot \text{m}$$

ChatGPT Integration

- Ask ChatGPT to find the maximum bending moment and its location for a simply supported beam with a point load or UDL.
- Use ChatGPT to explain how support types (fixed, pinned, cantilever) affect where the maximum moment occurs.
- Have ChatGPT compare section sizes needed for steel and aluminum beams under the same bending moment.

PROMPT 88

Analyze internal forces in a simple frame structure using equilibrium and section cuts.

Objective

To understand and apply equilibrium and section cut methods for analyzing internal forces (axial force, shear force, and bending moment) in a simple frame structure.

Activity

Frames are common structural systems used in mechanical and civil engineering to support loads. Internal force analysis in frames involves isolating sections of the structure, applying equilibrium equations ($\Sigma F_x = 0$, $\Sigma F_y = 0$, $\Sigma M = 0$), and solving for unknown forces and moments. This process provides insights into how frames resist applied loads and transfer forces through their members. In this prompt, you will learn how to analyze internal forces in a simple rectangular frame under combined horizontal and vertical loading.

Numerical Example

Analyze the internal forces at a section cut in a rectangular frame. The frame has

- width (w) = 6 m, height (h) = 4 m
- a horizontal load (H = 500 N) applied at the top-right corner
- a vertical load (P = 1000 N) applied at the top-left corner

The frame is fixed at the bottom corners.

Determine the internal axial force, shear force, and bending moment at a section cut located at $x = 2$ m from the left vertical member.

Solution

Step 1: Determine the reactions at the supports

1. Find the sum of the vertical forces ($\Sigma F_y = 0$):

$$R_{A_y} + R_{B_y} = P$$

By substituting, we obtain

$$R_{A_y} + R_{B_y} = 1000 \text{ N}$$

2. Determine the sum of the horizontal forces ($\Sigma F_x = 0$):

$$R_{A_x} = H = 500 \text{ N}$$

3. Determine the sum of the moments about A ($\Sigma M_A = 0$):

$$R_{B_y} \cdot 6 = P \cdot \frac{w}{2}$$

By substituting, we obtain

$$R_{B_y} = \frac{3000}{6} = 500 \ N$$

4. Determine the reaction at A:

$$R_{A_y} = P - R_{B_y} = 1000 - 500 = 500 \ N$$

Step 2: Solve for a section cut at $x = 2$ m.

1. Determine the vertical equilibrium ($\Sigma F_y = 0$).
   Shear force V balances the vertical reactions and loads:

$$V = R_{A_y} = 500 \ N$$

2. Determine the horizontal equilibrium ($\Sigma F_x = 0$):
   Axial force N balances the horizontal reaction:

$$N = R_{A_x} = 500 \ N$$

3. Use the moment equilibrium ($\Sigma M = 0$):
   Solving for the bending moment at the section, we obtain

$$M = R_{A_y} \cdot x = 500 \cdot 2 = 1000 \ N$$

ChatGPT Integration

- Use ChatGPT to analyze internal forces for frames with multiple sections and varied loading conditions.
- Request step-by-step derivations of axial, shear, and bending forces for complex frame configurations.
- Explore additional examples, such as multi-bay or three-dimensional frames, to deepen your understanding.

PROMPT 89

Explain the role of internal forces in ensuring the stability and safety of mechanical systems.

Objective

To understand the role of internal forces (axial force, shear force, and bending moment) in maintaining the stability and safety of mechanical systems under various loading conditions.

Activity

Internal forces are critical for maintaining the structural integrity and safety of mechanical systems. These forces, which arise due to external loads, help ensure that components remain stable and operate within their design limits. Axial forces influence elongation or compression, shear forces resist sliding between layers, and bending moments cause curvatures that must be controlled to prevent failure. Engineers use these internal force analyses to design systems capable of withstanding operational stresses, enhancing both safety and performance.

This prompt focuses on identifying internal force contributions in typical systems like beams, trusses, and frames. It also discusses how failure modes, such as buckling, shear failure, or bending failure, can occur if forces exceed the system's capacity.

Numerical Example

A steel beam of length $L = 6$ m is subjected to a uniformly distributed load $w = 400$ N/m over its entire length. The beam is simply supported at both ends.

1. Determine the maximum internal forces (shear force and bending moment).
2. Discuss how these forces affect the beam's stability and safety.

Solution

Step 1: Find the reactions at the supports

1. Find the vertical reactions (R_A and R_B).
   Use the equilibrium:
   - Sum of the vertical forces:

   $$R_A + R_B = w \cdot L$$

   By substituting w = 400 N/m and L = 6 m, we obtain

   $$R_A + R_B = 400 \times 6 = 2400\,\text{N}$$

   - Symmetry of the load:

   $$R_A = R_B = \frac{2400}{2} = 1200\ \text{N}$$

Step 2: Perform the shear force and bending moment calculations.

1. Solve the shear force equation.

   For any section at distance x from the left support,

   $$V(x) = R_A - w \cdot x$$

   By substituting R_A = 1200 and w = 400, we obtain

   $$V(x) = 1200 - 400 \cdot x$$

2. Determine the maximum shear force.
   At x = 0,

   $$V(0) = 1200N$$

   At $x = L$ = 6 m,

   $$V(6) = 1200 - 400 \cdot 6 = -1200\ \text{N}$$

3. Solve the bending moment equation.
   The bending moment at any section is

   $$M(x) = R_A \cdot x - \frac{w \cdot x^2}{2}$$

   By substituting R_A = 1200 and w = 400, we obtain

   $$M(x) = 1200 \cdot x - 200 \cdot x^2$$

Step 3: Understand the stability and safety considerations.

1. Shear failure risk:

   The maximum shear force (V = 1200 N) must be less than the shear strength of the material to prevent sliding failure.

2. Bending failure risk:

   The maximum bending moment (M = 1800 N·m) should not exceed the beam's moment capacity, which depends on its cross-section and material properties.

3. Deflection and buckling:

   Excessive bending can lead to large deflections, affecting stability. If compressive forces are present, buckling may also occur, requiring additional design checks.

ChatGPT Integration

- Use ChatGPT to explore similar problems with different load distributions and support conditions.
- Ask ChatGPT to calculate maximum deflections to assess stability further.
- Request examples of combined loading scenarios to understand interactions between shear, bending, and axial forces.

PROMPT 90

Solve a real-world problem involving internal force analysis in a multi-span beam.

Objective

To apply principles of internal force analysis to a practical multi-span beam problem, focusing on determining shear forces, bending moments, and their locations under various loads.

Activity

Multi-span beams are common in bridges, floors, and industrial equipment. Their analysis involves identifying reactions, shear forces, and bending moments across multiple spans subjected to complex loading. This prompt demonstrates how to systematically approach these problems using equilibrium equations and moment continuity principles. You will solve a realistic scenario to understand how internal forces are distributed in multi-span systems and their implications for design.

Numerical Example

A continuous beam with three spans (AB, BC, CD) has the following properties:

- span lengths: $L_{AB} = L_{BC} = L_{CD} = 4\,\mathrm{m}$
- point load $P = 1000$ N applied at the center of AB
- uniformly distributed load $w = 500$ N/m over BC
- a vertical reaction at D

Determine

1. the reactions at the supports (R_A, R_B, R_C, R_D)
2. the maximum bending moment and shear force in each span

Solution

Step 1: Determine the reactions at the supports.

1. Find the sum of the vertical forces $(\Sigma F_y = 0)$:
$$R_A + R_B + R_C + R_D = P + w \cdot L_{BC}$$

By substituting, we obtain
$$R_A + R_B + R_C + R_D = 3000\,N$$

2. Solve for the moment equilibrium about A $(\Sigma MA = 0)$. Consider the spans and loads to obtain
$$R_B \times 4 + R_C \times 8 + R_D \times 12 = P \times 2 + w \cdot L_{BC} \cdot \frac{L_{BC}}{2} \cdot 6$$

By substituting in values and simplifying, we obtain
$$R_B \cdot 4 + R_C \cdot 8 + R_D \cdot 12 = 6000 \text{ N.m}$$

3. Find the symmetry of loading and spans.
By symmetry, $R_A = R_D$ and $R_B = R_C$. Solving for reactions, we obtain
$$R_A = R_D = 750 \text{ N}, R_B = R_C = 750 \text{ N}$$

Step 2: Determine the internal forces in each span.

1. Determine span AB:
   - Shear force (V):
$$V(x) = R_A - P \cdot \delta(x-2)$$

Maximum shear:
$$V_{max} = 750\,\text{N}$$

- Bending moment (M):
$$M(x) = R_A \cdot x$$

Maximum bending moment:
$$M_{max} = R_A \cdot 2 = 1500 \text{ N.m}$$

2. Determine span BC.
   - Shear force (V):

$$V(x) = R_B - \omega.x$$

Maximum shear:
$$V_{max} = 750\,\text{N}$$

- Bending moment (M):

$$M(x) = R_B \cdot x - \frac{w \cdot x^2}{2}$$

Maximum bending moment at $x = 2$ m:
$$M_{max} = 500 \text{ N.m}$$

3. Determine span CD.
   Symmetric to *AB*, with results:
   - $V_{max} = 750$ N
   - $M_{max} = 1500\,\text{N·m}$

ChatGPT Integration

- Ask ChatGPT to solve variations of this problem with additional loads or spans.
- Request guidance for analyzing multi-span beams with varying cross-sections.
- Use ChatGPT to explore interaction effects between spans under dynamic loading.

CHAPTER 10

REAL-WORLD APPLICATIONS IN STATICS

Prompt 91 Define stress and strain and explain their relationship to internal forces in static structures.

Prompt 92 Differentiate between normal stress, shear stress, and bending stress with practical examples.

Prompt 93 Derive the relationship between bending moment and normal stress in beams, and explain its significance in beam design.

Prompt 94 Solve a problem involving stress distribution in a beam subjected to combined loading.

Prompt 95 Explain the concept of torsional stress in shafts and its significance in mechanical design.

Prompt 96 Analyze a structural element subjected to combined stresses (e.g., axial, bending, and torsional) and determine critical stress points.

Prompt 97 Explain how internal force analysis informs the design of trusses and frames for varying load conditions.

Prompt 98 Discuss how stress analysis informs the design against failure, such as yielding and buckling.

Prompt 99 Discuss how stress analysis informs the design against failure, such as yielding, fatigue, and creep.

Prompt 100 Explore real-world applications where statics transitions into the strength of materials, such as bridge design, pressure vessels, or mechanical components.

PROMPT 91

Define stress and strain and explain their relationship to internal forces in static structures.

Objective

To understand the fundamental concepts of stress and strain, their mathematical definitions, and how they relate to internal forces in structural members under static equilibrium.

Activity

Stress (σ) is the internal force per unit area within a material, arising due to external loads or reactions. Strain (ϵ) is the measure of deformation or elongation that a material undergoes under stress. These concepts are interrelated and form the basis for analyzing how materials respond to forces. Stress can be classified as *normal stress* (caused by axial forces or bending moments) and *shear stress* (caused by transverse forces). Strain is categorized similarly as normal strain or shear strain. In this task, you will explore the mathematical definitions of stress and strain, their units, and their significance in static structures.

Numerical Example

A steel rod with a cross-sectional area of $A = 500$ mm² is subjected to an axial tensile load of $F = 100$ kN. The rod has an original length of $L_0 = 2$ m and elongates by $\Delta L = 1.5$ mm under the load.

1. Calculate the normal stress in the rod.
2. Determine the normal strain.
3. If the modulus of elasticity (E) of steel is 200 GPa, verify the strain using Hooke's Law.

Solution

1. Determine the normal stress.

   Use the definition of normal stress:

   $$\sigma = \frac{F}{A}$$

   Substitute the values to obtain

   $$\sigma = \frac{100 \times 10^3}{500 \times 10^{-6}} = 200 \text{ MPa}$$

2. Determine the normal strain.

   Use the definition of normal strain:

   $$\epsilon = \frac{\Delta L}{L_0}$$

   Substitute the values to obtain

   $$\epsilon = \frac{1.5 \times 10^{-3}}{2} = 0.00075 = 0.075\%$$

3. Verfiy using Hooke's Law.

   Hooke's Law is

   $$\sigma = E \cdot \epsilon$$

   Substitute the values to obtain

   $$\epsilon = \frac{200 \times 10^6}{200 \times 10^9} = 0.00075$$

ChatGPT Integration

- Ask ChatGPT to explain the difference between normal and shear stress with examples.
- Use ChatGPT to solve additional problems involving stress and strain in beams or shafts.
- Request an explanation of how material properties, such as modulus of elasticity, influence strain under different loading conditions.

PROMPT 92

Differentiate between normal stress, shear stress, and bending stress with practical examples.

Objective

To understand the differences between normal stress, shear stress, and bending stress in materials and their relevance to internal force analysis in static structures.

Activity

Stress is categorized based on the type of internal force causing it. *Normal stress* arises due to axial loads (tensile or compressive) or bending moments. *Shear stress* is caused by transverse forces or torques. *Bending stress* is a specific type of normal stress distributed across a beam's cross-section due to bending moments. Each stress type influences the design and analysis of mechanical components differently. This task involves distinguishing these stress types, identifying where they occur, and solving a numerical example.

Numerical Example

A rectangular beam of width b = 100 mm and height h = 200 mm is subjected to

1. a tensile axial force of F = 50 kN
2. a transverse shear force of V = 10 kN
3. a bending moment of M = 5 kN·m

Determine

1. the normal stress due to the axial force
2. the maximum shear stress due to the transverse force
3. the maximum bending stress in the beam

Solution

Step 1: Determine the normal stress (σ).

1. Use the definition of normal stress:

$$\sigma = \frac{F}{A}$$

2. Identify the area of the cross-section:

$$A = b \cdot h = 100 \times 200 = 20000 \text{ mm}^2 = 0.02 \text{ m}^2$$

3. Substitute the values to obtain

$$\sigma = \frac{50 \times 10^3}{0.02} = 2.5 \text{ MPa}$$

Step 2: Determine the shear stress (τ).

1. Use the definition of the maximum shear stress for a rectangular cross-section:

$$\tau_{max} = \frac{3}{2} \cdot \frac{V}{A}$$

2. Substitute the values to obtain

$$\tau_{max} = \frac{3}{2} \cdot \frac{10 \times 10^3}{0.02} = 750 \text{ kPa}$$

Step 3: Determine the bending stress (σ_b).

1. Use the definition of the maximum bending stress:

$$\sigma_b = \frac{M \cdot c}{I}$$

where

- c = distance from the neutral axis = $h/2$ = 200/2 = 100 mm = 0.1 m
- I = moment of inertia = b·h³/12

2. Calculate I:

$$I = \frac{b \cdot h^3}{12} = 66.67 \times 10^{-6} \text{ m}^4$$

3. Substitute values into σ_b:

$$\sigma_b = \frac{M \cdot c}{I} = \frac{5 \times 10^3 \times 0.1}{66.67 \times 10^{-6}} = 7.5 \text{ MPa}$$

ChatGPT Integration

- Use ChatGPT to explore problems involving combined stresses in beams and shafts.
- Ask for clarification on how stress distributions vary across different cross-sectional shapes.
- Request derivations of shear and bending stress formulas for non-rectangular sections.

PROMPT 93

Derive the relationship between bending moment and normal stress in beams and explain its significance in beam design.

Objective

To derive the formula connecting the bending moment in a beam to the normal stress induced in its cross-section, highlighting its importance in designing safe and efficient beams.

Activity

When a beam is subjected to bending, normal stresses develop due to the internal bending moment. These stresses vary linearly across the cross-section, reaching a maximum at the extreme fibers and zero at the neutral axis. The relationship between the bending moment (M) and the resulting normal stress (σ) is fundamental for beam design, ensuring that the material strength is not exceeded. This task involves deriving the equation $\sigma = M \cdot c / I$ and understanding its role in assessing beam performance.

Solution

Step 1: Identify the assumptions for the derivation.

1. The beam is *prismatic* (uniform cross-section along its length).
2. The material is *linearly elastic* and follows Hooke's Law.
3. Plane sections remain *plane* after bending (Bernoulli's hypothesis).

Step 2: Determine the strain distribution in the cross-section.
Normal strain (ϵ) at a distance y from the neutral axis is proportional to the curvature (κ):

$$\epsilon = -y \cdot \kappa$$

the negative sign in the equation $\epsilon = -y \cdot \kappa$ originates from the convention of how strain and stress are defined relative to the neutral axis in a bending beam. The neutral axis is the location in the cross-section where there is no strain or stress during bending. It separates the region in tension from the region in compression. The coordinate yyy measures the perpendicular distance from the neutral axis:

- $y > 0$: above the neutral axis
- $y < 0$: below the neutral axis

When a beam bends,

- the material above the neutral axis is *compressed* (negative strain)
- the material below the neutral axis is *stretched* (positive strain)

To reflect this, the strain is defined as $\epsilon = -y \cdot \kappa$. For $y > 0$: Compression ($\epsilon < 0$) and for $y < 0$, Tension ($\epsilon > 0$).

Step 3: Find the stress distribution in the cross-section.

Hooke's Law relates stress to strain: $\sigma = E \cdot \epsilon$.

By substituting, we obtain

$$\sigma = -E \cdot y \cdot \kappa$$

Step 4: Relate the curvature to the bending moment.

1. The bending moment (M) is the result of the internal stresses distributed over the cross-sectional area:

$$M = \int_A \sigma \cdot y \, dA$$

By substituting $\sigma = -E \cdot y \cdot \kappa$ and considering the fact that E and κ are constants, we find

$$M = -E \cdot \kappa \int_A y^2 \, dA$$

2. The term $\int_A y^2 \, dA$ is the *moment of inertia* (I) about the neutral axis:

$$M = -E \cdot \kappa \cdot I$$

3. Solve for the curvature (κ):

$$\kappa = -\frac{M}{E \cdot I}$$

Step 5: Determine the maximum normal stress.

1. Normal stress at the outermost fiber ($y = c$):

$$\sigma = -E \cdot c \cdot \kappa$$

2. Substituting $\kappa = -\dfrac{M}{E \cdot I}$:

$$\sigma = \frac{M \cdot c}{I}$$

Significance in Beam Design

1. The equation $\sigma = \dfrac{M \cdot c}{I}$ shows the following:

   - Bending Moment (M): A higher moment increases the stress.
   - Distance from Neutral Axis (c): Stress is at the maximum at the outermost fibers.
   - Moment of Inertia (I): A larger value of I reduces stress, highlighting the importance of cross-sectional design.

2. Engineers use this relationship to do the following:

   - ensure that the material's maximum stress does not exceed its yield strength
   - optimize the cross-sectional shape for strength and weight efficiency

ChatGPT Integration

- Use ChatGPT to derive similar relationships for beams with non-standard cross-sections.
- Request examples of beam design based on the bending stress formula.
- Ask for a step-by-step analysis of stress distribution for different loading and support conditions.

PROMPT 94

Solve a problem involving stress distribution in a beam subjected to combined loading.

Objective

To analyze stress distribution in a beam under combined axial force, bending moment, and transverse shear force, demonstrating the superposition of stress components.

Activity

Combined loading scenarios are common in mechanical systems, where beams experience axial forces, bending moments, and transverse shear forces simultaneously. Each load type contributes to the overall stress distribution in the beam. This task involves calculating the axial, bending, and shear stresses at specific points in a beam's cross-section and visualizing the combined stress state.

Numerical Example

A rectangular beam with a width b = 120 mm and height h = 240 mm is subjected to

1. an axial tensile force F = 80 kN
2. a bending moment M = 4 kN·m, counterclockwise
3. a transverse shear force V = 12 kN

Determine the combined stress at the following points on the cross-section:

1. Point A: top edge, directly above the neutral axis
2. Point B: bottom edge, directly below the neutral axis
3. Point C: neutral axis, at the center

Solution

Step 1: Calculate the axial stress (σ_{axial}).

1. Use the formula:

$$\sigma_{axial} = \frac{F}{A}$$

2. Solve for the cross-sectional area:

$$A = b \cdot h = 120 \times 240 = 28800 \text{ mm}^2 = 0.0288 \text{ m}^2$$

3. Substitute values to obtain

$$\sigma_{axial} = \frac{80 \times 10^3}{0.0288} = 2777.78 \text{ kPa} = 2.78 \text{ MPa}$$

Step 2: Calculate bending stress (σ_b).

1. Use the formula:

$$\sigma_b = \frac{M \cdot c}{I}$$

where
- c = h/2 = 240/2 = 120 mm = 0.12 m.
- Moment of inertia I = b·h³/12.

2. Calculate I:

$$I = \frac{b \cdot h^3}{12} = 1.3824 \times 10^{-4} \text{ m}^4$$

3. Substitute values to obtain

$$\sigma_b = \frac{4 \times 10^3 \cdot 0.12}{1.3824 \times 10^{-4}} = 3.472 \text{ MPa}$$

Step 3: Calculate the shear stress (τ).

1. Use the formula:

$$\tau = \frac{V \cdot Q}{I \cdot b}$$

where
- $Q = A' \cdot y'$, the first moment of area for the section above or below the neutral axis.
- $A' = b \cdot h/2$, and $y' = h/4$.

2. Calculate Q:

$$Q = b \cdot \frac{h}{2} \cdot \frac{h}{4} = 8.64 \times 10^{-4} \text{ m}^3$$

3. Substitute values to obtain

$$\tau = \frac{12 \times 10^3 \times 8.64 \times 10^{-4}}{1.3824 \times 10^{-4} \times 0.12} = 62.5 \text{ kPa}$$

Step 4: Find the combined stresses.
- Point A (top edge):

$$\sigma_A = \sigma_{axial} - \sigma_b = -0.692 \text{ MPa}$$

- Point B (bottom edge):

$$\sigma_B = \sigma_{axial} + \sigma_b = 6.252 \text{ MPa (tension)}$$

- Point C (neutral axis):

$$\sigma_C = \sigma_{axial} = 2.78 \text{ MPa}, \quad \tau_C = 62.5 \text{ kPa}$$

ChatGPT Integration

- Use ChatGPT to calculate combined stresses for beams with different cross-sections.
- Explore how stress distributions vary for non-uniform loading conditions.
- Ask for examples involving three-dimensional stress analysis.

PROMPT 95

Explain the concept of torsional stress in shafts and its significance in mechanical design.

Objective

To understand torsional stress, how it develops in circular shafts under applied torques, and its importance in the design of mechanical components such as drive shafts, gears, and axles.

Activity

Torsional stress occurs in circular shafts subjected to applied torques, causing angular deformation or twisting. The stress varies linearly with the radial distance from the shaft's center, reaching a maximum at the outer surface. This concept is vital for mechanical design, as shafts must be designed to resist torsional stress without exceeding their material strength. You will explore the formula for torsional stress, its derivation, and its application in evaluating the safety and performance of shafts.

Numerical Example

A solid circular steel shaft with a diameter d = 50 mm is subjected to a torque T = 1200 N·m. The shaft's material has a shear modulus G = 80 GPa and allowable shear stress τ_{allow} = 50 MPa.
Determine

1. the maximum shear stress in the shaft
2. the angle of twist (θ) over a length L = 2 m
3. whether the shaft is safe under the applied torque

Solution

Step 1: Determine the maximum shear stress (τ_{max}).

1. Use the formula for torsional stress:

$$\tau = \frac{T \cdot r}{J}$$

where

- T: applied torque (1200 N·m)
- r: radius of the shaft ($r = 50/2 = 25$ mm $= 0.025$ m)
- J: polar moment of inertia for a circular shaft

2. Determine the polar moment of inertia:

$$J = \frac{\pi \cdot d^4}{32}$$

By substituting $d = 0.05$ m, we obtain

$$J = \frac{\pi \cdot (0.05)^4}{32} = 6.136 \times 10^{-7}\ \text{m}^4$$

3. Substitute this into the torsional stress formula:

$$\tau_{max} = \frac{T \cdot r}{J} = \frac{1200 \cdot 0.025}{6.136 \times 10^{-7}} = 48.88\ \text{MPa}$$

Step 2: Determine the angle of twist (θ).

1. Use the formula for the angle of twist:

$$\theta = \frac{T \cdot L}{J \cdot G}$$

where

- $T = 1200$ N·m
- $L = 2$ m
- $G = 80$ GPa
- $J = 6.136 \times 10^{-7} \text{m}^4$

2. Substitute values to obtain

$$\theta = \frac{1200 \cdot 2}{6.136 \times 10^{-7} \cdot 80 \times 10^9} = 0.0489\ \text{radians}$$

Step 3: Perform a safety check with the Factor of Safety.

The *Factor of Safety* (FoS) ensures that a design is sufficiently robust to withstand uncertainties in loading, material properties, and environmental conditions. It is calculated as follows:

$$FoS = \frac{\tau_{allow}}{\tau_{max}}$$

If we substitute in values, we obtain

$$FoS = \frac{50}{48.88} = 1.02$$

Since $\tau_{max} < \tau_{allow}$, the shaft is safe under the applied torque. However, adhering to standard factors of safety is essential in actual design to account for uncertainties and ensure reliability.

ChatGPT Integration

- Use ChatGPT to analyze torsional stresses for hollow or composite shafts.
- Ask ChatGPT for comparisons of torsional stress in materials with different properties (e.g., steel vs. aluminum).
- Explore the impact of shaft geometry (e.g., diameter and length) on the torsional stress and angle of twist.

PROMPT 96

Analyze a structural element subjected to combined stresses (e.g., axial, bending, and torsional) and determine critical stress points.

Objective

To understand how combined stresses (axial, bending, and torsional) interact in structural elements and determine the critical stress points for safe and efficient design.

Activity

Structural elements often experience multiple types of stresses simultaneously, such as axial, bending, and torsional stresses. These stresses must be combined using superposition principles to evaluate the critical stress points. This analysis is crucial for determining whether the material can safely support the combined loading without failure. In this task, you will calculate combined stresses in a shaft and assess its safety under given loading conditions.

Numerical Example

A circular steel shaft has a diameter d = 60 mm and a length L = 3 m. It is subjected to

1. an axial tensile force F = 50 kN
2. a bending moment M = 800 N·m
3. a torque T = 600 N·m

The material has an allowable normal stress σ_{allow} = 120 MPa and an allowable shear stress τ_{allow} = 60 MPa. Determine

1. the normal stresses due to axial and bending loads
2. the shear stress due to torsion
3. the maximum principal stress and maximum shear stress
4. whether the shaft is safe under the combined loading

Solution

Step 1: Normal stress due to axial force (σ_{axial}).

1. Use the formula:

$$\sigma_{axial} = \frac{F}{A}$$

2. Find the cross-sectional area:

$$A = \frac{\pi \cdot d^2}{4} = \frac{\pi (0.06)^2}{4} = 2.827 \times 10^{-3} \, m^2$$

3. Substitute values to obtain

$$\sigma_{axial} = \frac{50 \times 10^3}{2.827 \times 10^{-3}} = 17.7 \ \text{MPa}$$

Step 2: Determine the bending stress (σ_b).

1. Use the formula:

$$\sigma_b = \frac{M.c}{I}$$

where c = d/2 = 0.03 m and I = π·d⁴/64.

2. Find the moment of inertia:

$$I = \frac{\pi \cdot d^4}{64} = 6.362 \times 10^{-7} \text{ m}^4$$

3. Substitute the values to obtain

$$\sigma_b = \frac{800 \times 0.03}{6.362 \times 10^{-7}} = 37.73 \text{ MPa}$$

Step 3: Determine the shear stress due to torsion (τ).

1. Use the formula:

$$\tau = \frac{T.r}{J}$$

where $J = 2 \cdot I$.

2. Calculate J:

$$J = 2 \cdot I = 1.273 \times 10^{-6} \text{ m}^4$$

3. Substitute values to obtain

$$\tau = \frac{600 \times 0.03}{1.273 \times 10^{-6}} = 14.14 \text{ MPa}$$

Step 4: Determine the maximum principal stress.

1. Use the formula:

$$\sigma_{max} = \frac{\sigma_x + \sigma_y}{2} + \sqrt{\left(\frac{\sigma_x - \sigma_y}{2}\right)^2 + \tau^2}$$

Here, $\sigma_x = \sigma_{axial} + \sigma_b$ and $\sigma_y = 0$.

2. Substitute values to obtain

$$\sigma_{max} = \frac{17.7 + 37.73}{2} + \sqrt{\left(\frac{17.7 - 37.73}{2}\right)^2 + 14.14^2}$$

Simplify:

$$\sigma_{max} = 27.715 + \sqrt{100.3 + 199.94} = 45.04 \text{ MPa}$$

Step 5: Determine the maximum shear stress

1. Use the formula:

$$\tau_{max} = \sqrt{\tau^2 + \left(\frac{\sigma_x - \sigma_y}{2}\right)^2}$$

2. Substitute the values to obtain

$$\tau_{max} = \sqrt{199.94 + 100.3} = 17.33 \text{ MPa}$$

Step 6: Perform a safety check.

1. Normal stress:

$$\text{FoS}_6 = \frac{120}{45.1} \approx 2.66$$

The normal stress has a safety margin with an FoS of approximately 2.66. This must be checked against the standards.

2. Shear stress:

$$\text{FoS}_\delta = \frac{60}{17.4} \approx 3.45$$

The shear stress also has a safety margin with an FoS of approximately 3.45. This must be checked against the standards.

ChatGPT Integration

- Request a step-by-step explanation of the combined stress analysis for different geometries.
- Explore alternative materials to improve the safety factor under combined loading.
- Solve similar problems for hollow or composite shafts.

PROMPT 97

Explain how internal force analysis informs the design of trusses and frames for varying load conditions.

Objective

To understand how analyzing internal forces in trusses and frames enables the efficient design of structural systems that can withstand varying loads without failure.

Activity

Internal force analysis is essential for determining how forces are distributed among the members of a truss or frame. This analysis helps identify whether members are under tension or compression and ensures that their material and cross-sectional properties are adequate to support the applied loads. Engineers use methods like the method of joints, method of sections, and graphical methods to calculate internal forces. By understanding load paths and stress distributions, designers can optimize material use and ensure safety.

Numerical Example

A simple truss consists of three members: AB, BC, and AC. The truss is supported at A (pinned) and C (roller). A vertical load $P = 20$ kN is applied at joint B. The geometry of the truss is as follows:

- AB and BC: 3 m each
- AC: hypotenuse (4.24 m) of a right triangle

Determine the internal forces in all members using the *method of joints* and specify whether the members are in tension or compression.

Solution

Step 1: Identify the reaction forces.

1. Free body diagram:
   - The truss is in static equilibrium.
   - Vertical reaction at A: R_A
   - Horizontal reaction at A: H_A
   - Vertical reaction at C: R_C

2. Equilibrium equations:
   - $\Sigma F_x = 0$: $H_A = 0$
   - $\Sigma F_y = 0$: $R_A + R_C = P = 20$ kN
   - $\Sigma M_A = 0$:
   - Taking the moments about A:
   $$RC \times 3 = 20 \times 3 \Rightarrow RC = 20 \text{ kN}$$
   - Substitute R_C into $\Sigma F_y = 0$: $R_A = 0$

Step 2: Use the method of joints (joint C)

1. Free body diagram of joint C:
   - Forces: T_{AC} (along AC), T_{BC} (along BC), and RC = 20 kN.
2. Equilibrium equations at C:
   - $\Sigma F_x = 0$:

$$T_{AC} \cdot \frac{3}{4.24} = T_{BC}$$

   - $\Sigma F_y = 0$:

$$T_{AC} \cdot \frac{4}{4.24} = 20 \text{ kN}$$

3. Solve for T_{AC}:

$$T_{AC} = \frac{20 \times 4.24}{4} = 21.2 \text{ kN}$$

4. Solve for T_{BC}:

$$T_{BC} = 21.2 \cdot \frac{3}{4.24} = 15 \text{ kN}$$

Step 3: Use the method of joints (joint B).

1. Free body diagram of joint B:
   - Forces: T_{AB} (along AB), T_{BC} = 15 kN, and vertical load $P = 20$ kN
2. Equilibrium equations at B:
   - $\Sigma F_x = 0$: $T_{AB} = T_{BC}$
   - $\Sigma F_y = 0$: $T_{AB} \cdot \frac{4}{4.24} = 20$
3. Solve for T_{AB}:

$$T_{AB} = 20 \cdot \frac{4.24}{4} = 21.2 \text{ kN}$$

Significance in Design

1. Material selection: Members under compression need to resist buckling, while tensile members must have adequate strength.
2. Cross-sectional area: It ensures each member has sufficient area to prevent failure under the calculated internal forces.

3. Load optimization: Internal force analysis helps identify over-stressed members, guiding reinforcement or redesign.

PROMPT 98

Discuss how stress analysis informs the design against failure, such as yielding and buckling.

Objective

To understand how stress analysis helps identify potential failure modes, including yielding and buckling, and how it guides engineers to design components that meet safety and performance requirements.

Activity

Stress analysis is a cornerstone of mechanical design, ensuring that structures and components operate safely under applied loads. Yielding occurs when the material exceeds its elastic limit, while buckling is a stability failure common in slender members under compressive loads. Engineers use analytical tools like stress-strain relationships, failure theories, and critical load calculations to prevent these failure modes. This task involves understanding yielding and buckling, exploring their differences, and analyzing how stress analysis aids in preventing failure.

Numerical Example

A slender steel column with a circular cross-section has

- Length $L = 3$ m
- Diameter $d = 100$ mm
- Material properties: Yield strength $\sigma_y = 250$ MPa, Young's modulus $E = 200$ GPa

The column is fixed at one end and free at the other (cantilevered). Determine

1. the critical load (P_{cr}) for buckling
2. the maximum compressive stress under a load $P = 60$ kN
3. whether the column is safe against both buckling and yielding

Solution

Step 1: Determine the critical load for buckling (P_{cr}).

1. Use the formula for the buckling load (for a cantilevered column):

$$P_{cr} = \frac{\pi^2 \cdot E \cdot I}{(2L)^2}$$

where
- $I = \pi \cdot d^4/64$: moment of inertia
- $2L = 2 \times 3 = 6$ m: effective length for a cantilever

2. Calculate I:

$$I = \frac{\pi \cdot (0.1)^4}{64} = 4.91 \times 10^{-6}\ \text{m}^4$$

3. Substitute into P_{cr}:

$$P_{cr} = \frac{\pi^2 \cdot 200 \times 10^9 \times 4.91 \times 10^{-6}}{6^2} = 269.22\ \text{kN}$$

Step 2: Determine the maximum compressive stress (σ_{max}).

1. Use the formula for compressive stress:

$$\sigma_{max} = \frac{P}{A}$$

where $A = \pi \cdot d^2/4$.

2. Calculate A:

$$A = \frac{\pi \cdot (0.1)^2}{4} = 7.854 \times 10^{-3}\ \text{m}^2$$

3. Substitute this into σ_{max}:

$$\sigma_{max} = \frac{60 \times 10^3}{7.854 \times 10^{-3}} = 7.64\ \text{MPa}$$

Step 3: Perform the safety check.

1. Determine the buckling safety:
Compare $P = 60$ kN to $P_{cr} = 269.22$ kN:

$$\text{FoS}_{buckling} = \frac{P_{cr}}{P} = \frac{269.22}{60} = 4.5$$

The buckling limit is considered safe, as $FoS_{buckling} > 1.0$. However, to meet practical design standards, the FoS must be considered.

2. Determine the yielding safety.

Compare $\sigma_{max} = 7.6 \, \text{MPa}$ to $\sigma_y = 250 \, \text{MPa}$:

$$FoS_{yield} = \frac{\sigma_y}{\sigma_{max}} > 1$$

Yielding is considered safe. However, to meet practical design standards, the FoS must be considered.

ChatGPT Integration

- Ask ChatGPT to analyze buckling in non-standard geometries (e.g., I-beams and hollow sections).
- Request comparisons of critical loads for different boundary conditions.
- Use ChatGPT to explore advanced failure theories like Von Mises or Mohr-Coulomb for combined loading scenarios.

PROMPT 99

Discuss how stress analysis informs the design against failure, such as yielding, fatigue, and creep.

Objective

To understand how stress analysis addresses failure modes like yielding, fatigue, and creep, ensuring that materials and structures meet performance and safety standards over their operational lifespan.

Activity

Stress analysis is critical in designing structures and components to prevent failure due to yielding (exceeding the elastic limit), fatigue (progressive failure under cyclic loading), and creep (time-dependent deformation under constant load). By combining stress-strain relationships, material properties, and failure theories, engineers can predict these failure modes and design against them. This involves choosing suitable materials, applying appropriate factors of safety, and performing life cycle analysis.

Numerical Example

A steel beam with a rectangular cross-section ($b = 100$ mm, $h = 200$ mm) is subjected to

1. a constant tensile load $P = 150$ kN
2. cyclic stress with an amplitude $\sigma_a = 50$ MPa
3. operating temperature $T = 450°C$

Material properties:

- Yield strength σ_y: 250 MPa
- Endurance limit σ_e: 90 MPa
- Creep coefficient $C = 1.5 \times 10^{-12}$, h^{-1} MPa^{-n} and creep exponent $n = 5$

Determine

1. safety against yielding
2. safety against fatigue
3. estimate the creep strain after $t = 10000$ hours

Solution

Step 1: Determine the safety against yielding.

1. Use the formula for the tensile stress:

$$\sigma = P / A$$

where $A = b \cdot h = 0.1 \times 0.2 = 0.02$ m^2.

2. Substitute values:

$$\sigma = \frac{150 \times 10^3}{0.02} = 7.5 \text{ MPa}$$

3. Perform the safety check:

$$\text{FoS}_{\text{yield}} = \frac{\sigma_y}{\sigma} = \frac{250}{7.5} = 33.33$$

Conclusion: The beam seems safe against yielding. However, the standard FoS must be considered for practical cases.

Step 2: Determine the safety against fatigue.

1. Use the formula for the fatigue stress range:

$$\sigma_{range} = 2\sigma_a = 2 \times 50 = 100 \text{ MPa}$$

2. Compare this with the endurance limit:

$$\text{FoS}_{fatigue} = \frac{\sigma_e}{\sigma_{range}} = \frac{90}{100} = 0.9$$

Conclusion: The beam is **not safe** against fatigue and requires redesign or load reduction.

Step 3: Estimate the creep strain.

1. Use the creep strain formula:

$$\epsilon_{creep} = C \cdot \sigma^n \cdot t$$

where
- $\sigma = 7.5 \text{ MPa}$
- $t = 10000 \text{ hours} = 3.6 \times 10^7 \text{ seconds}$

2. Substitute the values to obtain

$$\epsilon_{creep} = 1.5 \times 10^{-12} \cdot (7.5)^5 \times 10^4 = 3.56 \times 10^{-4}$$

Conclusion: The creep strain is minimal and does not pose a significant risk. However, standard limits must be checked for practical cases.

ChatGPT Integration

- Use ChatGPT to explore failure theories like Von Mises or Tresca for combined stresses.
- Ask for examples of fatigue life predictions using S-N curves.
- Request creep strain calculations for different materials and operating conditions.

PROMPT 100

Explore real-world applications where statics transitions into the strength of materials, such as bridge design, pressure vessels, or mechanical components.

Objective

To illustrate how principles of statics provide the foundation for the strength of materials in solving real-world engineering problems, ensuring the safety, efficiency, and reliability of various structures and mechanical systems.

Activity

Statics serves as the starting point for analyzing forces, moments, and reactions in engineering structures. However, real-world applications often require extending these analyses to include material behavior under stresses and strains. This transition to the strength of materials enables engineers to predict failure modes, optimize material usage, and ensure long-term performance. Examples include analyzing the stress distribution in bridges, determining the wall thickness of pressure vessels, and evaluating the torsional stress in shafts.

Numerical Example

Design a cylindrical pressure vessel with the following specifications:

- internal pressure $P = 5$ MPa
- internal diameter $D = 2$ m
- allowable tensile stress of the material $\sigma_{allow} = 100$ MPa

Determine

1. the required wall thickness to withstand the internal pressure
2. the hoop and longitudinal stresses in the vessel
3. safety factor for the design

Solution

Step 1: Determine the required wall thickness.

1. Use the formula for the hoop stress (σ_h):

$$\sigma_h = \frac{P \cdot r}{t}$$

Rearranging for t, we obtain

$$t = \frac{P \cdot r}{\sigma_{allow}}$$

where $r = D/2 = 1$ m.

2. Substitute values to obtain

$$t = \frac{5 \times 10^6 \times 1}{100 \times 10^6} = 0.05 \text{ m} = 50 \text{ mm}$$

Step 2: Find the hoop and longitudinal stresses.

1. Hoop stress (σh):

$$\sigma_h = \frac{P \cdot r}{t} = 100 \text{ MPa}$$

2. Longitudinal stress (σl):

$$\sigma_1 = \frac{P \cdot r}{2t} = 50 \text{ MPa}$$

Step 3: Determine the Factor of Safety.

1. Use the formula:

$$\text{FoS} = \frac{\sigma_{\text{allow}}}{\sigma_h}$$

2. Substitute values to obtain

$$\text{FoS} = \frac{100}{100} = 1$$

Interpretation: The design meets the minimum safety requirement. To increase safety, consider increasing the wall thickness or using a stronger material. However, the standard limits must be checked against for practical cases. Please note that the solutions assumes thin-wall cylinderical

Significance in Design

1. Bridge design: Analyze force distribution in trusses and beams, then transition to stress and strain analysis for material selection and sizing.

2. Pressure vessels: Use statics to calculate internal forces, then apply material strength to determine wall thickness and safety.

3. Mechanical components: Evaluate torsional, bending, and axial stresses in shafts, gears, and machine elements.

ChatGPT Integration

- Explore advanced applications like thermal stresses in pressure vessels.
- Simulate dynamic loading scenarios for bridges using ChatGPT.
- Request the optimization of material used for multi-functional mechanical components.

Statics Wizard Tutor App

This GPT App is designed to complement **Part 1** of this book (Statics) by providing an interactive and immersive learning experience. Users can explore their preferred chapters and prompts from the book's comprehensive list, enabling them to research deeper into specific topics of interest. The app features tools for solving numerical examples, step-by-step walkthroughs for complex problems, and instant feedback to enhance understanding. Additionally, it supports interactive visualizations, personalized suggestions based on user progress, and detailed explanations to bridge theoretical concepts with practical applications. Whether you are reviewing key concepts, solving practice problems, or seeking clarification, the app serves as a dynamic extension of the book to support a hands-on approach to mastering statics.

Users can obtain access and run the Statics Wizard Tutor App directly through OpenAI GPTs. To get started, when available, navigate to the GPT Apps section within OpenAI's platform and search for "Statics Wizard Tutor App." Once launched, the app's user-friendly interface allows you to select chapters, prompts, or specific topics from Part 1 of this book. Simply input your questions or problems, and the app will provide tailored guidance, step-by-step solutions, and explanations to enhance your understanding of statics.

FIGURE 10.1 A logo generated using Logo Creator App, OpenAI Store

PART 2: DYNAMICS

INTRODUCTION TO DYNAMICS

Dynamics is a fundamental branch of mechanics that examines the motion of bodies and the forces that cause this motion. Unlike statics, which focuses on equilibrium, dynamics explores how objects move under the influence of external forces, making it essential for understanding real-world systems in motion. Engineers rely on dynamics to design and analyze machinery, vehicles, robotics, and other systems where motion and time-dependent behaviors play a critical role. From predicting the trajectory of a satellite to optimizing the motion of mechanical linkages, dynamics forms the backbone of modern engineering analysis.

This part of the book introduces the foundational concepts of dynamics in a structured manner, covering key principles such as kinematics, kinetics, energy methods, impulse-momentum principles, and the dynamics of rigid bodies in both planar and three-dimensional systems. Each chapter builds on the preceding one, offering a progressive understanding of the subject with interactive prompts, real-world applications, and numerical examples to deepen your learning experience.

Part 2 is divided into ten chapters, continuing from Part 1:

Chapter 11   Kinematics of Particles

Chapter 12   Kinetics of Particles: Force and Acceleration

Chapter 13   Kinetics of Particles: Work and Energy

Each chapter includes approximately ten interconnected prompts designed to guide readers in exploring the subject matter in greater depth. These prompts incorporate numerical examples, practical scenarios, and exploratory questions, ensuring a comprehensive and engaging approach to mastering dynamics.

CHAPTER 11

KINEMATICS OF PARTICLES

Prompt 101 Define kinematics and its importance in analyzing particle motion.

Prompt 102 Derive the equations of motion for a particle under uniform acceleration.

Prompt 103 Explain the difference between rectilinear and curvilinear motion, with examples.

Prompt 104 Solve a problem involving the motion of a particle with constant velocity.

Prompt 105 Analyze the motion of a particle in a straight-line using velocity-time and displacement-time graphs.

Prompt 106 Discuss the concept of relative velocity in one-dimensional motion and solve a related problem.

Prompt 107 Explain projectile motion and derive the expressions for range, maximum height, and time of flight.

Prompt 108 Solve a problem involving a particle's motion in two dimensions with constant acceleration.

Prompt 109 Discuss motion in polar coordinates and its application to curved paths.

Prompt 110 Solve a problem involving the velocity and acceleration components of a particle moving in a circular path.

PROMPT 101

Define kinematics and its importance in analyzing particle motion.

Objective

To understand the concept of kinematics and its significance in describing and analyzing the motion of particles without considering the forces causing the motion.

Activity

Kinematics is the branch of mechanics that focuses on describing motion through parameters such as displacement, velocity, acceleration, and time. Unlike kinetics, kinematics does not consider the forces or masses involved. It provides the foundation for analyzing particle motion, which is critical in various applications, such as predicting the trajectory of a projectile, determining the speed of a vehicle, or analyzing the motion of a robotic arm.

In engineering, kinematics allows for precise modeling of systems where motion is a key factor. For example, understanding the kinematics of machinery components ensures smooth operation and proper alignment in systems like gears, linkages, and cams.

Numerical Example

A particle moves along a straight line with a constant acceleration of 2 m/s^2. Its initial velocity is 5 m/s, and its initial position is 0 m. Calculate

1. the particle's position after 10 s
2. its velocity at the same time

Solution

Step 1: Determine the position after 10 seconds.

1. Use the kinematic equation for displacement:

$$s = ut + \frac{1}{2}at^2$$

2. Substitute the values to obtain

$$s = 50 + \frac{1}{2}(2)(100) = 50 + 100 = 150 \text{ m}$$

Step 2: Find the velocity after 10 seconds.

1. Use the kinematic equation for velocity:

$$v = u + at$$

2. Substitute the values to obtain

$$v = 5 + (2)(10) = 25 \text{ m/s}$$

ChatGPT Integration

- Use ChatGPT to derive kinematic equations step by step from first principles.
- Request ChatGPT solve problems involving different initial conditions or accelerations.
- Ask ChatGPT for examples of real-world systems where kinematics plays a key role, such as vehicle motion or projectile trajectories.

PROMPT 102

Derive the equations of motion for a particle under uniform acceleration.

Objective

To derive the fundamental equations of motion that describe the displacement, velocity, and acceleration of a particle under constant acceleration.

Activity

Uniform acceleration describes a motion where the rate of change of velocity remains constant. This type of motion is frequently encountered in physics and engineering, such as in free-falling objects, uniformly accelerating vehicles, or conveyor belt systems. Deriving the equations of motion allows for the prediction of a particle's behavior at any given time or location along its trajectory.

The equations of motion are derived from the basic definitions of velocity and acceleration, integrated over time. These equations form the basis for solving practical problems in kinematics.

Derivation of the Equations of Motion

1. First equation: $v = u + at$

- Definition of acceleration:

$$a = \frac{dv}{dt}$$

- Rearrange and integrate:

$$\int_{u}^{v} dv = \int_{0}^{t} a\, dt$$

- Result:

$$v = u + at$$

2. Second equation: $s = ut + \frac{1}{2}at^2$

- Velocity as the rate of change of displacement:

$$v = \frac{ds}{dt}$$

- Substitute $v = u + at$ into the equation:

$$\frac{ds}{dt} = u + at$$

Rearrange and integrate:

$$\int_{0}^{s} ds = \int_{0}^{t} (u + at)\, dt$$

- Solve:

$$s = ut + \frac{1}{2}at^2$$

3. Third equation: $v^2 = u^2 + 2as$

- Start with the chain rule for acceleration:

$$a = v\frac{dv}{ds}$$

- Rearrange and integrate: $\int_{u}^{v} v\, dv = \int_{0}^{s} a\, ds \Rightarrow \frac{v^2}{2} - \frac{u^2}{2} = as$
- Solve:

$$v^2 = u^2 + 2as$$

Final Equations of Motion

1. $v = u + at$

2. $s = ut + \dfrac{1}{2}at^2$

3. $v^2 = u^2 + 2as$

ChatGPT Integration

▪ Ask ChatGPT to explain the derivation step by step.

▪ Request real-world examples where these equations apply, such as braking distances or projectile motion.

▪ Use ChatGPT to simulate the motion of a particle under various acceleration conditions.

PROMPT 103

Explain the difference between rectilinear and curvilinear motion, with examples.

Objective

To understand the distinction between rectilinear and curvilinear motion and their applications in analyzing particle trajectories.

Activity

Rectilinear motion occurs when a particle moves along a straight path, whereas *curvilinear motion* describes movement along a curved trajectory. Rectilinear motion is simpler, as the position, velocity, and acceleration are all aligned along a single direction. In contrast, curvilinear motion requires analyzing components of velocity and acceleration in multiple directions.

For example, the motion of a car on a straight road is rectilinear, while the motion of a roller coaster or a satellite in orbit is curvilinear. These concepts are crucial for understanding real-world systems such as the trajectory of projectiles, the paths of robotic arms, and the movement of fluids along curved surfaces.

Numerical Example

A particle moves along a curved path defined by $x = 3t^2$ and $y = 2t^3$, where t is in seconds, x and y are in meters. Determine

1. the particle's velocity components at $t = 2$ s
2. the magnitude of the velocity at $t = 2$ s
3. the direction of the velocity at $t = 2$ s (angle with respect to the x-axis)

Solution

Step 1: Determine the velocity components.

1. Use the formulas for the velocity components:

$$v_x = \frac{dx}{dt}, \quad v_y = \frac{dy}{dt}$$

2. Derive v_x and v_y:

$$v_x = \frac{d}{dt}\left(3t^2\right) = 6t, \quad v_y = \frac{d}{dt}\left(2t^3\right) = 6t^2$$

3. Substitute $t = 2$ s into the formulas:

$$v_x = 6(2) = 12 \text{ m/s}, \quad v_y = 6\left(2^2\right) = 24 \text{ m/s}$$

Step 2: Determine the magnitude of velocity.

1. Use the formula for the magnitude:

$$v = \sqrt{v_x^2 + v_y^2}$$

2. Substitute values to obtain

$$v = \sqrt{12^2 + 24^2} = \sqrt{720} = 26.83 \text{ m/s}$$

Step 3: Find the direction of the Velocity.

1. Use the formula for the angle:

$$\theta = \tan^{-1}\left(\frac{v_y}{v_x}\right)$$

2. Substitute values to obtain

$$\theta = \tan^{-1}\left(\frac{24}{12}\right) = 63.43°$$

ChatGPT Integration

- Ask ChatGPT to differentiate between rectilinear and curvilinear motion with additional real-world examples.
- Request ChatGPT solve problems involving other parametric paths for curvilinear motion.
- Use ChatGPT to simulate velocity and acceleration components for specified trajectories.

PROMPT 104

Solve a problem involving the motion of a particle with constant velocity.

Objective

To understand how to analyze particle motion with constant velocity and calculate displacement over a given time interval.

Activity

Constant velocity motion is the simplest type of motion in kinematics. In this case, the particle covers equal distances in equal intervals of time. The velocity remains unchanged in magnitude and direction, meaning there is no acceleration. Examples include a car cruising at a constant speed or a conveyor belt moving at a steady rate.

Understanding this concept is critical for simplifying more complex systems, such as determining relative motion or initial conditions for accelerated motion.

Numerical Example

A particle moves in a straight line with a constant velocity of 15 m/s.

1. Calculate the displacement of the particle after 8 seconds.
2. Determine how long it will take for the particle to cover a distance of 300 m.

Solution

Step 1: Calculate the displacement after 8 seconds.

1. Use the formula for the displacement:

$$s = v \cdot t$$

2. Substitute values:

$$s = 15 \cdot 8 = 120 \text{ m}$$

Step 2: Determine the time to cover 300 meters.

1. Rearrange the formula:

$$t = \frac{s}{v}$$

2. Substitute values:

$$t = \frac{300}{15} = 20 \text{ seconds}$$

ChatGPT Integration

- Ask ChatGPT to provide examples of systems with constant velocity motion, such as conveyor belts or ships in calm water.
- Request ChatGPT simulate similar problems with varying values of velocity and displacement.
- Use ChatGPT to explore scenarios involving relative motion between two objects moving at constant velocities.

PROMPT 105

Analyze the motion of a particle in a straight-line using velocity-time and displacement-time graphs.

Objective

To understand how to represent and interpret particle motion in a straight line using velocity-time and displacement-time graphs, and to extract key motion characteristics.

Activity

Graphical analysis provides a powerful tool for understanding the motion of particles. A velocity-time graph reveals the particle's velocity changes and can be used to calculate displacement via the area under the curve. A displacement-time graph shows how a particle's position changes over time, allowing for the determination of velocity by analyzing the slope of the curve.

These techniques are widely used in engineering applications, such as analyzing braking systems, conveyor belts, and robotic arms, where motion must be precisely controlled and monitored.

Numerical Example

A particle starts from rest and accelerates uniformly at 2 m/s² for 5 seconds. It then continues at a constant velocity for 3 seconds.

1. Plot the velocity-time graph and displacement-time graph for this motion.

2. Calculate the total displacement of the particle.

Solution

Step 1: Create the velocity-time graph.

1. Determine the velocities:
   From $v = u + at$, after $5\,\text{s}$,
   $$v = 0 + (2)(5) = 10 \text{ m/s}$$

2. Create the graph description:
   - From $t = 0\,\text{s}$ to $t = 5\,\text{s}$, the velocity increases linearly from 0 to $10\,\text{m/s}$.
   - From $t = 5\,\text{s}$ to $t = 8\,\text{s}$, the velocity remains constant at $10\,\text{m/s}$.

Step 2: Create the displacement-time graph.

1. Calculate displacements:
   - First segment (0s to 5s):
     Use $s = ut + \dfrac{1}{2}at^2$:
     $$s_1 = (0)(5) + \frac{1}{2}(2)(5^2) = 25 \text{ m}$$
   - Second segment (5s to 8s):
     Use $s = vt$:
     $$s_2 = (10)(3) = 30 \text{ m}$$

2. **Graph description:**
   - From $t = 0\,\text{s}$ to $t = 5\,\text{s}$, the displacement increases parabolically as the particle accelerates.

- From $t = 5\,\text{s}$ to $t = 8\,\text{s}$, the displacement increases linearly as the velocity remains constant.

Step 3: Find the total displacement.

1. Use the formula for the total displacement:

$$s_{\text{total}} = s_1 + s_2$$

2. Substitute values to obtain

$$s_{\text{total}} = 25 + 30 = 55 \text{ m}$$

ChatGPT Integration

- Ask ChatGPT to explain how to derive displacement from the velocity-time graph.
- Use ChatGPT to simulate motion for different acceleration values and durations.
- Request ChatGPT generate similar problems involving graphical motion analysis.

PROMPT 106

Discuss the concept of relative velocity in one-dimensional motion and solve a related problem.

Objective

To understand the concept of relative velocity in one-dimensional motion and learn how to calculate the relative motion of two particles or objects moving along the same or opposite directions.

Activity

Relative velocity describes the velocity of one object as observed from another moving object. It is particularly useful in understanding the motion of objects in the same frame of reference, such as cars on a highway or boats in a river. The relative velocity between two objects depends on their velocities and the direction of their motion.

For example, two vehicles moving in the same direction at different speeds will have a smaller relative velocity compared to vehicles moving toward each other. Engineers often use this concept in designing systems like conveyor belts, drones, or intersecting transportation systems.

Numerical Example

Two cars, A and B, are traveling on a straight road. Car A moves at a constant velocity of 20 m/s, and car B moves at 15 m/s:

1. Determine the relative velocity of A with respect to B when both cars are moving in the same direction.
2. Determine the relative velocity of A with respect to B if they move toward each other.

Solution

Step 1: Determine the relative velocity for the same direction.

1. Use the formula:

$$v_{rel} = v_A - v_B$$

2. Substitute values:

$$v_{rel} = 20 - 15 = 5 \text{ m/s}$$

Step 2: Determine the relative velocity for the opposite direction. Substitute values:

$$v_{rel} = 20 - (-15) = 35 \text{ m/s}$$

ChatGPT Integration

- Ask ChatGPT to explain the concept of relative velocity in multidimensional motion.
- Request ChatGPT solve problems involving relative velocity in different frames of reference.
- Use ChatGPT to explore applications of relative velocity in drone navigation or traffic flow analysis.

PROMPT 107

Explain projectile motion and derive the expressions for range, maximum height, and time of flight.

Objective

To understand the physics of projectile motion and derive closed-form expressions for range, maximum height, and time of flight of an object launched with an initial velocity at an angle.

Activity

Projectile motion is an important concept in dynamics and engineering, involving two-dimensional motion under the influence of gravity, with no air resistance.

This activity involves

1. decomposing the initial velocity into horizontal and vertical components
2. analyzing vertical motion using kinematic equations
3. deriving expressions for maximum height, total flight time, and horizontal range

Numerical Example

Derive the general equations for projectile motion given

- Initial velocity: v_0
- Launch angle: θ
- Acceleration due to gravity: g
- Neglect air resistance

Solution

1. Resolve the initial velocity into components.

$$v_{0x} = v_0 \cos\theta$$

$$v_{0y} = v_0 \sin\theta$$

2. Determine the time of flight.

   The total time of flight is based on the time taken to rise and fall vertically:

   $$t_{\text{flight}} = \frac{2v_0 \sin\theta}{g}$$

3. Determine the maximum height.

   The maximum height is reached when vertical velocity becomes zero:

   $$H = \frac{(v_0 \sin\theta)^2}{2g}$$

4. Determine the range of the projectile.

   The range is the horizontal distance traveled during the flight time:

   $$R = v_0 \cos\theta \cdot \left(\frac{2v_0 \sin\theta}{g}\right)$$

   $$R = \frac{v_0^2 \sin(2\theta)}{g}$$

ChatGPT Integration

You can use ChatGPT to

- derive projectile motion equations for different launch angles and elevations
- simulate projectile motion with and without air resistance
- visualize trajectory plots and analyze how changing parameters affect time, range, and height

PROMPT 108

Solve a problem involving a particle's motion in two dimensions with constant acceleration.

Objective

To analyze the motion of a particle in two dimensions under constant acceleration and calculate its position, velocity, and trajectory.

Activity

Two-dimensional motion under constant acceleration occurs in scenarios such as projectile motion or objects moving along inclined planes. By resolving the motion into horizontal and vertical components, we can use kinematic equations to describe the trajectory and calculate quantities like displacement, velocity, and time.

This concept is essential in engineering for analyzing paths of vehicles, projectiles, and objects subjected to gravity or other uniform forces.

Numerical Example

A particle is launched with an initial velocity of 25 m/s at an angle of 30° to the horizontal.

1. Calculate the particle's position (x, y) after 2 s.
2. Determine the particle's velocity components (v_x, v_y) at $t = 2$ s.
3. Find the total velocity magnitude and direction at $t = 2$ s.

Use $g = 9.81$ m/s².

Solution

Step 1: Find the horizontal and vertical motion components.

1. Solve for the horizontal velocity (u_x):

$$u_x = u \cos \theta = 25 \cos 30° = 25 \times 0.866 = 21.65 \text{ m/s}$$

2. Solve for the vertical velocity (u_y):

$$u_y = u \sin \theta = 25 \sin 30° = 25 \times 0.5 = 12.5 \text{ m/s}$$

Step 2: Position after 2 seconds.

1. Solve for the horizontal displacement (x):

$$x = u_x \cdot t = 21.65 \times 2 = 43.3 \text{ m}$$

2. Solve for the vertical displacement (y):

$$y = u_y \cdot t - \frac{1}{2} g t^2 = 12.5 \times 2 - \frac{1}{2} \times 9.81 (2^2) = 5.38 \text{ m}$$

Step 3: Determine the velocity components at 2 seconds.

1. Solve for the horizontal velocity (v_x):

$$v_x = u_x = 21.65 \text{ m/s}$$

2. Solve for the vertical velocity (v_y):

$$v_y = u_y - g \cdot t = 12.5 - 9.81 \times 2 = -7.12 \text{ m/s}$$

Step 4: Determine the total velocity and direction.

1. Find the magnitude of the velocity (v):

$$v = \sqrt{v_x^2 + v_y^2} = \sqrt{(21.65)^2 + (-7.12)^2} = \sqrt{519.41} = 22.79 \text{ m/s}$$

2. Find the direction of the velocity (θ):

$$\theta = \tan^{-1}\left(\frac{v_y}{v_x}\right) = \tan^{-1}\left(\frac{-7.12}{21.65}\right) = -18.17°$$

ChatGPT Integration

- Use ChatGPT to verify the calculations step by step.
- Ask ChatGPT to simulate similar two-dimensional motion problems with varying initial velocities and angles.
- Request ChatGPT explore the effect of air resistance on the motion for advanced analysis.

PROMPT 109

Discuss motion in polar coordinates and its application to curved paths.

Objective

To understand how to describe particle motion using polar coordinates and analyze velocity and acceleration components for curved paths.

Activity

Polar coordinates are used to describe motion in terms of radial (r) and angular (θ) components. This system is particularly useful for analyzing motion along curved paths, such as circular or spiral trajectories. The position, velocity, and acceleration of a particle can be expressed in terms of the radial (r) and transverse (θ) directions.

Applications of polar coordinates include analyzing the motion of pendulums, planetary orbits, and rotating machinery components, where the trajectory is naturally described in terms of angles and radii.

Important Equations in Polar Coordinates

1. **Position Vector:**

$$\vec{r} = r\hat{e}_r$$

2. **Velocity:**

$$\vec{v} = \dot{r}\hat{e}_r + r\dot{\theta}\hat{e}_\theta$$

3. **Acceleration:**

$$\vec{a} = \left(\ddot{r} - r\dot{\theta}^2\right)\hat{e}_r + \left(r\ddot{\theta} + 2\dot{r}\dot{\theta}\right)\hat{e}_\theta$$

Numerical Example

A particle moves along a spiral path described by $r = 2t$ and $\theta = kt^2$, where r is in meters, θ is in radians, t is in seconds, and k has the units of s^{-2}. Assume $k = 0.5\ s^{-2}$. Determine

1. the velocity components (v_r, v_θ) at $t = 2$ s
2. the acceleration components (a_r, a_θ) at $t = 2$s

Solution

Step 1: Determine the velocity components.

1. Radial velocity (v_r):

$$v_r = \dot{r} = \frac{d}{dt}(2t) = 2 \text{ m/s}$$

2. Transverse velocity (v_θ):

$$\dot{\theta} = \frac{d}{dt}(kt^2) = 2kt = 2(0.5)(2) = 2 \text{ rad/s}$$

Step 2: Deteremine the acceleration components.

1. Radial acceleration (a_r):

$$a_r = \ddot{r} - r\dot{\theta}^2 = \frac{d^2}{dt^2}(2t) - (2t)(2kt)^2$$

At $t = 2$s, $k = 0.5$ 1/s²:

$$a_r = \ddot{r} - r\dot{\theta}^2 = 0 - 8\ k^2 t^2, \quad a_r = -8(0.5)^2(2^2) = -8 \text{ m/s}^2$$

2. Transverse acceleration (a_θ):

Have $\theta = \dfrac{d}{dt}(2kt) = 2k$. At $t = 2s$ and $k = 0.5$ 1/s², we obtain

$$a_\theta = r\ddot\theta + 2\dot r\dot\theta = 4kt + 8kt = 12kt, \quad a_\theta = 12(0.5)(2) = 12 \text{ m/s}^2$$

ChatGPT Integration

▪ Ask ChatGPT to explain the derivation of motion equations in polar coordinates step by step.

▪ Request ChatGPT solve similar problems involving motion along circular or spiral paths.

▪ Use ChatGPT to visualize the trajectory of particles described in polar coordinates.

PROMPT 110

Solve a problem involving the velocity and acceleration components of a particle moving in a circular path.

Objective

To calculate the velocity and acceleration components of a particle moving along a circular path using radial and transverse components.

Activity

Motion along a circular path is commonly encountered in engineering systems such as gears, turbines, and robotic arms. The velocity and acceleration components are expressed in terms of radial (r) and angular (θ) coordinates. This problem will demonstrate how to calculate these components for a particle undergoing circular motion.

Numerical Example

A particle moves along a circular path of radius 5 m. Its angular velocity is given by $\omega = 3t$ rad/s, and its angular acceleration is $\alpha = 3$ rad/s². Determine

1. the tangential velocity (v_θ) and radial velocity (v_r) at $t = 2$ s

2. the radial acceleration (a_r) and tangential acceleration (a_θ) at $t = 2$ s

Solution

Step 1: Determine the velocity components.

1. Tangential velocity (v_θ):

   At $t = 2s$,

   $$v_\theta = r \cdot \omega, \quad v_\theta = 5(3 \times 2) = 30 \text{ m/s}$$

2. Radial velocity (v_r):

   For circular motion with constant radius $(r = 5 \text{ m})$, the radial velocity is zero:

   $$v_r = 0$$

Step 2: Find the acceleration components.

1. Radial acceleration (a_r):

   At $t = 2s$,

   $$a_r = -r \cdot \omega^2, \quad a_r = -5(3 \times 2)^2 = -180 \text{ m/s}^2$$

2. Tangential acceleration (a_θ):

   At $t = 2\,s$,

   $$a_\theta = r \cdot \alpha, \quad a_\theta = 5 \times 3 = 15 \text{ m/s}^2$$

ChatGPT Integration

- Use ChatGPT to derive the radial and tangential acceleration components step by step.
- Ask ChatGPT for applications of circular motion in real-world systems, such as rotating machinery or planetary orbits.
- Request ChatGPT simulate the motion of particles with varying angular velocities and accelerations.

CHAPTER 12

KINETICS OF PARTICLES: FORCE AND ACCELERATION

Prompt 111 Define the concept of force and its relationship to acceleration using Newton's second law.

Prompt 112 Solve a problem involving the motion of a particle subjected to a single constant force.

Prompt 113 Explain the concept of net force and its effect on a particle in dynamic equilibrium.

Prompt 114 Derive the equation of motion for a particle under the influence of multiple forces.

Prompt 115 Solve a problem involving a particle moving on an inclined plane with variable friction.

Prompt 116 Discuss the application of free-body diagrams in analyzing the kinetics of particles.

Prompt 117 Solve a problem involving the motion of a particle under the action of variable forces.

Prompt 118 Explain the concept of inertial and non-inertial frames of reference with examples.

Prompt 119 Discuss the role of force components in analyzing motion in two or three dimensions.

Prompt 120 Solve a problem involving the motion of a particle under centripetal force in a circular path.

PROMPT 111

Define the concept of force and its relationship to acceleration using Newton's second law.

Objective

To understand the concept of force, its relationship with acceleration, and how Newton's second law governs particle motion.

Activity

Force is a vector quantity that describes the interaction between objects. It causes a change in motion, as per Newton's second law, which states that the net force acting on a particle is directly proportional to the rate of change of its momentum. For most engineering applications, the relationship is simplified to

$$\vec{F}_{net} = m\vec{a}$$

This principle forms the foundation for analyzing dynamics in mechanical systems, from simple objects on inclined planes to complex motion in multi-body systems.

Numerical Example

A 10 kg box is subjected to a constant horizontal force of 50 N on a frictionless surface.

1. Calculate the acceleration of the box.
2. If the force is applied for 5 s, determine the velocity and displacement of the box at the end of this time.

Solution

Step 1: Calculate the acceleration.

1. Use Newton's second law:

$$\vec{F}_{net} = m\vec{a}$$

2. Rearrange to solve for the acceleration and solve:

$$\vec{a} = \frac{\vec{F}_{net}}{m} = \frac{50}{10} = 5 \, \text{m/s}^2$$

Step 2: Determine the velocity after 5 seconds.

1. Use the equation for velocity:

$$v = u + at$$

2. Substitute to obtain

$$v = 0 + (5)(5) = 25 \text{ m/s}$$

Step 3: Determine the displacement after 5 seconds.

1. Use the equation for the displacement:

$$s = ut + \frac{1}{2}at^2$$

2. Substitute to obtain

$$s = 0 + \frac{1}{2}(5)(25) = 62.5 \text{ m}$$

ChatGPT Integration

- Use ChatGPT to explain Newton's second law with additional examples from everyday life.
- Request ChatGPT solve problems involving forces acting on multiple objects or systems.
- Explore advanced topics, such as the effect of variable forces, using ChatGPT.

PROMPT 112

Solve a problem involving the motion of a particle subjected to a single constant force.

Objective

To calculate the acceleration, velocity, and displacement of a particle subjected to a constant force using Newton's second law.

Activity

When a particle is subjected to a constant force, its motion can be analyzed using Newton's second law, $\vec{F}_{net} = m\vec{a}$. The force generates constant acceleration, which allows us to apply the kinematic equations for uniform acceleration to compute velocity and displacement. This is widely applicable in analyzing vehicles, machinery, and projectiles.

Numerical Example

A 20 kg object is subjected to a horizontal force of 100 N on a surface with a frictional force of 20 N.

1. Calculate the acceleration of the object.
2. If the force is applied for 4 s4 \, \text{s}4s, determine the object's velocity and displacement.

Solution

Step 1: Calculate the acceleration.

1. Use Newton's second law:

$$\vec{F}_{net} = m\vec{a}$$

2. Perform the net force calculation:

$$\vec{F}_{net} = F_{applied} - F_{friction}$$

Substitute $F_{applied} = 100\,N$ and $F_{friction} = 20\,N$:

$$\vec{F}_{net} = 100 - 20 = 80\ N$$

3. Determine the acceleration:

$$a = \frac{80}{20} = 4\ m/s^2$$

Step 2: Determine the velocity after 4 seconds.

1. Use the equation for velocity:

$$v = u + at$$

2. By substituting the initial velocity u=0, we obtain

$$v = 0 + (4)(4) = 16\ m/s$$

Step 3: Determine the displacement after 4 seconds.

1. Use the equation for the displacement:

$$s = ut + \frac{1}{2}at^2$$

2. By substituting, we obtain

$$s = 0 + 1/2(4)(16) = 32 \text{ m}$$

ChatGPT Integration

- Use ChatGPT to simulate motion problems with different force magnitudes and frictional conditions.
- Ask ChatGPT to explain the impact of varying frictional forces on the acceleration and displacement.
- Explore the extension to systems with variable forces using ChatGPT.

PROMPT 113

Explain the concept of net force and its effect on a particle in dynamic equilibrium.

Objective

To understand how net force determines the motion of a particle and how equilibrium conditions apply to dynamic systems.

Activity

Net force ($\vec{F}net$) is the vector sum of all forces acting on a particle. In dynamic equilibrium, the net force is nonzero, resulting in a constant acceleration rather than zero velocity. Newton's second law governs the relationship between the net force, mass, and acceleration:

$$\vec{F}_{net} = m\vec{a}$$

Dynamic equilibrium is a key concept in systems where constant forces, such as thrust or friction, influence steady acceleration or deceleration. Examples include vehicles moving with uniform acceleration or objects sliding on inclined planes.

Numerical Example

A 5 kg block slides down an inclined plane making an angle of 30° with the horizontal. The coefficient of kinetic friction between the block and the plane is 0.2. Calculate

1. the net force acting on the block
2. the acceleration of the block

Solution

Step 1: Perform the free-body diagram analysis.

1. Identify the forces acting on the block:
   - Gravitational force: $F_g = mg$
   - Normal force: F_N
   - Frictional force: $F_f = \mu_k F_N$

2. Identify the components of the gravitational force:
   - Parallel to the plane:

$$F_g^{\parallel} = F_g \sin\theta$$

   - Perpendicular to the plane:

$$F_g^{\perp} = F_g \cos\theta$$

3. Determine the normal force:

$$F_N = F_g^{\perp} = mg\cos\theta$$

4. Determine the frictional force:

$$F_f = \mu_k F_N = \mu_k mg\cos\theta$$

Step 2: Determine the net force.

1. Identify the net force along the incline:

$$\vec{F}_{net} = F_g^{\parallel} - F_f$$

2. Substitute the components:

$$\vec{F}_{net} = mg\sin\theta - \mu_k\, mg\cos\theta$$

3. Perform the numerical substitutions:

$m = 5$ kg, g = 9.81 m/s^2, μk = 0.2, and θ = 30°

$$\vec{F}_{net} = (5)(9.81)(\sin 30°) - (0.2)(5)(9.81)(\cos 30°) = 24.525 - 8.496 = 16.029$$

Step 3: Determine the acceleration.

Use Newton's second law:

$$\vec{a} = \frac{\vec{F}_{net}}{m}$$

Substitute in $\vec{F}_{net}$ = 16.029 N and m = 5:

$$a = \frac{16.029}{5} = 3.206 \text{ m/s}^2$$

ChatGPT Integration

- Use ChatGPT to explain the decomposition of forces in dynamic equilibrium scenarios.
- Ask ChatGPT to simulate motion on inclined planes with different angles and friction coefficients.
- Explore how varying net forces affect acceleration and motion using ChatGPT.

PROMPT 114

Derive the equation of motion for a particle under the influence of multiple forces.

Objective

To understand how multiple forces acting on a particle are combined to derive its equation of motion using Newton's second law.

Activity

When multiple forces act on a particle, their vector sum determines the net force, which in turn dictates the particle's motion. Newton's second law can be written as

$$\vec{F}_{net} = \sum \vec{F}_i = m\vec{a}$$

where $\sum \vec{F}_i$ represents the sum of all forces acting on the particle. This derivation is critical for analyzing systems involving multiple forces, such as a block on an inclined plane with friction, tension in cables, or aerodynamic drag.

Numerical Example

A 10 kg object is subjected to three forces:

1. $\vec{F}_1 = 50\,N$ at $0°$(horizontal)
2. $\vec{F}_2 = 30\,N$ at $90°$ (vertical)
3. $\vec{F}_1 = 40\,N$ at $180°$(horizontal, opposite direction)

Determine

1. the net force acting on the object
2. the resulting acceleration of the object

Solution

Step 1: Resolve the forces into components.

Resolve each force into the horizontal (x) and vertical (y) components.

1. Horizontal components (F_x):

$$F_{1x} = F_1 \cos 0° = 50 \text{ N}, \quad F_{2x} = F_2 \cos 90° = 0, \quad F_{3x} = F_3 \cos 180° = -40 \text{ N}$$

Net horizontal force:
$$F_x = F_{1x} + F_{2x} + F_{3x} = 50 + 0 - 40 = 10 \text{ N}$$

2. Vertical components (F_y):

$$F_{1y} = F_1 \sin 0° = 0, \quad F_{2y} = F_2 \sin 90° = 30 \text{ N}, \quad F_{3y} = F_3 \sin 180° = 0$$

Net vertical force:
$$F_y = F_{1y} + F_{2y} + F_{3y} = 0 + 30 + 0 = 30 \text{ N}$$

Step 2: Calculate the net force.

1. Find the magnitude of the net force:

$$F_{net} = \sqrt{F_x^2 + F_y^2}$$

Through substitution, we obtain
$$F_{net} = \sqrt{10^2 + 30^2} = \sqrt{100 + 900} = \sqrt{1000} = 31.62 \text{ N}$$

2. Determine the direction of the net force (θ):

$$\theta = \tan^{-1}\left(\frac{F_y}{F_x}\right)$$

Through substitution, we obtain
$$\theta = \tan^{-1}\left(\frac{30}{10}\right) = 71.57°$$

Step 3: Calculate the acceleration.

Use Newton's second law:

$$a = \frac{31.62}{10} = 3.162 \text{ m/s}^2$$

ChatGPT Integration

- Use ChatGPT to derive equations of motion for multiple force systems step by step.
- Request ChatGPT analyze similar problems involving inclined planes, pulleys, or drag forces.
- Ask ChatGPT for guidance on visualizing force diagrams for complex systems.

PROMPT 115

Solve a problem involving a particle moving on an inclined plane with variable friction.

Objective

To analyze the motion of a particle on an inclined plane when the coefficient of friction varies with velocity, applying Newton's second law.

Activity

Inclined planes with variable friction are common in real-world systems where friction changes based on surface properties or speed, such as braking systems or conveyor belts. This problem introduces a dynamic coefficient of friction:

$$\mu_k = 0.2 + 0.05\,v$$

where v is the particle's velocity in m/s. The goal is to calculate acceleration under these varying conditions.

Numerical Example

A 10 kg block slides down an inclined plane at 30°. The coefficient of kinetic friction is given by $\mu_k = 0.2 + 0.05\,v$, where v is the velocity in m/s. Determine the block's acceleration when its velocity is 4 m/s.

Solution

Step 1: Calculate the forces.

1. Identify the gravitational force (F_g).

$$F_g = mg$$

Through substitution, we obtain

$$F_g = 10 \times 9.81 = 98.1 \text{ N}$$

2. Identify the components of the gravitational force.
   - Parallel to the plane:

$$F_g^{\parallel} = F_g \sin\theta = 98.1 \times \sin 30° = 49.05 \text{ N}$$

   - Perpendicular to the plane:

$$F_g^{\perp} = F_g \cos\theta = 98.1 \times \cos 30° = 84.96 \text{ N}$$

3. Identify the frictional force (F_f).

   The coefficient of friction is velocity-dependent.
   Through substitution $(v = 4 \text{ m/s})$, we obtain

$$\mu_k = 0.2 + 0.05 \times 4 = 0.4$$

   Solve for the frictional force:

$$F_f = \mu_k F_g^{\perp} = 0.4 \times 84.96 = 33.98 \text{ N}$$

Step 2: Determine the net force.

1. Identify the net force along the incline:

$$\vec{F}_{net} = F_g^{\parallel} - F_f$$

Through substitution, we obtain

$$\vec{F}_{net} = 49.05 - 33.98 = 15.07 \text{ N}$$

Step 3: Find the acceleration.

Use Newton's second law:

$$a = \frac{15.07}{10} = 1.51 \text{ m/s}^2$$

ChatGPT Integration

- Use ChatGPT to explore how variable friction impacts motion in other systems, such as braking vehicles or conveyor belts.
- Request ChatGPT analyze how changes in the velocity affect acceleration in real-time simulations.
- Experiment with different coefficients of friction to observe their influence on net force and acceleration.

PROMPT 116

Discuss the application of free-body diagrams in analyzing the kinetics of particles.

Objective

To understand the role of free-body diagrams (FBDs) in solving kinetics problems by visually representing forces and moments acting on a particle.

Activity

A free-body diagram is a simplified visual representation of a particle or body, isolating it to show all forces and moments acting upon it. FBDs help engineers and physicists identify the net force acting on a particle and are essential for applying Newton's second law effectively. By resolving forces into components and eliminating irrelevant details, FBDs provide clarity in solving problems involving motion, equilibrium, and force analysis.

Numerical Example

A 5 kg block is being pulled up a 25° incline by a rope with a tension $T = 50$ N. The coefficient of kinetic friction between the block and the incline is $\mu_k = 0.1$. Draw the free-body diagram and calculate

1. the net force acting on the block
2. the acceleration of the block

Solution

Step 1: Create the free-body diagram.

The FBD includes

1. gravitational force (F_g) acting vertically downward
2. tension force (T) acting along the incline upward
3. frictional force (F_f) opposing motion, acting along the incline downward
4. normal force (F_N) perpendicular to the incline

Step 2: Resolve the forces.

1. Use the gravitational force (F_g):

$$F_g = mg$$

Through substitution, we obtain

$$F_g = 5 \times 9.81 = 49.05 \text{ N}$$

2. Use the components of the gravitational force:
   - Parallel to the incline:

$$F_g^{\parallel} = F_g \sin\theta$$

   - Perpendicular to the incline:

$$F_g^{\perp} = F_g \cos\theta$$

Through substitution $(\theta = 25°)$, we obtain

$$F_g^{\parallel} = 49.05 \times \sin 25° = 20.73 \text{ N}, \quad F_g^{\perp} = 49.05 \times \cos 25° = 44.45 \text{ N}$$

3. Determine the frictional force (F_f):

$$F_f = \mu_k F_N$$

The normal force is $F_f = \mu_k F_g^{\perp}$, so we find

$$F_f = 0.1 \times 44.45 = 4.44 \text{ N}$$

Step 3: Determine the net force.

1. Identify the net force along the incline:

$$\vec{F}_{net} = T - F_g^{\parallel} - F_f$$

2. Through substitution, we obtain

$$\vec{F}_{net} = 50 - 20.73 - 4.44 = 24.83 \text{ N}$$

Step 4: Determine the acceleration.

1. Use Newton's second law:

$$\vec{a} = \frac{\vec{F}_{net}}{m}$$

2. Through substitution, we obtain

$$a = \frac{24.83}{5} = 4.97 \text{ m/s}^2$$

ChatGPT Integration

- Use ChatGPT to create free-body diagrams for various scenarios.
- Ask ChatGPT to simulate changes in incline angles, friction coefficients, or applied forces.
- Request explanations for resolving forces and their applications in engineering problems.

PROMPT 117

Solve a problem involving the motion of a particle under the action of variable forces.

Objective

To analyze the motion of a particle under the influence of a variable force, using integration to determine velocity and displacement as functions of time.

Activity

Variable forces occur in many engineering scenarios, such as spring systems, aerodynamic drag, or magnetic forces. Unlike constant forces, the net force varies with time, position, or velocity. Analyzing these requires applying Newton's second law and solving differential equations or integrating force expressions. This problem will explore how to find the velocity and displacement of a particle under a time-dependent force.

Numerical Example

A particle of mass 2 kg is acted upon by a force $F(x) = -kx^2$, where $k = 3$ N/m² and x is the position in meters. The particle starts from rest at $x = 4$ m.

1. Determine its velocity as a function of position.
2. Find the particle's position when its velocity reaches 2 m/s.

Solution

Step 1: Write the equation of motion.

1. Use Newton's second law:

$$\vec{F}_{net} = m\vec{a} \quad \Rightarrow \quad F(x) = m\frac{d^2x}{dt^2}$$

2. Substitute $F(x) = -kx^2$:

$$-kx^2 = m\frac{d^2x}{dt^2}$$

3. Relate the acceleration to the velocity:

$$a = \frac{d^2x}{dt^2} = \frac{dv}{dt} = \frac{dv}{dx}\cdot\frac{dx}{dt} = v\frac{dv}{dx}$$

4. Through substitution, we obtain

$$-kx^2 = m \cdot v\frac{dv}{dx}$$

Step 2: Determine the velocity as a function of position.

1. Rearrange the equation and integrate:

$$\int v\, dv = -\frac{k}{m}\int x^2\, dx$$

The following are the solutions to the integrals:

$$\frac{v^2}{2} = -\frac{k}{m}\cdot\frac{x^3}{3} + C$$

2. Apply the initial condition ($v = 0$ when $x = 4$):

$$C = \frac{k}{m}\times\frac{64}{3}$$

3. Determine the final velocity equation:

$$v^2 = \frac{2k}{m}\left(\frac{64 - x^3}{3}\right)$$

Step 3: Determine the position when velocity is 2 m/s.

Set $v = 2$ m/s in the velocity equation and solve for x:

$$4 = \frac{6}{2}\left(\frac{64 - x^3}{3}\right) \quad \Rightarrow \quad x = 3.91\,\text{m}$$

Among the three solutions, the acceptable one is $x \approx 3.915$ m.

ChatGPT Integration

- Use ChatGPT to analyze other position-dependent forces, such as gravitational forces in orbital mechanics or restoring forces in springs.
- Simulate problems with variable initial conditions or non-linear force relationships.
- Request step-by-step guidance for similar advanced mechanics problems.

PROMPT 118

Explain the concept of inertial and non-inertial frames of reference with examples.

Objective

To understand the distinction between inertial and non-inertial frames of reference, including the forces and principles applicable in each case, and to analyze their significance in solving mechanics problems.

Activity

Inertial and non-inertial frames of reference are essential concepts in classical mechanics. An *inertial frame of reference* is one in which Newton's laws of motion hold true without modification. In contrast, a *non-inertial frame of reference* is an accelerating or rotating frame where fictitious forces, such as the Coriolis force or centrifugal force, must be introduced to explain motion.

In daily life, inertial frames are often approximated by stationary or uniformly moving systems (e.g., a train moving at a constant velocity), while non-inertial frames include accelerating vehicles or rotating systems like carousels. The concept is crucial in analyzing relative motion and solving problems involving forces and accelerations.

Numerical Example

A passenger is sitting in a car accelerating at $a = 3$ m/s^2. The passenger observes a ball of mass 0.5 kg hanging from the ceiling of the car, making an angle θ with the vertical.

1. Determine the angle θ of the string with the vertical in the car's frame (non-inertial).
2. Find the tension in the string.

Solution

Step 1: Analyze the forces in the non-inertial frame.
In the car's non-inertial frame

1. the ball appears to be stationary relative to the observer
2. a fictitious force, F_f, acts horizontally opposite to the car's acceleration:

$$F_f = ma$$

Step 2: Resolve the forces.

The forces acting on the ball are

1. tension T in the string, is resolved into the
    - horizontal component: $T\sin\theta$
    - vertical component: $T\cos\theta$
2. the gravitational force F_g = mg acting downward
3. fictitious force $F_f = ma$ acting horizontally

In equilibrium,

$$T\cos\theta = mg, \quad T\sin\theta = ma$$

Step 3: Determine θ.

Divide the two equations and substitute to obtain

$$\tan\theta = \frac{a}{g} = \frac{3}{9.81} = 0.306$$

Solve for θ:

$$\theta = \tan^{-1}(0.306) = 17°$$

Step 4: Find the tension.

$$T = \frac{mg}{\cos\theta}$$

By substituting, we obtain

$$T = \frac{0.5 \times 9.81}{0.956} \approx 5.13 \text{ N}$$

ChatGPT Integration

* Use ChatGPT to simulate scenarios involving other fictitious forces, such as centrifugal or Coriolis forces.
* Ask for step-by-step guidance on solving problems in non-inertial frames.
* Explore examples of inertial and non-inertial frames in everyday situations, such as elevators or rotating systems.

PROMPT 119

Discuss the role of force components in analyzing motion in two or three dimensions.

Objective

To understand how forces in two or three dimensions can be resolved into components, enabling a systematic approach to analyzing motion using Newton's laws.

Activity

Force components are vital in mechanics to break down a force into manageable parts along chosen axes, simplifying calculations. In 2D problems, forces are often resolved into horizontal and vertical components. In 3D problems, forces involve components along x, y, and z-axes. These components are combined using vector addition to determine net forces, accelerations, and trajectories.

Applications include projectile motion, inclined planes, and forces on particles in systems such as pendulums or spaceships. Understanding these components is essential for solving problems involving equilibrium or dynamics in complex systems.

Numerical Example

A particle of mass $m = 5$ kg is subjected to three forces:

- $\vec{F_1} = 50$ N at $0°$, to the horizontal

  $\vec{F_2} = 30$ N at $90°$

- $\vec{F_3} = 40$ N at $180°$

Determine

1. the net force acting on the particle
2. the resulting acceleration

Solution

Step 1: Resolve the forces into components.

1. Define the force components along the x-axis.

$$F_{1x} = F_1 \cos 0° = 50 \text{ N}, \quad F_{2x} = F_2 \cos 90° = 0, \quad F_{3x} = F_3 \cos 180° = -40 \text{ N}$$

$$F_x = F_{1x} + F_{2x} + F_{3x} = 50 + 0 - 40 = 10 \text{ N}$$

2. Define the force components along the y-axis.

$$F_{1y} = F_1 \sin 0° = 0, \quad F_{2y} = F_2 \sin 90° = 30 \text{ N}, \quad F_{3y} = F_3 \sin 180° = 0$$

$$F_y = F_{1y} + F_{2y} + F_{3y} = 0 + 30 + 0 = 30 \text{ N}$$

Step 2: Determine the net force.

Define the magnitude of the net force:

$$F_{net} = \sqrt{F_x^2 + F_y^2}$$

By substituting values, we obtain

$$F_{net} = \sqrt{10^2 + 30^2} = \sqrt{100 + 900} = \sqrt{1000} = 31.62 \text{ N}$$

Determine the direction of the net force:

$$\theta = \tan^{-1}\left(\frac{F_y}{F_x}\right)$$

By substituting values, we obtain

$$\theta = \tan^{-1}\left(\frac{30}{10}\right) = \tan^{-1}(3) \approx 71.57°$$

Step 3: Determine the resulting acceleration.

Use Newton's second law:

$$a = \frac{F_{net}}{m}$$

Substitute $F_{net} = 31.62$ N and $m = 5$ kg to obtain

$$a = \frac{31.62}{5} = 6.32 \text{ m/s}^2$$

ChatGPT Integration

- Use ChatGPT to solve problems involving forces in 3D systems.
- Request detailed examples involving inclined planes or systems with multiple forces.
- Experiment with varying force magnitudes and angles to analyze their effects on motion.

PROMPT 120

Solve a problem involving the motion of a particle under centripetal force in a circular path.

Objective

To analyze the motion of a particle in uniform circular motion, derive the relationship between centripetal force, velocity, and radius, and solve a problem involving these parameters.

Activity

Centripetal force is the inward force required to keep a particle moving in a circular path. It is directly proportional to the square of the particle's velocity and inversely proportional to the radius of the path. Real-world applications include satellite orbits, vehicle turns, and rotating machinery.

Understanding centripetal force enables engineers to design safe curves for roads, calculate the forces on rotating components, and predict the behavior of orbiting bodies.

Numerical Example

A car of mass 1500 kg moves around a circular track of radius 50 m at a constant speed of 20 m/s.

1. Determine the centripetal force acting on the car.
2. Calculate the coefficient of friction required between the tires and the road to maintain this motion.

Solution

Step 1: Write the formula for the centripetal force.
The centripetal force is given by

$$F_c = \frac{mv^2}{r}$$

where
- $m = 1500$ kg, the mass of the car
- $v = 20$ m/s, the speed
- $r = 50$ m, the radius

Step 2: Calculate the centripetal force.

Substitute the given values into the formula:

$$F_c = \frac{1500 \times 20^2}{50} = 12 \text{ kN}$$

Step 3: Relate the friction to the centripetal force.
For the car to maintain its circular motion, the force of friction must equal the centripetal force. The frictional force is given by

$$F_f = \mu N$$

where
- μ is the coefficient of friction
- $N = mg$ is the normal force

Substitute $N = mg$:

$$F_f = \mu \, mg$$

Equate the frictional force to the centripetal force and solve for ì :

$$\mu = \frac{F_c}{mg}$$

Step 4: Calculate the coefficient of friction.

Substitute in $F_c = 12000$ N, $m = 1500$ kg, and $g = 9.81$ m/s²:

$$\mu = \frac{12000}{1500 \times 9.81} = 0.816$$

ChatGPT Integration

- Use ChatGPT to explore the effects of varying the radius or speed on the centripetal force.
- Ask for examples involving banked curves or non-uniform circular motion.
- Simulate scenarios for real-world applications like roller coasters or centrifuges.

CHAPTER 13

KINETICS OF PARTICLES: WORK AND ENERGY

Prompt 121 Define the concepts of work and energy in the context of particle motion. Explain the work-energy principle with examples.

Prompt 122 Derive the expression for the work done by a force acting on a particle along a straight path.

Prompt 123 Solve a problem involving the work done by a variable force on a particle.

Prompt 124 Discuss the concept of kinetic energy and its relationship with work.

Prompt 125 Explain the potential energy function for conservative forces and its significance in mechanical systems.

Prompt 126 Solve a problem involving the conservation of mechanical energy in a system with no non-conservative forces.

Prompt 127 Discuss the role of power in particle motion and derive its mathematical expression.

Prompt 128 Solve a problem involving the instantaneous power delivered to a particle by a varying force.

Prompt 129 Explain the concept of work done by non-conservative forces and how it modifies the work-energy principle.

Prompt 130 Solve a multi-step problem involving the combined effects of conservative and non-conservative forces.

PROMPT 121

Define the concepts of work and energy in the context of particle motion. Explain the work-energy principle with examples.

Objective

To introduce the concepts of work and energy, explain their significance in particle motion, and understand the work-energy principle through practical applications.

Activity

Work and energy are foundational concepts in dynamics. *Work* is the transfer of energy through force applied over a displacement. It quantifies the effort exerted on a particle or body. *Energy* is the ability of a system to perform work, existing in various forms such as kinetic and potential energy. The work-energy principle states that the net work done on a particle equals its change in kinetic energy.

For example, lifting an object against gravity involves work, while the kinetic energy of a moving car increases due to the engine's work against resistive forces.

Numerical Example

A block of mass 5 kg is pushed along a frictionless horizontal surface by a constant force $F = 20$ N over a distance of 4 m.

1. Calculate the work done by the force.
2. Determine the block's final velocity if it starts from rest.

Solution

Step 1: Calculate the work done.

The work W done by a constant force is given by

$$W = F \cdot d \cdot \cos\theta$$

where

- $F = 20$ N
- $d = 4$ m
- $\theta = 0°$ (force is in the direction of motion)

Substitute the values to obtain

$$W = 20 \times 4 \cdot \cos 0° = 20 \times 4 \times 1 = 80\,J$$

The work done by the force is 80 J.

Step 2: Apply the work-energy principle.

The work-energy principle states

$$W = \Delta KE = \frac{1}{2}mv^2 - \frac{1}{2}mu^2$$

where

- $m = 5$ kg
- $u = 0$ m/s (starts from rest)
- $W = 80$ J

Substitute the values to obtain

$$80 = \frac{1}{2} \times 5 \times v^2$$

Simplify and solve for v:

$$v = \sqrt{32} = 5.66 \text{ m/s}$$

The block's final velocity is approximately 5.66 m/s.

ChatGPT Integration

- Ask ChatGPT to explain the work-energy principle with additional examples, such as lifting objects or accelerating cars.
- Explore variations of this problem by changing the force, angle, or distance.
- Use ChatGPT to calculate work in scenarios with multiple forces or resistive forces.

PROMPT 122

Derive the expression for the work done by a force acting on a particle along a straight path.

Objective

To derive the mathematical expression for the work done by a force acting on a particle, including variable forces, and understand the principles underlying the calculation.

Activity

The work done by a force is a measure of energy transferred to or from a particle as it moves along a path. For a constant force, the calculation is straightforward, but for a variable force, integration is required. This derivation is fundamental in analyzing energy transformations in mechanical systems.

Derivation

Step 1: Find the work done by a constant force.

The work W done by a constant force F over a displacement d is

$$W = F \cdot d \cdot \cos\theta$$

where

- F is the magnitude of the force
- d is the displacement
- θ is the angle between the force and displacement vectors

Step 2: Find the work done by a variable force.

For a variable force $\vec{F}(x)$, the displacement is divided into infinitesimally small segments dx. The work done over each segment is:

$$dW = F(x) \cdot dx$$

The total work done as the particle moves from x_1 to x_2 is the integral of dW:

$$W = \int_{x_1}^{x_2} F(x)\, dx$$

Step 3: Find the work in vector form.

In three-dimensional space, the work done by a force $\vec{F}$ as the particle moves along a path is given by the line integral:

$$W = \int_{r_1}^{r_2} \vec{F} \cdot d\vec{r}$$

where

- $\vec{F}$ is the force vector
- $d\vec{r}$ is the infinitesimal displacement vector

Numerical Example

A particle moves along the x-axis under a variable force $F(x) = 5x^2$ N. Find the work done by the force as the particle moves from $x = 1$ m to $x = 3$ m.

Solution

Step 1: Write the expression for work.

The work is given by

$$W = \int_{x_1}^{x_2} F(x)\, dx$$

Substitute $F(x) = 5x^2$ and solve the integral:

$$\int 5x^2\, dx = \left[\frac{5x^3}{3} \right]_1^3 = 43.33\, J$$

ChatGPT Integration

- Ask ChatGPT to derive work expressions for forces in other coordinate systems (e.g., polar coordinates).
- Use ChatGPT to practice line integrals for forces acting in 2D or 3D.
- Explore problems where $F(x)$ is non-linear or includes trigonometric functions.

PROMPT 123

Solve a problem involving the work done by a variable force on a particle

Objective

To solve a numerical example involving the work done by a variable force on a particle, emphasizing the integration approach.

Activity

Work done by a variable force is calculated by integrating the force over the displacement. This method is crucial in systems where forces depend on position, such as springs or gravitational fields.

Numerical Example

A particle is subjected to a force $F(x) = 6x^2$ N along the x-axis. Find the work done as the particle moves from $x = 1$ to $x = 4$.

Solution

Step 1: Write the expression for work.

The work done is given by

$$W = \int_{x_1}^{x_2} F(x)\,dx$$

Step 2: Solve the integral.
Substitute $F(x) = 6x^2$:

$$W = \int_1^4 6x^2\,dx = \left[2x^3 \right]_1^4 = 126\,J$$

ChatGPT Integration

- Use ChatGPT to explore other examples of variable forces and their work calculations.
- Ask ChatGPT to demonstrate work done by forces with exponential or trigonometric dependencies.
- Request ChatGPT provide practice problems involving variable force integration.

PROMPT 124

Discuss the concept of kinetic energy and its relationship with work.

Objective

To explain the concept of kinetic energy and its connection with the work-energy principle, highlighting its importance in mechanical analysis.

Activity

Kinetic energy is the energy associated with a particle's motion. According to the work-energy theorem, the net work done on a particle equals the change in its kinetic energy. This principle is widely used in analyzing motion and energy transformations in mechanical systems. Understanding this relationship is crucial for solving a wide range of engineering problems.

Numerical Example

A block of mass $m = 5$ kg is moving along a frictionless horizontal surface with an initial velocity $u = 2$ m/s. A constant force $F = 10$ N is applied over a distance $d = 4$ m. Determine the final velocity v of the block using the work-energy theorem.

Solution

Step 1: Write the work-energy theorem.

The work-energy theorem states

$$W = \Delta KE = \frac{1}{2}mv^2 - \frac{1}{2}mu^2$$

Step 2: Calculate the work done.

The work done by the force is

$$W = F \cdot d$$

Substitute the values to obtain

$$W = 10 \times 4 = 40 \text{ J}$$

Step 3: Apply the work-energy theorem.

From the work-energy theorem,

$$40 = \frac{1}{2} \times 5 \times v^2 - \frac{1}{2} \times 5 \times 2^2$$

Simplify:

$$v^2 = \frac{50}{2.5} = 20$$

$$v = \sqrt{20} \approx 4.47 \, \text{m/s}$$

ChatGPT Integration

You can use ChatGPT to

- explore scenarios involving non-constant forces and their effects on kinetic energy
- derive the work-energy theorem for systems with rotational motion or variable mass
- generate additional practice problems involving energy transformations and their applications in engineering systems

PROMPT 125

Explain the potential energy function for conservative forces and its significance in mechanical systems.

Objective

To define the potential energy function for conservative forces, emphasizing its role in energy conservation and mechanical system analysis.

Activity

Potential energy represents stored energy in a system due to its position or configuration in a force field.

Conservative forces, such as gravitational and elastic (spring) forces, possess potential energy functions that allow engineers to apply the principle of energy conservation instead of directly analyzing force and motion.

This is fundamental in analyzing systems like pendulums, springs, and planetary motion, where the total mechanical energy remains constant in the absence of non-conservative forces.

Numerical Example

A spring with a stiffness constant $k = 200 \, \text{N} / \text{m}$ is compressed by a distance $x = 0.1 \, \text{m}$.

Determine the potential energy stored in the spring.

Solution

Step 1: Define the potential energy function.

The potential energy stored in a spring (Hooke's Law) is

$$U = \frac{1}{2} k x^2$$

Step 2: Substitute known values.

Substitute $k = 200 \, \dfrac{\text{N}}{\text{m}}$ and $x = 0.1 \, \text{m}$

Step 3: Simplify.

$$U = \frac{1}{2} \times 200 \times 0.01 = 1 \, \text{J}$$

ChatGPT Integration

You can use ChatGPT to

- derive potential energy functions for other conservative forces, such as gravitational and electrostatic forces
- request practice problems involving energy conservation in systems with springs, pendulums, and projectiles
- explain the relationship between potential energy, force, and system stability, including the use of derivatives to identify equilibrium points

PROMPT 126

Solve a problem involving the conservation of mechanical energy in a system with no non-conservative forces.

Objective

To demonstrate the application of energy conservation in an idealized system, assuming no energy dissipation due to non-conservative forces such as friction or air resistance.

Activity

Energy conservation is a powerful tool in analyzing mechanical systems. By equating the total mechanical energy at two different positions, we can solve for unknown variables without explicitly using Newton's laws or resolving forces. This principle is widely used in modeling pendulums, roller coasters, and free-fall problems, where mechanical energy is conserved in the absence of non-conservative forces.

Numerical Example

A block of mass $m = 2$ kg slides down a frictionless inclined plane of height $h = 5$ m. Determine the velocity of the block at the bottom of the incline.

Solution

Step 1: Write the conservation of mechanical energy equation.

$$E_{top} = E_{bottom}$$

Expand the energy terms:

$$U_{top} + KE_{top} = U_{bottom} + KE_{bottom}$$

Step 2: Simplify for the given system.

At the top of the incline, we have

$$U_{top} = mgh, \quad KE_{top} = 0$$

At the bottom of the incline, we have

$$U_{bottom} = 0, \quad KE_{bottom} = \frac{1}{2}mv^2$$

Step 3: Solve for the velocity.

Substitute into the conservation equation to obtain

$$mgh = \frac{1}{2}mv^2$$

Cancel m from both sides and solve for v:

$$v^2 = 2 \times 9.81 \times 5 = 98.1 \Rightarrow v = \sqrt{98.1} \approx 9.9 \text{ m/s}$$

ChatGPT Integration

You can use ChatGPT to

- verify energy conservation in systems involving pendulums, sliding blocks, or rotating bodies
- analyze real-world cases where friction or air resistance modifies the ideal conservation equations
- explore variations of this problem, such as changing incline angles, initial speeds, or adding spring forces

PROMPT 127

Discuss the role of power in particle motion and derive its mathematical expression.

Objective

To explore the concept of power as the rate of work done, derive its mathematical relationship with force and velocity, and highlight its significance in engineering systems.

Activity

Power quantifies the rate at which work is performed or energy is transferred. In mechanical systems, power is directly related to force and velocity. Understanding power is crucial in optimizing energy usage in machines and vehicles, as well as in analyzing their efficiency. Engineers use power calculations to design engines, transmissions, and control systems for performance and energy efficiency.

Numerical Example

A car engine applies a constant force of $F = 1000$ N to move the car at a constant velocity of

$v = 20$ m/s. Calculate the power delivered by the engine.

Solution

Step 1: Define power.

Power is defined as the rate of doing work:

$$P = \frac{dW}{dt}$$

Step 2: Express work in terms of force and displacement.

Work is defined as

$$W = F \cdot d$$

For constant velocity v, displacement is $d = v \cdot t$, so

$$P = F \cdot \frac{d}{dt}(vt)$$

$$P = F \cdot v$$

Step 3: Solve for the given example.

Substitute the known values to obtain

$$P = 1000 \times 20 = 20000 \text{ W}$$

ChatGPT Integration

You can use ChatGPT to

- explore power calculations for systems with *non-constant forces* or *velocities*
- solve problems involving *rotational power* and *torque*
- compare power outputs of different *mechanical systems*, such as engines, motors, and turbines

PROMPT 128

Solve a problem involving the instantaneous power delivered to a particle by a varying force.

Objective

To calculate the instantaneous power delivered to a particle under a non-constant force, highlighting the importance of real-time energy analysis in dynamic systems.

Activity

Instantaneous power provides insight into the dynamic performance of systems where forces or velocities vary with time. It is especially important in applications such as electric motors, turbines, and regenerative braking systems, where performance and energy efficiency depend on time-dependent interactions between force and motion.

This activity involves

1. defining and applying the formula for instantaneous power
2. substituting time-dependent expressions for force and velocity
3. evaluating the power at a specific time

Numerical Example

A particle experiences a force

$$F(t) = 10e^{-t} \text{ N}$$

while moving with a velocity

$$v(t) = 2t \text{ m/s}$$

Find the instantaneous power delivered to the particle at $t = 2\,\text{s}$.

Solution

Step 1: Define the instantaneous power.

The instantaneous power is the product of force and velocity:

$$P(t) = F(t) \cdot v(t)$$

Step 2: Substitute in the force and velocity functions.

Substitute $F(t) = 10e^{-t}$ and $v(t) = 2t$:

$$P(t) = \left(10e^{-t}\right) \cdot (2t)$$

Step 3: Evaluate at $t = 2$ s.

$$P(2) = \left(10e^{-2}\right) \times (4) = 10 \times 0.135 \times 4 = 5.4 \text{ W}$$

ChatGPT Integration

You can use ChatGPT to

- solve similar problems involving time-varying forces and velocities
- analyze real-world examples of instantaneous power in systems like electric motors and turbines
- explore how instantaneous power analysis contributes to energy efficiency in dynamic systems

PROMPT 129

Explain the concept of work done by non-conservative forces and how it modifies the work-energy principle.

Objective

To introduce the concept of non-conservative forces, such as friction and air resistance, and discuss their effects on mechanical energy transformations.

Activity

Non-conservative forces dissipate mechanical energy as heat, sound, or deformation, altering the total energy balance of a system. This modified principle is essential for accurately analyzing real-world systems where energy is not conserved due to friction or other dissipative effects. Engineers use this principle to design efficient machines and predict motion outcomes in systems like sliding blocks, vehicle brakes, or industrial processes.

Numerical Example

A block of mass $m = 10$ kg slides down a rough inclined plane of height $h = 5$ m and length $L = 10$ m. The coefficient of kinetic friction between the block and the incline is $\mu_k = 0.2$. Determine the velocity of the block at the bottom of the incline.

Solution

Step 1: Write the modified work-energy principle.

$$W_{nc} = \Delta KE + \Delta U$$

where

- W_{nc} is the work done by non-conservative forces
- ΔKE is the change in kinetic energy
- ΔU is the change in potential energy

Step 2: Determine the work done by friction.

Friction performs negative work, opposing motion:

$$W_{friction} = -f_k \cdot L$$

Substitute $f_k = \mu k \cdot N$, where $N = mg \cos\theta$:

$$W_{friction} = -\mu_k \cdot (mg \cos\theta) \cdot L$$

Step 3: Determine the energy conservation with non-conservative work.

Substitute the above into the energy balance to obtain

$$-\mu_k (mg \cos\theta) \cdot L = \frac{1}{2} mv^2 - mgh$$

Solve for v^2:

$$v^2 = 2gh - 2\mu_k gL \cos\theta$$

Step 4: Evaluate known values.

Given

- $\mu_k = 0.2$
- $g = 9.81$ m/s^2
- $h = 5$ m
- $L = 10$ m

Calculate the angle factor:

$$\cos\theta = \frac{h}{L} = \frac{5}{10} = 0.5$$

Substitute into the expression for $v^2 v^\wedge 2 v^2$:

$$v^2 = 2(9.81)(5) - 2(0.2)(9.81)(10)(0.5) = 78.48$$

$$v = 8.86 \, \text{m/s}$$

ChatGPT Integration

You can use ChatGPT to

- analyze systems with non-conservative forces, such as friction, drag, or damping
- request step-by-step solutions for energy-based problems involving energy loss
- explore the impact of different coefficients of friction or inclinations on energy dissipation and final velocity

PROMPT 130

Solve a multi-step problem involving the combined effects of conservative and non-conservative forces.

Objective

To solve a multi-step problem that incorporates both conservative and non-conservative forces, highlighting their combined effects on the energy transformations within a system.

Activity

The analysis of systems with both conservative and non-conservative forces involves accounting for both energy conservation and dissipation. This approach is essential in understanding real-world mechanical systems such as sliding blocks, pendulums with air resistance, and vehicles moving on uneven terrain.

Numerical Example

A block of mass $m = 5 \, kg$ is attached to a spring with stiffness constant $k = 100 \, \text{N/m}$. The block is initially compressed by $x = 0.2 \, \text{m}$ and then released. It slides on a rough horizontal surface with a coefficient of kinetic friction $\mu_k = 0.1$.

Determine the velocity of the block after it has traveled $d = 0.5$ m on the rough surface.

Solution

Step 1: Write the total energy equation.

The total mechanical energy includes contributions from spring potential energy, kinetic energy, and work done by friction:

$$U_{spring} + KE_{initial} + W_{nc} = KE_{final}$$

Step 2: Calculate the initial spring potential energy.

The potential energy stored in the spring is

$$U_{spring} = \frac{1}{2}kx^2 = \frac{1}{2} \times 100(0.2)^2 = 2\,J$$

Step 3: Calculate the work done by friction.

Work done by friction is

$$W_{friction} = -f_k \cdot d$$

Where friction force $f_k = \mu_k\ N$, and $N = mg$:

$$W_{friction} = -\mu_k \cdot m \cdot g \cdot d$$

$$W_{friction} = -0.1 \cdot (5 \cdot 9.81) \cdot 0.5 = -2.45\,J$$

Step 4: Apply the energy equation.

Use the total energy balance:

$$2 + 0 - 2.45 = \frac{1}{2}mv^2$$

This leads to a nonphysical (negative) kinetic energy, which means the block does not reach the full 0.5 m travel. Let's recalculate using energy balance up to the point where spring energy is completely dissipated by friction.

Corrected Approach: Find Distance the Block Can Travel Before Stopping

Set

$$U_{spring} = -W_{friction}$$

$$2 = \mu k \cdot m \cdot g \cdot d$$

Solve for d:

$$d = \frac{2}{\mu_k \cdot m \cdot g} = \frac{2}{0.1 \cdot 5 \cdot 9.81} \approx 0.407 \text{ m}$$

Interpretation

Since the block can only travel 0.407 m before stopping, it never reaches the full 0.5 m distance. Therefore, the final velocity at 0.5 m is not defined — the block has already stopped due to friction.

ChatGPT Integration

You can use ChatGPT to

- analyze energy transformations in systems with both conservative and non-conservative forces
- solve problems involving varying coefficients of friction or nonlinear spring behavior
- generate visualizations of energy input, dissipation, and motion in real-world mechanical systems

CHAPTER 14

KINETICS OF PARTICLES: IMPULSE AND MOMENTUM

Prompt 131 Define the concepts of impulse and momentum and explain their relationship.

Prompt 132 Derive the impulse-momentum equation for a particle.

Prompt 133 Solve a problem involving the impulse applied to a particle and its resulting change in momentum.

Prompt 134 Explain the principle of conservation of linear momentum and its applications.

Prompt 135 Compare and contrast elastic and inelastic collisions, providing examples.

Prompt 136 Discuss the concept of angular momentum for a particle and derive its mathematical expression.

Prompt 137 Solve a problem involving the angular impulse-momentum principle.

Prompt 138 Explain the relationship between external forces and the change in linear momentum of a system.

Prompt 139 Discuss the impact of impulsive forces in mechanical systems and their significance.

Prompt 140 Solve a real-world problem involving a collision or impact between two particles using the impulse-momentum principle.

PROMPT 131

Define the concepts of impulse and momentum and explain their relationship.

Objective

To introduce the fundamental concepts of impulse and momentum, explain their mathematical definitions, and establish their relationship in mechanical systems.

Activity

Impulse and momentum are fundamental principles in dynamics. Momentum represents the quantity of motion a particle possesses, while impulse describes the effect of a force applied over a time interval. The impulse-momentum theorem relates these two concepts and is essential in analyzing collisions, braking forces, and impact scenarios in engineering applications.

Numerical Example

A 2 kg object is moving with an initial velocity of 3 m/s. A force of 10 N is applied to the object for 4 s. Determine the final velocity of the object after the force is applied.

Solution

Step 1: Define the impulse-momentum theorem.

The impulse-momentum theorem states

$$J = \Delta p = mv_f - mv_i$$

Step 2: Compute the impulse.

Impulse is given by the integral of force over time:

$$J = \int_{t_1}^{t_2} F \, dt$$

For a constant force, this simplifies to

$$J = F \cdot \Delta t$$

By substituting $F = 10$ N and $\Delta t = 4$ s, we obtain

$$J = 10 \times 4 = 40 \text{ N.s}$$

Step 3: Apply the impulse-momentum theorem.

Use the impulse-momentum equation:

$$40 = 2v_f - 2(3)$$

Solve for v_f:

$$v_f = \frac{46}{2} = 23 \text{ m/s}$$

ChatGPT Integration

- Use ChatGPT to analyze impulse-momentum relationships in different scenarios such as vehicle braking or collisions.
- Ask ChatGPT to derive variations of the impulse equation for non-constant forces.
- Request ChatGPT generate visualizations of impulse effects on moving objects.

PROMPT 132

Derive the impulse-momentum equation for a particle.

Objective

To derive the impulse-momentum equation by integrating Newton's second law, demonstrating how force applied over time changes a particle's momentum.

Activity

Impulse and momentum are closely related concepts in dynamics. Newton's second law states that the rate of change of momentum is equal to the applied force. By integrating this equation over a time interval, we derive the impulse-momentum theorem. This derivation is crucial for understanding collisions, force interactions, and impact dynamics in engineering applications.

Derivation

Step 1: Start with Newton's second law.

Newton's second law states that the force applied to a particle equals the time rate of change of its momentum:

$$F = \frac{dp}{dt}$$

Step 2: Express the momentum in terms of the mass and velocity. Momentum is defined as

$$p = m v$$

For a constant mass system, the time derivative of momentum is

$$\frac{dp}{dt} = m\frac{dv}{dt} = ma$$

Step 3: Integrate both sides over a time interval. Impulse is defined as the integral of force over time:

$$J = \int_{t_1}^{t_2} F\, dt$$

Substituting $F = \dfrac{dp}{dt}$, we obtain

$$\int_{t_1}^{t_2} F\, dt = \int_{p_1}^{p_2} dp$$

Evaluate the integral:

$$J = p_2 - p_1 = mv_f - mv_i$$

Step 4: Solve the final impulse-momentum equation.

Thus, the impulse-momentum equation is

$$J = \Delta p = mv_f - mv_i$$

ChatGPT Integration

- Use ChatGPT to derive variations of the impulse equation for systems with varying mass.
- Ask ChatGPT to explain impulse in the context of real-world applications such as airbags or sports collisions.
- Request ChatGPT provide step-by-step solutions for complex impulse-momentum problems.

PROMPT 133

Solve a problem involving the impulse applied to a particle and its resulting change in momentum.

Objective

To apply the impulse-momentum theorem in solving a practical problem, demonstrating how impulse affects a particle's velocity.

Activity

Impulse represents the effect of a force acting over a time interval, resulting in a change in momentum. By applying the impulse-momentum theorem, we can determine how an external force alters an object's motion. This principle is widely used in engineering applications such as vehicle crashes, sports, and force analysis in mechanical systems.

Numerical Example

A 5 kg ball is initially moving at 4 m/s in the positive x-direction. A force of 15 N is applied in the same direction for 3 seconds. Determine the final velocity of the ball.

Solution

Step 1: Write the impulse-momentum equation.

The impulse-momentum theorem states

$$J = \Delta p = mv_f - mv_i$$

Step 2: Compute the impulse.

Impulse is given by the integral of force over time:

$$J = F \cdot \Delta t$$

For a constant force, this simplifies to

$$J = 15 \times 3 = 45 \text{ Ns}$$

Substituting F = 15 N and Δt = 3 s.

Step 3: Apply the impulse-momentum theorem.

Using the impulse-momentum equation:

$$45 = 5v_f - 5(4)$$

$$v_f = \frac{65}{5} = 13 \text{ m/s}$$

ChatGPT Integration

- Use ChatGPT to explore impulse-momentum problems involving forces at different angles.
- Ask ChatGPT to generate real-world scenarios where impulse is crucial, such as car crashes or bouncing balls.
- Request ChatGPT visualize the effect of impulse on objects with varying masses and forces.

PROMPT 134

Explain the principle of conservation of linear momentum and its applications.

Objective

To explain the principle of conservation of linear momentum, derive its mathematical form, and discuss its applications in engineering and physics.

Activity

The principle of conservation of linear momentum states that in the absence of external forces, the total momentum of a system remains constant. This principle is fundamental in analyzing collisions, rocket propulsion, and multi-body interactions in mechanical systems. Engineers apply this concept in designing safety mechanisms, impact absorption systems, and analyzing motion in vehicles and machinery.

Derivation and Explanation

Step 1: Define the linear momentum.

Linear momentum is defined as

$$p = m\,v$$

For a system of particles, the total momentum is

$$P_{total} = \sum m_i v_i$$

Step 2: Apply Newton's second law.

Newton's second law states

$$F_{net} = \frac{dP}{dt}$$

If no external forces act on the system, then

$$F_{net} = 0 \Rightarrow \frac{dP}{dt} = 0$$

$$P = \text{constant}$$

This shows that in the absence of external forces, momentum remains conserved.

Numerical Example

Two ice skaters, one with a mass of 60 kg and the other with a mass of 40 kg, push off each other while initially at rest. The 60 kg skater moves at 2 m/s after the push. Determine the velocity of the 40 kg skater.

Solution

Step 1: Apply the principle of the conservation of the linear momentum. The total initial momentum is

$$P_{initial} = 0$$

After they push off, we find

$$m_1 v_1 + m_2 v_2 = 0$$

Substituting values gives us

$$(60)(2) + (40) v_2 = 0$$

Solving for v_2:

$$v_2 = -3 \text{ m/s}$$

The negative sign indicates that the second skater moves in the opposite direction.

ChatGPT Integration

- Use ChatGPT to simulate conservation of momentum in multi-body systems.
- Ask ChatGPT to solve real-world momentum problems, such as rocket propulsion.
- Request ChatGPT visualize elastic and inelastic collisions using physics simulations.

PROMPT 135

Compare and contrast elastic and inelastic collisions, providing examples.

Objective

To explain the key differences between elastic and inelastic collisions, their governing principles, and provide real-world examples demonstrating their effects in mechanical systems.

Activity

Collisions are categorized based on how they conserve energy and momentum. In elastic collisions, both kinetic energy and momentum are conserved. In inelastic collisions, momentum is conserved, but kinetic energy is not, as some energy is converted into deformation, heat, or sound. Understanding these principles is crucial in engineering applications such as crash safety analysis, material testing, and mechanical impact dynamics.

TABLE 14.1 Comparison of elastic and inelastic collisions

Property	Elastic Collision	Inelastic Collision
Momentum	Conserved	Conserved
Kinetic Energy	Conserved	Not conserved (partially lost)
Deformation	Objects do not permanently deform	Objects may deform or stick together
Example	Billiard balls colliding	A car crash with crumpling of the frame

Numerical Example

A 2 kg ball moving at 5 m/s collides elastically with a 4 kg ball that is initially at rest. Determine their velocities after the collision.

Solution

Step 1: Apply the principle of the conservation of momentum.

For a two-body system,

$$m_1 v_{1i} + m_2 v_{2i} = m_1 v_{1f} + m_2 v_{2f}$$

Substituting given values, we obtain

$$(2)(5) + (4)(0) = (2) v_{1f} + (4) v_{2f}$$

$$2 v_{1f} + 4 v_{2f} = 10$$

Step 2: Apply the principle of the conservation of kinetic energy.

For elastic collisions, we find

$$\frac{1}{2}m_1 v_{1i}^2 + \frac{1}{2}m_2 v_{2i}^2 = \frac{1}{2}m_1 v_{1f}^2 + \frac{1}{2}m_2 v_{2f}^2$$

Substituting given values, we obtain

$$\frac{1}{2}(2)(5)^2 + \frac{1}{2}(4)(0)^2 = \frac{1}{2}(2)v_{1f}^2 + \frac{1}{2}(4)v_{2f}^2$$

$$v_{1f}^2 + 2v_{2f}^2 = 25$$

Step 3: Solve for the final velocities.

Solving these two equations simultaneously gives

$$v_{1f} = -1.67 \text{ m/s}, \quad v_{2f} = 3.33 \text{ m/s}$$

ChatGPT Integration

- Use ChatGPT to simulate elastic and inelastic collisions using interactive animations.
- Ask ChatGPT to analyze how energy loss varies between different materials during collisions.
- Request ChatGPT solve real-world collision problems in crash simulations and sports physics.

PROMPT 136

Discuss the concept of angular momentum for a particle and derive its mathematical expression.

Objective

To introduce the concept of angular momentum, derive its mathematical expression, and explain its significance in rotational motion.

Activity

Angular momentum is a fundamental quantity in rotational dynamics, analogous to linear momentum in translational motion. It is essential for analyzing the motion of rotating bodies, gyroscopic effects, and orbital mechanics. The conservation of angular momentum is widely applied in engineering and physics, from turbine design to spaceflight trajectory adjustments.

Derivation

Step 1: Define the angular momentum.

Angular momentum for a particle is defined as the cross product of its position vector and linear momentum.

$$\vec{L} = \vec{r} \times \vec{p}$$

Since linear momentum is given by $\vec{p} = m\vec{v}$, angular momentum can be expressed as

$$\vec{L} = \vec{r} \times m\vec{v}$$

Step 2: Express the angular momentum in terms of perpendicular components.
The magnitude of angular momentum is

$$L = mrv \sin\theta$$

where θ is the angle between the position vector and the velocity vector. For a particle in circular motion, velocity is related to angular velocity as

$$v = r\omega$$

Substituting into the equation for angular momentum, we find

$$L = mr^2\omega$$

which is the fundamental expression for angular momentum of a particle in circular motion.

Step 3: Relate the angular momentum to the torque.

The time derivative of angular momentum gives the net external torque acting on the particle.

$$\frac{d\vec{L}}{dt} = \vec{r} \times \vec{F} = \vec{\tau}$$

which states that the rate of change of angular momentum is equal to the applied torque.

Numerical Example

A particle of mass 3 kg moves in a circular path of radius 2 m with an angular velocity of 4 rad/s. Determine its angular momentum.

Solution

Step 1: Use the angular momentum formula.

From the derived equation, we find
$$L = mr^2 \omega$$

Step 2: Substitute the given values.
$$L = (3)(2^2)(4) = 48 \text{ kg.m}^2/\text{s}$$

ChatGPT Integration

- Use ChatGPT to derive the angular momentum expression for rigid bodies and extended systems.
- Ask ChatGPT to analyze real-world applications of angular momentum conservation, such as figure skating and satellite motion.
- Request ChatGPT solve numerical problems involving torque and angular impulse.

PROMPT 137

Solve a problem involving the angular impulse-momentum principle.

Objective

To apply the angular impulse-momentum principle in solving a practical problem, demonstrating how torque applied over time affects a particle's angular motion.

Activity

Angular impulse is the product of torque and the time over which it acts, leading to a change in angular momentum. This principle is essential for analyzing rotational motion in mechanical systems such as rotating wheels, turbines, and space vehicle attitude control. Understanding the angular impulse-momentum relationship allows engineers to design braking systems, robotic arms, and energy transfer mechanisms.

Numerical Example

A 2 kg disc of radius 0.5 m is initially rotating at 10 rad/s. A constant torque of 4 N·m is applied to the disc for 3 seconds. Determine the final angular velocity of the disc.

Solution

Step 1: Write the angular impulse-momentum equation.

The angular impulse-momentum principle states

For constant torque, this simplifies to

$$\Delta L = \tau \cdot \Delta t$$

Since angular momentum is defined as $L = I\omega$, we can rewrite the equation as

$$I\omega_f - I\omega_i = \tau \cdot \Delta t$$

Step 2: Compute the moment of inertia.

For a solid disc, the moment of inertia about its center is

$$I = \frac{1}{2}mr^2$$

By substituting $m = 2$ kg and $r = 0.5$ m, we obtain

$$I = \frac{1}{2}(2)(0.5)^2 = 0.25 \text{ kg} \cdot \text{m}^2$$

Step 3: Solve for the final angular velocity.

Use the impulse-momentum equation:

$$(0.25)\omega_f - (0.25)(10) = (4)(3)$$

Solve for ω_f:

$$\omega_f = \frac{14.5}{0.25} = 58 \text{ rad/s}$$

ChatGPT Integration

- Use ChatGPT to analyze how varying torque affects angular motion over different time intervals.
- Ask ChatGPT to simulate real-world applications of the angular impulse-momentum principle, such as flywheels and braking systems.
- Request ChatGPT provide step-by-step solutions to rotational motion problems involving external torques.

PROMPT 138

Explain the relationship between external forces and the change in linear momentum of a system.

Objective

To explain how external forces influence the linear momentum of a system, derive the governing equations, and discuss applications in engineering dynamics.

Activity

Linear momentum changes when an external force is applied to a system. Newton's second law states that the time rate of change of momentum is equal to the net external force. This principle is essential for analyzing vehicle acceleration, rocket propulsion, and collision forces in mechanical systems. Understanding how external forces alter momentum helps engineers design systems that control motion efficiently, such as braking systems, robotic manipulators, and safety mechanisms.

Derivation and Explanation

Step 1: Express Newton's second law in terms of momentum.

Newton's second law is given by

$$F_{net} = ma$$

Since acceleration is the time derivative of velocity, we obtain

$$F_{net} = m\frac{dv}{dt}$$

Rewrite the formula using the definition of linear momentum, $p = mv$:

$$F_{net} = \frac{dp}{dt}$$

This equation shows that the net external force acting on a system equals the rate of change of momentum.

Step 2: Integrate over a time interval.

By integrating both sides over a time interval t_1 to t_2, we obtain

$$\int_{t_1}^{t_2} F_{net}\, dt = \int_{p_1}^{p_2} dp$$

Evaluate the integral:

$$J = p_2 - p_1 = mv_f - mv_i$$

This is the impulse-momentum theorem, which states that the impulse exerted on a system equals the change in its linear momentum.

Numerical Example

A 1500 kg car accelerates from 10 m/s to 25 m/s over 5 seconds due to a constant applied force. Determine the force exerted on the car.

Solution

Step 1: Apply the momentum change equation.

$$F_{net} \cdot \Delta t = mv_f - mv_i$$

Step 2: Substitute the given values.

$$F_{net} \cdot 5 = (1500)(25) - (1500)(10)$$

$$F_{net} = \frac{22500}{5} = 4500 \text{ N}$$

ChatGPT Integration

- Use ChatGPT to analyze external force effects on multi-body systems, such as vehicle towing.
- Ask ChatGPT to simulate momentum changes in real-world scenarios like rocket propulsion or braking systems.
- Request ChatGPT derive momentum equations for varying mass systems, such as fuel-burning rockets.

PROMPT 139

Discuss the impact of impulsive forces in mechanical systems and their significance.

Objective

To explain the concept of impulsive forces, discuss their effects in mechanical systems, and analyze their significance in real-world applications such as collisions, crash safety, and high-impact machinery.

Activity

An impulsive force is a force that acts over a very short time interval but produces a significant change in momentum. These forces appear in systems involving sudden impacts, such as vehicle crashes, hammering, or bouncing balls. Engineers design protective mechanisms such as airbags, shock absorbers, and energy-absorbing materials to minimize the harmful effects of impulsive forces. Understanding the impulse-momentum principle helps in analyzing the effects of sudden forces in mechanical and structural systems.

Definition and Mathematical Representation

Step 1: Define the impulsive force.

An *impulsive force* is one that acts for a short duration but results in a large change in momentum. The mathematical definition follows from the impulse-momentum theorem:

$$J = \int_{t_1}^{t_2} F\, dt$$

For an impulsive force, the force is very large but acts over a very small time interval. In many cases, we approximate the impulse as:

$$J = F_{avg} \cdot \Delta t$$

where F_{avg} is the average force exerted over the short time interval.

Numerical Example

A 0.2 kg baseball moving at 40 m/s is struck by a bat, reversing its direction and sending it back at 50 m/s. The bat is in contact with the ball for 0.005 s. Determine the average impulsive force exerted by the bat.

Solution

Step 1: Apply the impulse-momentum theorem.

$$J = \Delta p = mv_f - mv_i$$

Step 2: Compute the impulse.

By substituting the given values, we obtain

$$J = (0.2)(-50) - (0.2)(40) = -18 \text{ N.s}$$

The negative sign indicates that the impulse acted in the opposite direction of the ball's initial motion.

Step 3: Compute the average impulsive force.

Use the impulse approximation formula:

$$J = F_{avg} \cdot \Delta t$$

By substituting values, we obtain

$$F_{avg} = \frac{-18}{0.005} = -3600 \text{ N}$$

The large negative force magnitude reflects the strong impact force applied by the bat.

ChatGPT Integration

- Use ChatGPT to analyze impulsive forces in safety systems such as airbags and helmets.
- Ask ChatGPT to generate simulations of impact forces in various collision scenarios.
- Request ChatGPT derive variations of the impulse formula for forces that vary over time.

PROMPT 140

Solve a real-world problem involving a collision or impact between two particles using the impulse-momentum principle.

Objective

To apply the impulse-momentum principle in solving a real-world collision problem, demonstrating how impulse and momentum conservation determine the outcome of an impact.

Activity

Collisions occur frequently in engineering applications, from vehicle crashes to industrial machinery and sports physics. The impulse-momentum principle helps analyze forces exerted during impact and determine final velocities of colliding bodies. By solving a real-world problem, we can understand how impulse transfers momentum between objects and how impact forces are distributed.

Numerical Example

A 1200 kg car moving at 15 m/s collides head-on with a 1000 kg stationary car. After the collision, the 1200 kg car moves at 5 m/s in the same direction. Determine the velocity of the 1000 kg car immediately after the collision, assuming momentum is conserved.

Solution

Step 1: Apply the principle of the conservation of momentum.
The total momentum before the collision is equal to the total momentum after the collision:

$$m_1 v_{1i} + m_2 v_{2i} = m_1 v_{1f} + m_2 v_{2f}$$

Step 2: Substitute the given values into the formula.

$$(1200)(15) + (1000)(0) = (1200)(5) + (1000) v_{2f}$$

Step 3: Solve for the final velocity.

$$v_{2f} = \frac{12000}{1000} = 12 \, \text{m/s}$$

ChatGPT Integration

- Use ChatGPT to analyze vehicle collision scenarios with varying masses and velocities.
- Ask ChatGPT to solve impact problems involving angular momentum and rotational motion.
- Request ChatGPT simulate collision effects using impulse-momentum principles.

CHAPTER 15

PLANAR KINEMATICS OF RIGID BODIES

Prompt 141 Define the concepts of translation, rotation, and general planar motion of rigid bodies.

Prompt 142 Derive the velocity equations for different points on a rigid body in planar motion.

Prompt 143 Explain the instantaneous center of zero velocity (ICZV) and its applications.

Prompt 144 Solve a problem involving the velocity analysis of a rigid body using the ICZV method.

Prompt 145 Derive the acceleration equations for different points on a rigid body in planar motion.

Prompt 146 Discuss the Coriolis acceleration and solve a problem involving relative motion in rotating frames.

Prompt 147 Apply vector analysis to determine the velocity and acceleration of a point on a rigid body.

Prompt 148 Solve a problem involving motion of a rigid body with combined translation and rotation.

Prompt 149 Discuss the significance of angular velocity and angular acceleration in planar motion.

Prompt 150 Analyze a real-world engineering application of planar kinematics in mechanical systems.

PROMPT 141

Define the concepts of translation, rotation, and general planar motion of rigid bodies.

Objective

To define the fundamental motion types of rigid bodies in a plane, distinguishing between translation, rotation, and general planar motion, and explaining their significance in mechanical systems.

Activity

The motion of a rigid body in a plane can be classified into three primary types: translation, rotation about a fixed axis, and general planar motion.

- Translation occurs when all points on the body move in parallel paths.
- Rotation about a fixed axis occurs when all points move in circular paths about a fixed axis perpendicular to the plane of motion.
- General planar motion is a combination of translation and rotation.

Understanding these motion types is essential in analyzing mechanisms such as linkages, gears, and robotic arms, where rigid bodies undergo complex motions.

Definition and Mathematical Representation

Step 1: Define translation.

A rigid body undergoes pure translation when every point on the body moves with the same velocity and acceleration. The displacement of any point is given by

$$\vec{r}(t) = \vec{r_0} + \vec{v}t + \frac{1}{2}\vec{a}t^2$$

Velocity and acceleration are given by

$$\vec{v} = \frac{d\vec{r}}{dt}, \quad \vec{a} = \frac{d\vec{v}}{dt}$$

Step 2: Define the rotation about a fixed axis.

A rigid body undergoes pure rotation when all points move in circular paths about a fixed axis. The angular displacement, velocity, and acceleration are given by

$$\theta = \theta_0 + \omega t + \frac{1}{2}\alpha t^2$$

$$\omega = \frac{d\theta}{dt}, \quad \alpha = \frac{d\omega}{dt}$$

Velocity of a point at a distance r from the rotation axis is

$$v = r\omega$$

Acceleration is given by

$$a_t = r\alpha, \quad a_n = r\omega^2$$

where a_t is the tangential acceleration and a_n is the normal (centripetal) acceleration.

Step 3: Define general planar motion.

A rigid body undergoes general planar motion when it has both translational and rotational components. The velocity of any point on the rigid body is

$$\vec{v}_P = \vec{v}_G + \omega \times \vec{r}_{P/G}$$

Acceleration is

$$\vec{a}_P = \vec{a}_G + \alpha \times \vec{r}_{P/G} - \omega^2 \vec{r}_{P/G}$$

where G represents the center of mass of the rigid body.

Numerical Example

A rigid bar of length 2 meters is rotating about a fixed axis at one end with an angular velocity of 4 rad/s and an angular acceleration of 2 rad/s². Determine the velocity and acceleration of the free end of the bar.

Solution

Step 1: Compute the velocity.

Use $v = r\omega$, with $r = 2$ m and $\omega = 4$ rad/s to obtain

$$v = (2)(4) = 8\,\text{m/s}$$

Step 2: Compute the acceleration.

Tangential acceleration is given by

$$a_t = (2)(2) = 4\,\text{m/s}^2$$

By substituting in $r = 2$ m and $\alpha = 2$ rad/s^2, we obtain

$$a_t = (2)(2) = 4\,\text{m/s}^2$$

Normal (centripetal) acceleration is

$$a_n = r\omega^2$$

By substituting in $\omega = 4$ rad/s, we find

$$a_n = (2)(4)^2 = 32\,\text{m/s}^2$$

The magnitude of total acceleration

$$a = \sqrt{a_t^2 + a_n^2} = \sqrt{4^2 + 32^2} = \sqrt{16 + 1024} = \sqrt{1040} = 32.25\,\text{m/s}^2$$

ChatGPT Integration

- Use ChatGPT to analyze planar motion in linkages and robotic arms.
- Ask ChatGPT to simulate general planar motion in mechanical systems.
- Request ChatGPT provide step-by-step explanations for velocity and acceleration analysis in rotating bodies.

PROMPT 142

Derive the velocity equations for different points on a rigid body in planar motion.

Objective

To derive the velocity equations for different points on a rigid body undergoing planar motion, distinguishing between pure translation, pure rotation, and general motion.

Activity

In *planar motion*, the velocity of different points on a rigid body depends on whether the motion is pure translation, pure rotation, or general planar motion (a combination of translation and rotation). A velocity analysis is fundamental in mechanical design, robotics, and linkages, where engineers determine the motion of various components.

Derivation of the Velocity Equations

Step 1: Identify the velocity in the pure translation.

In pure translation, all points on the rigid body move with the same velocity:

$$\vec{v}_P = \vec{v}_G$$

where $\vec{v}_P$ is the velocity of any point and $\vec{v}_G$ is the velocity of the center of mass.

Step 2: Identify the velocity in pure rotation about a fixed axis.

For pure rotation about a fixed axis, every point moves in a circular path, and its velocity is given by

$$\vec{v}_P = \omega \times \vec{r}_P$$

Since velocity is always perpendicular to the radius vector $\vec{r}_P$, its magnitude is

$$\omega \times r_P$$

Step 3: Identify the velocity in general planar motion.

For a rigid body undergoing both translation and rotation, the velocity of any point P is given by

$$\vec{v}_P = \vec{v}_G + \omega \times \vec{r}_{P/G}$$

where

- $\vec{v}_P$ is the velocity of point P
- $\vec{v}_G$ is the velocity of the center of mass
- $\omega \times \vec{r}_{P/G}$ represents the velocity due to rotation about G

Numerical Example

A rigid rod of length 3 meters rotates about a fixed point at one end with an angular velocity of 5 rad/s. Determine the velocity of the other end of the rod.

Solution

Step 1: Use the velocity equation for the rotation.

$$v_P = r_P \omega$$

Step 2: Substitute in the given values.

$$v_P = (3)(5) = 15m/s$$

ChatGPT Integration

- Use ChatGPT to derive velocity equations for complex linkages in mechanical systems.
- Ask ChatGPT to provide graphical representations of velocity vectors in planar motion.
- Request ChatGPT analyze velocity changes in robotic arm movement and machine components.

PROMPT 143

Explain the instantaneous center of zero velocity (ICZV) and its applications.

Objective

To introduce the concept of the instantaneous center of zero velocity (ICZV), explain how it is used in analyzing planar motion of rigid bodies, and discuss its significance in mechanical systems.

Activity

The ICZV is a point in a rigid body undergoing planar motion where the instantaneous velocity is zero. It is a crucial tool in kinematics for simplifying velocity analysis, particularly in mechanisms such as linkages, gears, and rolling motion. By locating the ICZV, engineers can determine velocity distributions across a rigid body without using vector calculus.

Definition and Mathematical Representation

Step 1: Define the ICZV concept.

For a rigid body undergoing general planar motion, every point moves in a combination of translation and rotation. The ICZV is the point where the velocity is momentarily zero, meaning the body appears to be rotating about that point.

Step 2: Define the velocity relationship using ICZV.

The velocity of any point on a rigid body is related to the ICZV by

$$\vec{v}_P = \omega \times \vec{r}_{P/IC}$$

where

- $\vec{v}_P$ is the velocity of point P
- ω is the angular velocity of the body
- $\vec{r}_{P/IC}$ is the position vector from the ICZV to point P

By determining the ICZV location, engineers can solve velocity problems efficiently without needing complex vector equations.

Numerical Example

A rigid bar of length 3 meters rotates with an angular velocity of 6 rad/s. If one end moves with a velocity of 12 m/s, determine the location of the instantaneous center of zero velocity.

Solution

Step 1: Use the ICZV velocity equation.

$$\vec{r}_{P/IC} = \frac{\vec{v}_P}{\omega}$$

Step 2: Substitute in the given values.

$$\vec{r}_{P/IC} = \frac{12}{6} = 2 \text{ m}$$

This means the instantaneous center of zero velocity is located 2 meters from the point where the velocity was given.

ChatGPT Integration

- Use ChatGPT to determine the ICZV in different linkages and mechanisms.
- Ask ChatGPT to simulate velocity distributions in rotating and translating systems.
- Request ChatGPT generate visual representations of ICZV applications in mechanical linkages.

PROMPT 144

Solve a problem involving the velocity analysis of a rigid body using the ICZV method.

Objective

To apply the instantaneous center of zero velocity (ICZV) method in solving a velocity analysis problem for a rigid body in planar motion.

Activity

The ICZV is a powerful tool in kinematics, simplifying velocity analysis for rigid bodies undergoing planar motion. By identifying the location of the ICZV, we can determine the velocities of different points on the body without needing vector equations or differentiation. This method is widely used in analyzing mechanisms such as linkages, crankshafts, and robotic arms.

Numerical Example

A rigid rod of length 4 meters is rotating about a point A with an angular velocity of 6 rad/s counterclockwise. The velocity of point B, which is at the other end of the rod, is 8 m/s. Determine the location of the instantaneous center of zero velocity and the velocity of a point located 1.5 meters from point A along the rod.

Solution

Step 1: Apply the ICZV velocity equation.

For any point P on the rigid body, velocity is given by

$$\vec{v}_P = \omega \times \vec{r}_{P/IC}$$

Step 2: Locate the instantaneous center of zero velocity.

By substituting in values for point B, we obtain

$$r_{B/IC} = \frac{8}{6} = 1.33\,\text{m}$$

This means that the ICZV is located 1.33 meters from point B along the rod. Since the total rod length is 4 meters, the ICZV is positioned at

$$4 - 1.33 = 2.67\,\text{m from point A}$$

Step 3: Compute the velocity of a point 1.5 meters from A.

Use the velocity equation for ICZV:

$$v_P = \omega \cdot r_{P/IC}$$

By substituting $r_{P/IC} = 2.67 - 1.5 = 1.17$ m, we find

$$v_P = (6)(1.17) = 7.02 \, \text{m/s}$$

ChatGPT Integration

- Use ChatGPT to locate the ICZV for complex mechanical linkages and robotic arms.
- Ask ChatGPT to generate graphical representations of ICZV applications in engineering.
- Request ChatGPT analyze real-world velocity distribution in systems using ICZV methods.

PROMPT 145

Derive the acceleration equations for different points on a rigid body in planar motion.

Objective

To derive the acceleration equations for different points on a rigid body undergoing planar motion, distinguishing between translational, rotational, and general motion cases.

Activity

Acceleration analysis is fundamental in studying the motion of rigid bodies in mechanisms, linkages, and robotic systems. A rigid body in planar motion experiences both linear acceleration and rotational acceleration, which combine to determine the acceleration at any given point. Understanding how acceleration propagates across a rigid body is essential for engineering applications such as machine dynamics and vibration analysis.

Derivation of Acceleration Equations

Step 1: Identify the acceleration in pure translation.

If a rigid body is in pure translation, all points on the body experience the same acceleration:

$$\vec{a}_p = \vec{a}_G$$

where $\vec{a}_p$ is the acceleration of any point on the body and $\vec{a}_G$ is the acceleration of the center of mass.

Step 2: Identify the acceleration in pure rotation about a fixed axis.

For pure rotation about a fixed axis, acceleration consists of tangential and normal (centripetal) components. The velocity of any point at distance r from the rotation axis is

$$a_t = r\alpha$$

where α is the angular acceleration.

The normal (centripetal) acceleration is given by

$$a_n = r\omega^2$$

The total acceleration is the vector sum of these components.

Step 3: Identify the acceleration in general planar motion.

For a rigid body undergoing both translation and rotation, the acceleration of any point P is given by

$$\vec{a}_P = \vec{a}_G + \vec{\alpha} \times \vec{r}_{P/G} - \omega^2 \vec{r}_{P/G}$$

Numerical Example

A rigid rod of length 2 meters is rotating about a fixed axis with an angular velocity of 5 rad/s and an angular acceleration of 3 rad/s². Determine the acceleration of the free end of the rod.

Solution

Step 1: Compute the tangential acceleration.

$$a_t = r\alpha$$

By substituting $r = 2$ m and $\alpha = 3$ rad/s², we find

$$a_t = (2)(3) = 6\,\text{m/s}^2$$

Step 2: Compute the normal acceleration.

$$a_n = r\omega^2$$

By substituting $\omega = 5$ rad/s, we obtain

$$a_n = (2)(5)^2 = 50\,\text{m/s}^2$$

ChatGPT Integration

- Use ChatGPT to analyze acceleration in multi-body systems such as linkages and gears.
- Ask ChatGPT to generate vector diagrams illustrating acceleration distributions in rigid bodies.
- Request ChatGPT simulate real-world applications of acceleration in mechanical systems.

PROMPT 146

Discuss the Coriolis acceleration and solve a problem involving relative motion in rotating frames.

Objective

To explain the concept of Coriolis acceleration, derive its mathematical expression, and apply it to solve a problem involving relative motion in a rotating reference frame.

Activity

The Coriolis acceleration appears when an object moves within a rotating reference frame. This effect is crucial in analyzing rotating machinery, planetary motion, and navigation systems. Understanding the Coriolis acceleration is essential for applications such as weather pattern analysis, missile trajectory corrections, and robotic systems that operate in rotating environments.

Derivation of the Coriolis Acceleration

Step 1: Define the total acceleration in the rotating frames.

The absolute acceleration of a point P moving in a rotating reference frame is given by

$$\vec{a}_P = \vec{a}_O + \vec{\alpha} \times \vec{r}_{P/O} + \vec{\omega} \times (\vec{\omega} \times \vec{r}_{P/O}) + 2\vec{\omega} \times \vec{v}_{rel} + \vec{a}_{rel}$$

where

- $\vec{a}_P$ is the absolute acceleration of point P
- $\vec{a}_O$ is the acceleration of the rotating frame's origin
- $\vec{\alpha}$ is the angular acceleration of the rotating frame
- $\vec{\omega}$ is the angular velocity of the rotating frame
- $\vec{r}_{P/O}$ is the position vector of point P relative to the rotating frame's origin
- $\vec{v}_{rel}$ is the velocity of point P relative to the rotating frame
- $\vec{a}_{rel}$ is the acceleration of P relative to the rotating frame

The term $2\vec{\omega} \times \vec{v}_{rel}$ is called the Coriolis acceleration $\vec{a}_C$, and it arises due to motion relative to a rotating reference frame. The term $\vec{\omega} \times (\vec{\omega} \times \vec{r}_{P/O})$ represents centripetal acceleration, and $\vec{\alpha} \times \vec{r}_{P/O}$ is the Euler acceleration.

The magnitude of the Coriolis acceleration is

$$|\vec{a}_C| = 2\omega v_{rel} \sin\theta$$

where θ is the angle between vectors $\vec{\omega}$ and $\vec{v}_{rel}$.

Numerical Example

A small block (point P) moves inside a rotating reference frame. The reference frame has the following characteristics:

- The origin O of the rotating frame accelerates with respect to an inertial frame with $\vec{a}_O = (0.5, -1, 0)\,\text{m/s}^2$.
- The frame rotates about the z-axis with a constant angular velocity $\vec{\omega} = (0, 0, 2)\,\text{rad/s}$.
- The position of point P relative to the origin O is $\vec{r}_{P/O} = (3, 4, 0)\,\text{m}$.
- In the rotating frame, point P has a velocity $\vec{v}_{rel} = (1, 0, 0)\,\text{m/s}$.
- In the rotating frame, point P experiences an acceleration $\vec{a}_{rel} = (0, 2, 0)\,\text{m/s}^2$.

Solution

Step 1: Use the Euler acceleration.

Use the formula

$$\vec{\alpha} \times \vec{r}_{P/O} = 0$$

since $\vec{\alpha} = \dfrac{d\omega}{dt} = 0$.

Step 2. Identify the centripetal acceleration.

This term is

$$\vec{\omega} \times (\vec{\omega} \times \vec{r}_{P/O}) = (0,0,2) \times \underbrace{[(0,0,2) \times (3,4,0)]}_{= (-8,6,0)} = (-12,-16,0)$$

Step 3. Define the Coriolis acceleration.

This term is

$$2\vec{\omega} \times \vec{v}_{\text{rel}} = 2(0,0,2) \times (1,0,0) = (0,4,0)$$

Step 4. Sum all the contributions.

Now, add up all the terms:

$$\vec{a}_P = \vec{a}_O + \vec{\alpha} \times \vec{r}_{P/O} + \vec{\omega} \times (\vec{\omega} \times \vec{r}_{P/O}) + 2\vec{\omega} \times \vec{v}_{\text{rel}} + \vec{a}_{\text{rel}} = (0.5,-1,0) + (0,0,0) +$$

$$(-12,-16,0) + (0,4,0) + (0,2,0) = (0.5-12,-1-16+4+2,0) =$$

$$(-11.5,-11,0) \ \text{m/s}^2$$

ChatGPT Integration

- Use ChatGPT to analyze absolute acceleration in non-inertial frames such as rotating spacecraft.
- Ask ChatGPT to simulate the effects of Coriolis acceleration on projectile motion in rotating environments.
- Request ChatGPT solve advanced problems involving varying angular velocity and acceleration.

PROMPT 147

Apply vector analysis to determine the velocity and acceleration of a point on a rigid body.

Objective

To apply vector analysis to determine the velocity and acceleration of a specific point on a rigid body undergoing planar motion, using both translational and rotational motion components.

Activity

The velocity and acceleration of any point on a rigid body can be determined using vector analysis, which accounts for both translational motion of the body's center and rotational motion about the center. This method is widely used in analyzing linkages, gears, and robotic arms, where velocity and acceleration information is crucial for dynamic performance and control.

Velocity Analysis Using Vector Formulation

For a rigid body undergoing general planar motion, the velocity of any point P is given by

$$\vec{v}_P = \vec{v}_G + \vec{\omega} \times \vec{r}_{P/G}$$

where

- $\vec{v}_P$ is the velocity of point P
- $\vec{v}_G$ is the velocity of the center of mass
- $\vec{\omega}$ is the angular velocity vector
- $\vec{r}_{P/G}$ is the position vector of P relative to G

The velocity of any point can be obtained using vector cross products rather than direct scalar equations, which provides a general approach applicable to various orientations of motion.

Acceleration Analysis Using Vector Formulation

The absolute acceleration of a point P is given by

$$\vec{a}_P = \vec{a}_G + \vec{\alpha} \times \vec{r}_{P/G} - \vec{\omega} \times (\vec{\omega} \times \vec{r}_{P/G})$$

where

- $\vec{a}_P$ is the acceleration of point P
- $\vec{a}_G$ is the acceleration of the center of mass
- $\vec{\alpha}$ is the angular acceleration vector
- $\vec{\omega}$ is the angular velocity vector
- $\vec{r}_{P/G}$ is the position vector from G to P

The acceleration equation includes both tangential and normal (centripetal) acceleration components due to rotation.

Numerical Example

A rigid bar of length 3 meters rotates counterclockwise with an angular velocity of 4 rad/s and an angular acceleration of 2 rad/s². The velocity of its center G is 3 m/s to the right. Determine

1. the velocity of the free end P
2. the acceleration of P

Assume:

Bar lies in the xy-plane.

Point G is the center, so the distance from G to the free end P is 1.5m.

Solution

Step 1: Compute the velocity of point P.

Use the velocity equation:

$$\vec{v}_P = \vec{v}_G + \vec{\omega} \times \vec{r}_{P/G}$$

By substituting the given values, we obtain

$$\vec{v}_P = \left(3\hat{i}\right) + \left(4\hat{k} \times 1.5\hat{i}\right) = 3\hat{i} + 6\hat{j} = (3,6,0)$$

Step 2: Compute the acceleration of point P.

Use the acceleration equation:

$$\vec{a}_P = \vec{a}_G + \vec{\alpha} \times \vec{r}_{P/G} - \vec{\omega} \times \left(\vec{\omega} \times \vec{r}_{P/G}\right)$$

By substituting the values, we obtain

$$\vec{a}_P = (0) + \left(2\hat{k} \times 1.5\hat{i}\right) - \left(4\hat{k} \times \left(4\hat{k} \times 1.5\hat{i}\right)\right) = 24\hat{i} + 3\hat{j} \text{ m/s}^2$$

ChatGPT Integration

- Use ChatGPT to solve velocity and acceleration problems for multi-link mechanisms.
- Ask ChatGPT to simulate vector analysis for different rigid body motions.
- Request ChatGPT generate step-by-step derivations for acceleration equations in rotating systems.

PROMPT 148

Solve a problem involving motion of a rigid body with combined translation and rotation.

Objective

To apply kinematic equations to analyze the motion of a rigid body undergoing both translation and rotation, demonstrating how velocity and acceleration vary at different points on the body.

Activity

Rigid bodies in general planar motion experience both translational motion of their center of mass and rotational motion about the center. This type of motion is common in mechanical linkages, robotic arms, and vehicle dynamics, where understanding how different points move is critical for system performance and safety.

Numerical Example

A rigid bar of length 2 meters moves in a plane. The center of mass G has a velocity of 3 m/s in the positive x-direction and an angular velocity of 4 rad/s counterclockwise. Determine the velocity and acceleration of the bar's free end P, which is 2 meters away from G.

Solution

Step 1: Compute the velocity of point P.

Use the velocity equation for general planar motion:

$$\vec{v}_P = \vec{v}_G + \vec{\omega} \times \vec{r}_{P/G}$$

where

- $\vec{v}_G = 3\hat{i}$ (velocity of the center of mass)
- $\vec{\omega} = 4\hat{k}$ rad/s (counterclockwise rotation)
- $\vec{r}_{P/G} = 2\hat{j}$ m (distance from G)

Apply the cross product to obtain

$$\vec{v}_P = \left(3\hat{i}\right) + \left(4\hat{k}\right) \times \left(2\hat{j}\right) = -5\hat{i} \text{ m/s}$$

The velocity of P is 5 m/s in the negative x-direction.

Step 2: Compute the acceleration of point P.

Use the acceleration equation for general planar motion:

$$\vec{a}_P = \vec{a}_G + \vec{\alpha} \times \vec{r}_{P/G} - \vec{\omega} \times \left(\vec{\omega} \times \vec{r}_{P/G} \right)$$

Given

- $\vec{a}_G = 0$ (since G moves with constant velocity)
- $\vec{\alpha} = 0$ (since no angular acceleration is specified)

Compute the centripetal acceleration term:

$$\vec{\omega} \times \left(\vec{\omega} \times \vec{r}_{P/G} \right) = \left(4\hat{k} \right) \times \left(\left(4\hat{k} \right) \times \left(2\hat{j} \right) \right) = \left(4\hat{k} \right) \times \left(-8\hat{i} \right) = 32\hat{j} \, \text{m/s}^2$$

ChatGPT Integration

- Use ChatGPT to analyze motion in multi-link mechanisms where the translation and rotation occur simultaneously.
- Ask ChatGPT to simulate velocity and acceleration fields in rigid body motion.
- Request ChatGPT provide detailed step-by-step solutions for complex motion analysis problems.

PROMPT 149

Discuss the significance of angular velocity and angular acceleration in planar motion.

Objective

To explain the role of angular velocity and angular acceleration in planar motion, derive their mathematical relationships, and discuss their significance in engineering applications.

Activity

Angular velocity and angular acceleration are fundamental quantities in rotational motion. These parameters describe how fast a rigid body rotates and how its rotational speed changes over time. They are essential for analyzing mechanisms such as gears, flywheels, robotic arms, and rotating machinery, where precise control of motion is critical.

Definition and Mathematical Representation

Step 1: Define the angular velocity.

Angular velocity $\vec{\omega}$ represents the rate of change of angular position with respect to time:

$$\vec{\omega} = \frac{d\theta}{dt}\hat{k}$$

where

- θ is the angular displacement
- $\vec{\omega}$ is the angular velocity vector
- $\hat{k}$ is the unit vector normal to the plane of motion

The instantaneous velocity of a point P on a rigid body at a distance r from the axis of rotation is given by

$$\vec{v}_P = \vec{\omega} \times \vec{r}_{P/O}$$

The magnitude of the velocity is

$$v_P = r\omega$$

Step 2: Define the angular acceleration.

Angular acceleration $\vec{\alpha}$ represents the rate of the change of the angular velocity:

$$\vec{\alpha} = \frac{d\vec{\omega}}{dt}$$

The instantaneous acceleration of a point P on a rotating rigid body is given by

$$\vec{a}_P = \vec{\alpha} \times \vec{r}_{P/O} - \vec{\omega} \times \left(\vec{\omega} \times \vec{r}_{P/O}\right)$$

By expanding the terms, we obtain the tangential and normal acceleration components:

$$a_t = r\alpha, \quad a_n = r\omega^2$$

where

- a_t is the tangential acceleration
- a_n is the normal (centripetal) acceleration

Numerical Example

A rigid disc of radius 0.5 meters rotates counterclockwise with an angular velocity of 6 rad/s and an angular acceleration of 2 rad/s². Determine the velocity and acceleration of a point located at the rim of the disc.

Solution

Step 1: Compute the velocity of the point.

Use the velocity formula:

$$v = r\omega$$

By substituting r = 0.5 m and ω = 6 rad/s, we obtain

$$v = (0.5)(6) = 3\,\text{m/s}$$

Step 2: Compute the acceleration of the point.

The tangential acceleration is given by

$$a_t = r\alpha$$

By substituting r = 0.5 m and α = 2 rad/s², we obtain

$$a_t = (0.5)(2) = 1 \text{ m/s}^2$$

The normal acceleration is

$$a_n = r\omega^2$$

By substituting ω = 6 rad/s, we obtain

$$a_n = (0.5)(6)^2 = 18 \text{ m/s}^2$$

ChatGPT Integration

- Use ChatGPT to analyze the angular velocity and acceleration in mechanisms such as gears and robotic arms.
- Ask ChatGPT to generate vector field visualizations for angular motion in planar systems.
- Request ChatGPT provide detailed solutions for real-world problems involving angular motion.

PROMPT 150

Analyze a real-world engineering application of planar kinematics in mechanical systems.

Objective

To explore a real-world engineering application where planar kinematics principles are used to analyze the motion of mechanical systems, demonstrating how velocity and acceleration relationships influence system performance.

Activity

Planar kinematics is essential in analyzing the motion of linkages, robotic arms, gears, automotive suspensions, and industrial machinery. Understanding how different points on a rigid body move allows engineers to optimize system performance, reduce wear, and ensure smooth operation.

An important example is the four-bar linkage, commonly used in automobile suspensions, excavators, and machine presses. This mechanism exhibits both translation and rotation, making it a perfect case for applying planar kinematics principles. The following problem is a real-world example of a four-bar linkage in an excavator arm.

Numerical Example

An excavator arm consists of a four-bar linkage that moves a bucket. At a given instant,

- the main arm is rotating counterclockwise at 2 rad/s with an angular acceleration of 1.5 rad/s²
- the length of the arm is 1.8 meters
- the bucket is attached at the end of the arm

Determine

1. the velocity of the bucket
2. the acceleration of the bucket

Solution

Step 1: Compute the velocity of the bucket.

Use the velocity equation:

$$\vec{v}_P = \vec{v}_C + \vec{\omega} \times \vec{r}_{P/C}$$

where

- $\vec{v}_C = 0$ (assuming the pivot point is fixed)
- $\vec{\omega} = 2\hat{k}$ (counterclockwise)
- $\vec{r}_{P/C} = 1.8\hat{j}$ m

Compute the cross product:

$$\vec{v}_P = \left(2\hat{k}\right) \times \left(1.8\hat{j}\right) = -3.6\hat{i} \text{ m/s}$$

Step 2: Compute the acceleration of the bucket.

Use the acceleration equation:

$$\vec{a}_P = \vec{a}_C + \vec{\alpha} \times \vec{r}_{P/C} - \vec{\omega} \times \left(\vec{\omega} \times \vec{r}_{P/C}\right)$$

Given

- $\vec{a}_C = 0$
- $\vec{\alpha} = 1.5\hat{k}$ rad/s²
- $\vec{\omega} = 2\hat{k}$ rad/s
- $\vec{r}_{P/C} = 1.8\hat{j}$ m

Compute the tangential acceleration:

$$\vec{a}_t = \left(1.5\hat{k}\right) \times \left(1.8\hat{j}\right) = -2.7\hat{i}$$

Compute the normal acceleration:

$$\vec{a}_n = -\left(2\hat{k} \times \left(2\hat{k} \times 1.8\hat{j}\right)\right) = 7.2\hat{j}$$

Thus,

$$\vec{a}_P = -2.7\hat{i} + 7.2\hat{j} \text{ m/s}^2$$

ChatGPT Integration

- Use ChatGPT to analyze multi-link mechanisms such as excavator arms and suspension systems.

- Ask ChatGPT to simulate velocity and acceleration distributions in real-world machinery.

- Request ChatGPT provide an in-depth analysis of forces acting on linkages in dynamic motion.

CHAPTER 16

PLANAR KINETICS OF RIGID BODIES: FORCE AND ACCELERATION

Prompt 151 Explain the relationship between force, mass, and acceleration for a rigid body in planar motion.

Prompt 152 Derive the equations of motion for a rigid body using Newton's second law.

Prompt 153 Define the mass moment of inertia and explain its role in rotational motion.

Prompt 154 Derive the planar equations of motion using the mass center formulation.

Prompt 155 Solve a problem involving the equations of motion for a rigid body in translation.

Prompt 156 Derive and apply Euler's equation for rotational motion about a fixed axis.

Prompt 157 Solve a problem involving the dynamic equilibrium of a rotating rigid body.

Prompt 158 Discuss the concept of dynamic balancing in rotating machinery and its practical applications.

Prompt 159 Explain D'Alembert's principle and apply it to analyze planar motion of a rigid body.

Prompt 160 Solve a real-world problem involving force and acceleration analysis in a mechanical system.

PROMPT 151

Explain the relationship between force, mass, and acceleration for a rigid body in planar motion.

Objective

To explain how force, mass, and acceleration interact in the motion of a rigid body undergoing planar motion, including both translational and rotational effects.

Activity

In rigid body dynamics, the relationship between force, mass, and acceleration follows from Newton's second law, which states that the acceleration of an object is proportional to the net force acting on it and inversely proportional to its mass. However, for a rigid body, we extend this principle to include rotational motion, where torque plays a role analogous to force, and moment of inertia replaces mass.

Understanding these relationships is fundamental in analyzing machines, vehicles, and robotic systems, where forces and moments must be accounted for to predict motion.

Newton's Second Law for Planar Motion of a Rigid Body

A rigid body moving in a plane can have both translational motion and rotational motion. Newton's second law applies to both cases:

1. For translation (linear motion of the center of mass), the formula is

$$\sum \vec{F} = m\vec{a}_G$$

where

▪ $\sum \vec{F}$ is the sum of external forces acting on the body
▪ m is the mass of the rigid body
▪ $\vec{a}_G$ is the acceleration of the center of mass G

2. For the rotation about the center of mass, the formula is

$$\sum \vec{M}_G = I_G \vec{\alpha}$$

where

- $\sum \vec{M}_G$ is the sum of external moments about the center of mass G
- I_G is the mass moment of inertia about G
- $\vec{\alpha}$ is the angular acceleration

These two equations describe the planar motion of a rigid body by considering both linear and angular effects.

Numerical Example

A 5 kg uniform bar of length 2 meters is subjected to a horizontal force of 20 N applied at one end. The bar is free to rotate about its center. Determine

1. the acceleration of the center of mass
2. the angular acceleration of the bar

Solution

Step 1: Compute the acceleration of the center of mass.

Use Newton's second law for translation:

$$\sum \vec{F} = m\vec{a}_G$$

By substituting in the given values, we obtain

$$a_G = \frac{20}{5} = 4 \text{ m/s}^2$$

Step 2: Compute the angular acceleration.

Use Newton's second law for the rotation:

$$\sum \vec{M}_G = I_G \vec{\alpha}$$

For a uniform bar rotating about its center, the moment of inertia is

$$I_G = \frac{1}{12} m L^2$$

By substituting m = 5 kg and L = 2 m, we obtain

$$I_G = \frac{1}{12}(5)(2)^2 = \frac{20}{12} = 1.67 \text{ kg} \cdot \text{m}^2$$

The moment about the center of mass due to the applied force is

$$M_G = (20)(1) = 20 \text{ N} \cdot \text{m}$$

Solve for α:

$$|\vec{\alpha}| = \frac{M_G}{I_G} = \frac{20}{1.67} = 12 \text{ rad/s}^2$$

PROMPT 152

Derive the equations of motion for a rigid body using Newton's second law.

Objective

To derive the equations of motion governing the planar movement of a rigid body, using Newton's second law for both translation and rotation.

Activity

Newton's second law states that the sum of external forces equals mass times acceleration in translational motion, and the sum of external moments equals moment of inertia times angular acceleration in rotational motion. These fundamental relationships govern how rigid bodies move under the influence of external forces and torques.

These principles are crucial in analyzing mechanical linkages, automotive dynamics, robotic arms, and structural vibrations.

Derivation of the Equations of Motion

A rigid body undergoing planar motion can experience both translation and rotation.

1. The formula for the translational motion (Newton's second law in vector form) is

$$\sum \vec{F} = m\vec{a}_G$$

where

- $\sum \vec{F}$ is the sum of external forces acting on the body
- m is the mass of the rigid body
- $\vec{a}_G$ is the acceleration of the center of mass

2. The formula for the rotational motion (Newton's second law for rotations) is

$$\sum \vec{M}_G = I_C \vec{\alpha}$$

where

- $\sum \vec{M}_G$ is the sum of external moments about the center of mass GGG
- I_G is the mass moment of inertia about GGG
- $\vec{\alpha}$ is the angular acceleration

3. Determine the kinematic relationships (link linear and angular motion).

For the velocity of any point P on the rigid body, we use

$$\vec{v}_P = \vec{v}_G + \vec{\omega} \times \vec{r}_{P/G}$$

For the acceleration of any point PPP on the rigid body, we use

$$\vec{a}_P = \vec{a}_G + \vec{\alpha} \times \vec{r}_{P/G} - \vec{\omega} \times (\vec{\omega} \times \vec{r}_{P/G})$$

These equations completely describe the planar motion of a rigid body subjected to external forces and moments.

Numerical Example

A solid disc of mass 10 kg and radius 0.4 meters is rolling without slipping on a flat surface. A horizontal force of 50 N is applied tangentially to the disc at its edge. Determine

1. the acceleration of the center of mass
2. the angular acceleration of the disc

Solution

Step 1: Apply Newton's second law for translation.

$$\sum \vec{F} = m \vec{a}_G$$

By substituting in the given values, we obtain

$$a_G = \frac{50}{10} = 5 \text{ m/s}^2$$

Step 2: Apply Newton's second law for rotation.

For a solid disc rotating about its center, the moment of inertia is

$$I_G = \frac{1}{2}mR^2$$

By substituting $m = 10$ kg and $R = 0.4$ m, we obtain

$$I_G = \frac{1}{2}(10)(0.4)^2 = 0.8 \text{ kg} \cdot \text{m}^2$$

Since the force is applied tangentially at the rim, the moment is

$$M_G = (50)(0.4) = 20 \text{ N} \cdot \text{m}$$

Solve for α:

$$\alpha = \frac{20}{0.8} = 25 \text{ rad/s}^2$$

ChatGPT Integration

- Use ChatGPT to analyze force and acceleration relations in rolling motion and mechanical linkages.
- Ask ChatGPT to generate graphical representations of force and motion interactions in dynamic systems.
- Request ChatGPT provide alternative methods for deriving equations of motion in various mechanical systems.

PROMPT 153

Define the mass moment of inertia and explain its role in rotational motion.

Objective

To define the mass moment of inertia, explain its significance in rotational motion, and describe how it affects the dynamics of rigid bodies in planar motion.

Activity

The mass moment of inertia is a fundamental property that describes how mass is distributed in a rigid body relative to an axis of rotation. It plays a crucial role in Newton's second law for rotations, where torque is related to angular acceleration in the same way force is related to linear acceleration. Understanding moment of inertia is essential in engineering applications, such as rotating machinery, vehicle dynamics, and aerospace systems.

Definition and Mathematical Representation

The mass moment of inertia I of a rigid body about a given axis is defined as

$$I = \int r^2 \, dm$$

where

- I is the moment of inertia
- r is the perpendicular distance from the mass element dm to the axis of rotation
- the integral sums contributions from all mass elements in the body

This equation shows that mass farther from the axis of rotation contributes more to the moment of inertia than mass closer to the axis.

Parallel Axis Theorem

If the moment of inertia about the center of mass I_G is known, the moment of inertia about any parallel axis I_O can be found using

$$I_O = I_G + md^2$$

where

- I_O is the moment of inertia about the new axis
- I_G is the moment of inertia about the center of mass
- m is the total mass
- d is the perpendicular distance between the two axes

This theorem is useful for calculating moments of inertia for offset rotating components such as flywheels and gears.

Common Moments of Inertia

For standard rigid bodies, moments of inertia are derived using integration. Some common cases are as follows:

1. thin rod (about center):

$$I_G = \frac{1}{12}mL^2$$

2. thin rod (about end):

$$I_O = \frac{1}{3}mL^2$$

3. solid disc (about center):

$$I_G = \frac{1}{2}mR^2$$

4. solid sphere (about center):

$$I_G = \frac{2}{5}mR^2$$

Numerical Example

A solid disc of mass 8 kg and radius 0.3 meters is rotating about an axis through its center. Compute its moment of inertia and determine its angular acceleration if a torque of 12 N·m is applied.

Solution

Step 1: Compute the moment of inertia.

For a solid disc about its center, substitute $m = 8$ kg and $R = 0.3$ m to obtain

$$I_G = \frac{1}{2}(8)(0.3)^2 = 0.36\,\text{kg}\cdot\text{m}^2$$

Step 2: Compute the angular acceleration.

Use Newton's second law for rotation:

$$\sum \vec{M} = I_G\vec{\alpha}$$

By substituting in values, we obtain

$$\vec{\alpha} = \frac{12}{0.36} = 33.33 \, \text{rad/s}^2$$

ChatGPT Integration

- Use ChatGPT to derive the moments of inertia for non-standard shapes using integration.
- Ask ChatGPT to analyze the effect of moment of inertia on the stability of rotating objects.
- Request ChatGPT simulate how changes in moment of inertia influence rotational dynamics in mechanical systems.

PROMPT 154

Derive the planar equations of motion using the mass center formulation.

Objective

To derive the planar equations of motion for a rigid body using the mass center formulation, incorporating both linear and angular motion equations.

Activity

A rigid body in planar motion undergoes both translation and rotation. The mass center formulation simplifies the equations of motion by describing the body's motion relative to its center of mass (G). These equations are widely used in vehicle dynamics, robotic motion, and structural mechanics, where understanding force and moment interactions is crucial for stability and performance.

Derivation of the Planar Equations of Motion

A rigid body moving in a plane is subject to

1. external forces that cause translation
2. external moments (torques) that cause rotation

Step 1: Identify the translational motion using Newton's second law.

For translation, Newton's second law states

$$\sum \vec{F} = m\vec{a}_C$$

where

- $\sum \vec{F}$ is the sum of external forces acting on the rigid body
- m is the mass of the body
- $\vec{a}_C$ is the acceleration of the center of mass

This equation governs the linear motion of the center of mass.

Step 2: Determine the rotational motion using Newton's second law for rotations.

For the rotation about the center of mass, the equation of motion is given by

$$\sum \vec{M}_C = I_C \vec{\alpha}$$

where

- $\sum \vec{M}_C$ is the sum of external moments about the center of mass G
- I_C is the mass moment of inertia about G
- $\vec{\alpha}$ is the angular acceleration of the rigid body

This equation governs the rotational motion of the body about its center of mass.

Step 3: Combine the planar motion equations.

By combining both translational and rotational equations, we obtain the general planar equations of motion for a rigid body:

$$\sum F_x = ma_{Gx}, \quad \sum F_y = ma_{Gy}, \quad \sum M_G = I_G \alpha$$

These three equations completely describe the planar motion of a rigid body, allowing us to solve for unknown forces, accelerations, and torques.

Numerical Example

A rigid beam of mass 6 kg and length 1.5 meters is subjected to a horizontal force of 30 N at one end. The beam is free to rotate about its center. Determine

1. the acceleration of the center of mass
2. The angular acceleration of the beam

Solution

Step 1: Compute the acceleration of the center of mass.

Use Newton's second law for translation:

$$\sum \vec{F} = m\vec{a}_C$$

By substituting the given values, we obtain

$$a_C = \frac{30}{6} = 5 \text{ m/s}^2$$

Step 2: Compute the angular acceleration.

For a uniform beam rotating about its center, the moment of inertia is

$$I_C = \frac{1}{12} mL^2$$

By substituting $m = 6$ kg and $L = 1.5$ m, we obtain

$$I_C = \frac{1}{12}(6)(1.5)^2 = \frac{13.5}{12} = 1.125 \text{ kg} \cdot \text{m}^2$$

The moment about the center of mass due to the applied force is

$$M_C = (30)(0.75) = 22.5 \text{ N} \cdot \text{m}$$

Solve for α:

$$\alpha = \frac{22.5}{1.125} = 20 \text{ rad/s}^2$$

ChatGPT Integration

- Use ChatGPT to solve force and acceleration problems for real-world mechanical linkages.
- Ask ChatGPT to analyze how forces and moments influence motion in dynamic systems.
- Request ChatGPT derive equations of motion for customized mechanical systems.

PROMPT 155

Solve a problem involving the equations of motion for a rigid body in translation.

Objective

To apply the equations of motion to a rigid body undergoing pure translation, demonstrating how external forces influence acceleration.

Activity

A rigid body in pure translation moves such that every point on the body experiences the same linear acceleration. There is no rotation about the center of mass, so the equation of motion simplifies to Newton's second law for translation. This concept is critical in vehicle motion analysis, material handling systems, and robotic mechanisms.

Equations of the Motion for Translation

For a rigid body in pure translation, Newton's second law applies directly to the center of mass:

$$\sum \vec{F} = m\vec{a}_G$$

Since the body does not rotate, the moment equation is not needed:

$$\sum \vec{M}_G = 0$$

These equations govern rectilinear motion of rigid bodies without rotation.

Numerical Example

A crate of mass 50 kg is pushed along a horizontal surface by a force of 200 N. The coefficient of kinetic friction between the crate and the surface is 0.3. Determine

1. the acceleration of the crate
2. the normal force acting on the crate

Solution

Step 1: Compute the normal force.

The normal force N is determined from the vertical equilibrium equation. Since the only vertical forces are the weight W and the normal force N, we have

$$N = mg = (50)(9.81) = 490.5 \text{ N}$$

Step 2: Compute the friction force.

The kinetic friction force is given by

$$F_f = \mu_k N$$

By substituting $\mu_k = 0.3$ and $N = 490.5$:

$$F_f = (0.3)(490.5) = 147.15 \text{ N}$$

Step 3: Compute the acceleration.

Apply Newton's second law:

$$\sum F_x = ma_G$$

The net force in the horizontal direction is

$$a_G = \frac{200 - 147.15}{50} = 1.06 \text{ m/s}^2$$

ChatGPT Integration

▪ Use ChatGPT to analyze real-world problems involving friction forces and motion.

▪ Ask ChatGPT to generate force diagrams for translating rigid bodies.

▪ Request ChatGPT simulate the effects of varying friction on acceleration.

PROMPT 156

Derive and apply Euler's equation for rotational motion about a fixed axis.

Objective

To derive Euler's equation for rotational motion and apply it to solve a problem involving a rigid body rotating about a fixed axis.

Activity

Euler's equation describes how external moments cause angular acceleration in a rotating rigid body. It is the rotational analog of Newton's second law and is widely used in robotic arms, rotating machinery, vehicle dynamics, and aerospace systems. Understanding how torque and moment of inertia interact is essential for designing stable and efficient mechanical systems.

Derivation of Euler's Equation for a Rigid Body Rotating About a Fixed Axis

In the most general case, when a rigid body rotates freely in three dimensions, Euler's equations describe the angular motion about the principal axes as:

$$\sum M_i = I_i \dot{\omega}_i + (\text{cross-coupling terms})$$

The cross-coupling terms arise from the interaction between angular velocities about different axes and represent the internal gyroscopic effects. These terms vanish when the body rotates only about a single principal axis, simplifying the equation to the fixed-axis form.

For a rigid body rotating about a fixed axis, Newton's second law for rotational motion states

$$\sum \vec{M}_O = I_O \vec{\alpha}$$

where

- $\sum \vec{M}_O$ is the sum of external moments about the fixed axis passing through point O
- I_O is the moment of inertia of the body about the fixed axis
- $\vec{\alpha}$ is the angular acceleration

This equation states that the sum of torques acting on the body is proportional to its angular acceleration, similar to $\sum F = ma$ in linear motion.

Numerical Example

A solid disc of mass 10 kg and radius 0.5 meters is mounted on a fixed horizontal axis. A force of 30 N is applied tangentially at the rim of the disc. Determine

1. the moment of inertia of the disc about the axis
2. the angular acceleration of the disc

Solution

Step 1: Compute the moment of inertia.

For a solid disc rotating about its center, the moment of inertia is

$$I_O = \frac{1}{2}mR^2$$

By substituting m = 10 kg and R = 0.5 m, we obtain

$$I_O = \frac{1}{2}(10)(0.25) = 1.25 \text{ kg} \cdot \text{m}^2$$

Step 2: Compute the angular acceleration.

Use Euler's equation:

$$\sum \vec{M}_O = I_O \vec{\alpha}$$

Solve for α, using generated torque, $M_O = (30)(0.5) = 15 \text{ N} \cdot \text{m}$:

$$\alpha = \frac{15}{1.25} = 12 \text{ rad/s}^2$$

ChatGPT Integration

- Use ChatGPT to derive Euler's equation for complex rotating bodies.
- Ask ChatGPT to generate torque analysis diagrams for rotating systems.
- Request ChatGPT simulate the effect of varying torques on angular acceleration.

PROMPT 157

Solve a problem involving the dynamic equilibrium of a rotating rigid body.

Objective

To apply the equations of motion to a rigid body undergoing rotation, demonstrating how forces and moments interact to maintain dynamic equilibrium.

Activity

A rigid body in rotation must satisfy both translational and rotational equilibrium conditions. This concept is widely applied in rotor dynamics, vehicle suspensions, flywheels, and robotic mechanisms, where forces and moments must be carefully balanced to ensure stable operation.

Equations for Dynamic Equilibrium of a Rotating Rigid Body

For a rigid body undergoing planar motion, the equations of dynamic equilibrium are as follows:

1. force equilibrium in the x-direction:

$$\sum F_x = ma_{Gx}$$

2. force equilibrium in the y-direction:

$$\sum F_y = ma_{Gy}$$

3. moment equilibrium about the center of mass:

$$\sum M_G = I_G \alpha$$

These equations describe the linear and angular accelerations resulting from applied forces and moments.

Numerical Example

A uniform rod of mass 8 kg and length 2 meters is hinged at one end and is subjected to a force of 40 N applied at its free end in the horizontal direction. Determine

1. the acceleration of the center of mass
2. the angular acceleration of the rod

Solution

Step 1: Compute the acceleration of the center of mass.

The force equilibrium equation along the horizontal axis is

$$\sum F_x = ma_G$$

Since the force is applied at the free end, Newton's second law gives

$$a_G = \frac{40}{8} = 5 \text{ m/s}^2$$

Step 2: Compute the angular acceleration.

For a uniform rod rotating about a hinge, the moment of inertia is

$$I_O = \frac{1}{3} mL^2$$

By substituting $m = 8$ kg and $L = 2$ m, we obtain

$$I_O = \frac{1}{3}(8)(2)^2 = \frac{32}{3} \text{ kg} \cdot \text{m}^2$$

The moment about the hinge due to the applied force is

$$M_O = (40)(2) = 80 \text{ N} \cdot \text{m}$$

Use the moment equilibrium equation:

$$\alpha = \frac{80}{\left(\dfrac{32}{3}\right)} = 7.5 \text{ rad/s}^2$$

ChatGPT Integration

- Use ChatGPT to analyze equilibrium conditions in rotating systems.
- Ask ChatGPT to simulate force and moment interactions in robotic linkages.
- Request ChatGPT solve problems involving dynamic balancing of rotating rigid bodies.

PROMPT 158

Discuss the concept of dynamic balancing in rotating machinery and its practical applications.

Objective

To explain the concept of dynamic balancing, its significance in rotating machinery, and how it is applied in engineering systems to reduce vibrations and improve efficiency.

Activity

Dynamic balancing ensures that rotating components operate smoothly by eliminating unwanted vibrations caused by mass imbalances. It is a critical factor in automotive engines, turbines, industrial fans, robotic arms, and high-speed rotating equipment, where excessive vibrations can cause structural damage, reduced efficiency, and increased wear.

Definition of Dynamic Balancing

A rotating system is said to be dynamically balanced when the resultant of the inertial forces and moments acting on the system is zero.

A system is statically balanced if the net force due to mass distribution is zero. However, for high-speed rotating systems, dynamic balancing is required to eliminate both forces and moments.

For dynamic balancing, the following conditions must be satisfied:

$$\sum F_x = 0, \quad \sum F_y = 0, \quad \sum M_O = 0$$

Practical Applications of Dynamic Balancing

1. automobile wheels: ensures even weight distribution, preventing vibrations at high speeds
2. turbines and rotors: reduces fatigue failure due to oscillations in high-speed applications
3. machine tools and fans: enhances operational smoothness and increases bearing lifespan
4. aircraft propellers: prevents oscillations that can lead to severe structural damage

Numerical Example

A rigid disc of negligible mass has two attached masses of 3 kg and 5 kg placed at radii of 0.3 m and 0.2 m, respectively. The system rotates at 600 rpm. Determine

1. the required counterweight to achieve dynamic balance
2. the position where the counterweight should be placed

All three masses are assumed to lie along the same radial line or collinear, so vector summation reduces to scalar.

Solution

Step 1: Compute the mass moments.

The moment due to mass m_1 is

$$M_1 = m_1 r_1 = (3)(0.3) = 0.9 \text{ kg} \cdot \text{m}$$

The moment due to mass m_2 is

$$M_2 = m_2 r_2 = (5)(0.2) = 1.0 \text{ kg} \cdot \text{m}$$

Step 2: Compute the required counterweight.

To achieve dynamic balance, the counterweight m_c must satisfy

$$m_c r_c = M_1 + M_2$$

$$m_c = \frac{0.9 + 1.0}{0.25} = 7.6 \text{ kg}$$

Thus, a 7.6 kg counterweight should be placed at a radius of 0.25 meters to achieve dynamic balance.

ChatGPT Integration

- Use ChatGPT to analyze unbalanced rotating systems and recommend corrective actions.
- Ask ChatGPT to simulate the effects of unbalanced forces in rotating machinery.
- Request ChatGPT solve balancing problems for complex mechanical systems.

PROMPT 159

Explain D'Alembert's principle and apply it to analyze planar motion of a rigid body.

Objective

To explain D'Alembert's principle, derive its mathematical formulation, and apply it to solve a problem involving the planar motion of a rigid body.

Activity

D'Alembert's principle is a reformulation of Newton's second law that introduces the concept of inertial forces, transforming a dynamically accelerating system into a statistically equivalent system. This approach is widely used in engineering mechanics, vibration analysis, and vehicle dynamics, where solving motion problems using equilibrium conditions simplifies calculations.

Definition and Mathematical Formulation of D'Alembert's Principle

D'Alembert's principle states that the sum of real forces and inertial forces acting on a system is zero, allowing dynamic problems to be solved using equilibrium equations.

For a rigid body in planar motion, the equations of motion in terms of inertial forces are as follows:

1. Force Equilibrium for Translation:

$$\sum \vec{F} - m\vec{a}_C = 0$$

2. Moment Equilibrium for Rotation:

$$\sum \vec{M}_C - I_C \vec{\alpha} = 0$$

D'Alembert's principle treats acceleration as an additional force (inertial force), allowing the problem to be solved as a static equilibrium problem.

Numerical Example: Planar Motion of a Sliding Rod

This is an example of a problem involving the planar motion of a sliding rod. A rigid rod of mass 6 kg and length 1.5 meters slides without friction on a horizontal surface. A force of 50 N is applied at one end. Determine

1. the acceleration of the center of mass
2. the angular acceleration of the rod

Solution Using D'Alembert's Principle

Step 1: Apply the force equilibrium equation.

Use

$$\sum \vec{F} - m\vec{a}_G = 0$$

By substituting values, we obtain

$$a_G = \frac{50}{6} = 8.33 \text{ m/s}^2$$

Step 2: Apply the moment equilibrium equation.

For a uniform rod rotating about its center, the moment of inertia is

$$I_C = \frac{1}{12} mL^2$$

By substituting $m = 6$ kg and $L = 1.5$ m, we obtain

$$I_G = \frac{1}{12}(6)(1.5)^2 = \frac{13.5}{12} = 1.125 \text{ kg} \cdot \text{m}^2$$

Use

$$\sum \vec{M}_G - I_C\vec{\alpha} = 0$$

The applied force generates a moment (The force at one end causes rotation about the center of mass.):

$$M_G = (50)(0.75) = 37.5 \text{ N} \cdot \text{m}$$

Solve for α:

$$\alpha = \frac{37.5}{1.125} = 33.33 \text{ rad/s}^2$$

ChatGPT Integration

■ Use ChatGPT to analyze real-world problems using D'Alembert's principle.

- Ask ChatGPT to generate force diagrams for dynamically equivalent systems.
- Request ChatGPT apply D'Alembert's principle to multi-body mechanical systems.

PROMPT 160

Solve a real-world problem involving force and acceleration analysis in a mechanical system.

Objective

To apply the equations of motion to analyze the forces and accelerations in a real-world mechanical system, demonstrating the relationship between applied forces, inertia, and dynamic response.

Activity

Understanding force and acceleration interactions is essential for designing automobiles, conveyor systems, robotic arms, and industrial machinery. Engineers use Newton's second law, rotational dynamics, and force equilibrium to ensure efficient and stable operation of mechanical systems. The following example is about the force and acceleration in an elevator system.

Numerical Example

A 500 kg elevator cabin is lifted by a motor-driven pulley system with a tension force of 6000 N. The cabin moves vertically, and the friction in the pulley system is negligible. Determine

1. the acceleration of the elevator
2. the reaction force exerted by the passengers on the cabin floor if the total passenger mass is 150 kg

Solution

Step 1: Apply Newton's second law to the elevator system.

The forces acting on the elevator are

- Tension force $T = 6000$ N (upward).
- Weight of the elevator $W = mg = (500)(9.81) = 4905\,N$ (downward).

Apply Newton's second law:

$$\sum F_y = ma$$

$$T - mg = ma$$

By substituting the given values, we obtain

$$a = \frac{6000 - 4905}{500} = 2.19 \text{ m/s}^2$$

Step 2: Compute the reaction force on the passengers.

Each passenger experiences an apparent weight due to the acceleration of the elevator. The normal force Np exerted by the cabin floor is

$$N_p = m_p(g + a)$$

where $m_p = 150$ kg is the total passenger mass.

By substituting in values, we obtain

$$N_p = (150)(9.81 + 2.19) = 1800 N$$

ChatGPT Integration

- Use ChatGPT to analyze the force and acceleration in real-world mechanical systems.
- Ask ChatGPT to simulate the effects of acceleration on passengers in elevators, vehicles, and amusement rides.
- Request ChatGPT generate force diagrams for multi-body dynamic systems.

CHAPTER 17

PLANAR KINETICS OF RIGID BODIES: WORK AND ENERGY

Prompt 161 Explain the work-energy principle for a rigid body and its significance in planar motion analysis.

Prompt 162 Derive the general work-energy equation for a rigid body in planar motion.

Prompt 163 Solve a problem involving the work done by external forces on a rigid body.

Prompt 164 Explain the concept of kinetic energy for a rigid body and derive its mathematical expression.

Prompt 165 Apply the work-energy theorem to solve a problem involving rolling motion with and without slipping.

Prompt 166 Discuss the concept of potential energy in the context of rigid body motion and its practical applications.

Prompt 167 Solve a problem involving the conservation of mechanical energy for a rigid body in motion.

Prompt 168 Explain the role of power in planar kinetics and derive its mathematical formulation.

Prompt 169 Solve a problem involving instantaneous power delivered to a rigid body by external forces.

Prompt 170 Compare the force-based (Newtonian) and energy-based (work-energy) methods for solving a rigid body motion problem, analyzing their advantages and limitations.

PROMPT 161

Explain the work-energy principle for a rigid body and its significance in planar motion analysis.

Objective

To explain the work-energy principle as applied to rigid body motion, describing its significance in planar motion analysis and how it simplifies solving dynamic problems.

Activity

The work-energy principle states that the work done by external forces on a rigid body is equal to its change in kinetic energy. This principle allows engineers to analyze motion without directly solving force and acceleration equations, making it useful for problems involving variable forces, impact events, and rolling motion. It is widely applied in mechanical linkages, vehicle dynamics, crash simulations, and robotic motion analysis.

Mathematical Formulation of the Work-Energy Principle

For a rigid body in planar motion, the total work done by external forces equals the change in translational and rotational kinetic energy:

$$W_{net} = \Delta KE_{trans} + \Delta KE_{rot}$$

where

- W_{net} is the net work done by external forces
- KE_{trans} is the translational kinetic energy
- KE_{rot} is the rotational kinetic energy

Translational and Rotational Kinetic Energy Expressions

For a rigid body moving in a plane, the total kinetic energy consists of the following:

1. the translational kinetic energy (due to the motion of the center of mass):

$$KE_{trans} = \frac{1}{2}mv_G^2$$

2. the rotational kinetic energy (due to the rotation about the center of mass):

$$KE_{rot} = \frac{1}{2}I_C\omega^2$$

where

- m is the mass of the rigid body
- v_G is the velocity of the center of mass
- I_G is the mass moment of inertia about the center of mass
- ω is the angular velocity of the body

Thus, the total kinetic energy of the rigid body is

$$KE_{total} = \frac{1}{2}mv_G^2 + \frac{1}{2}I_C\omega^2$$

Numerical Example

The following is an example of work done on a rotating rod. A rigid rod of mass 5 kg and length 2 meters is rotating about its center. A force of 30 N is applied tangentially at one end of the rod, causing it to accelerate from rest to an angular velocity of 6 rad/s. Determine

1. the work done by the applied force
2. the final kinetic energy of the rod

Solution

Step 1: Compute the work done.

Since the force is applied tangentially at the end of the rod, the work done is

$$W = F \cdot d$$

The distance traveled by the point where the force is applied is the arc length:

$$d = R\theta = (1)(\theta)$$

Use the kinematic equation for rotational motion, and solve for θ.

$$\theta = \frac{\omega^2 - \omega_0^2}{2\alpha}$$

Since $\omega_0 = 0$ and $\alpha = \dfrac{\tau}{I_G}$, we find

$$\theta = \frac{(6)^2}{2\left(\dfrac{30}{\dfrac{1}{12}(5)(2)^2}\right)} = 3 \text{ rad}$$

Thus,

$$W = (30)(3) = 90 \text{ J}$$

Step 2: Compute the final kinetic energy.

For a rod rotating about its center, the moment of inertia is

$$I_G = \frac{1}{12}mL^2 = \frac{1}{12}(5)(2)^2 = 1.67 \text{ kg} \cdot \text{m}^2$$

The rotational kinetic energy is

$$KE_{rot} = \frac{1}{2}I_G\omega^2 = \frac{1}{2}(1.67)(6)^2 = 30 \text{ J}$$

The total kinetic energy is

$$KE_{total} = 30 \text{ J}$$

ChatGPT Integration

- Use ChatGPT to analyze work-energy interactions in dynamic systems.
- Ask ChatGPT to generate graphical representations of energy transformations in mechanical motion.
- Request ChatGPT simulate real-world applications of the work-energy principle in engineering.

PROMPT 162

Derive the general work-energy equation for a rigid body in planar motion.

Objective

To derive the general work-energy equation for a rigid body undergoing planar motion, explaining how external forces and torques influence kinetic energy changes.

Activity

The work-energy equation provides an alternative approach to solving rigid body dynamics problems by relating the work done by external forces and moments to the change in kinetic energy. This principle is particularly useful in mechanical linkages, rotating machinery, and vehicle dynamics, where forces vary with position.

Derivation of the General Work-Energy Equation

For a rigid body in planar motion, the total work done by external forces and moments is equal to the change in kinetic energy:

$$W_{net} = \Delta KE_{total}$$

where

- W_{net} is the net work done by external forces and torques
- KE_{total} is the total kinetic energy of the rigid body

Step 1: Define the kinetic energy components.

The total kinetic energy of a rigid body consists of the following:

1. the translational kinetic energy (due to motion of the center of mass):

$$KE_{trans} = \frac{1}{2}mv_G^2$$

2. the rotational kinetic energy (due to rotation about the center of mass):

$$KE_{rot} = \frac{1}{2}I_G\omega^2$$

Thus, the total kinetic energy is

$$KE_{total} = \frac{1}{2}mv_G^2 + \frac{1}{2}I_G\omega^2$$

Step 2: Identify the work done by external forces.

The work done by a force $\vec{F}$ over a displacement $d\vec{r}$ is

$$dW = \vec{F}\cdot d\vec{r}$$

The total work done by all external forces is

$$W_{net} = \int \sum \vec{F}\cdot d\vec{r}$$

Step 3: Determine the work done by external moments.

The work done by a moment M over an angular displacement $d\theta$ is

$$dW_{rot} = Md\theta$$

The total work done by all external moments is

$$W_{rot} = \int \sum M\, d\theta$$

Thus, the net work done on the rigid body is

$$W_{net} = W_{trans} + W_{rot}$$

Numerical Example

This problem is an example of work done on a rolling cylinder. A cylinder of mass 10 kg and radius 0.4 meters rolls without slipping down an inclined plane of 30°. If it starts from rest and travels 5 meters, determine

1. the work done by gravity
2. the final velocity of the cylinder

Solution

Step 1: Compute the work done by gravity.

The gravitational force acting parallel to the incline is

$$F = mg\sin\theta = (10)(9.81)\sin 30° = 49.05 \text{ N}$$

The work done by gravity is

$$W = Fd = (49.05)(5) = 245.25 \text{ J}$$

Step 2: Compute the final velocity using work-energy.

The initial kinetic energy is

$$KE_{\text{initial}} = 0$$

The final kinetic energy is

$$KE_{\text{final}} = \frac{1}{2}mv_G^2 + \frac{1}{2}I_G\omega^2$$

For a rolling cylinder, the moment of inertia about its center is

$$I_G = \frac{1}{2}mR^2 = \frac{1}{2}(10)(0.4)^2 = 0.8 \text{ kg} \cdot \text{m}^2$$

Since $v_G = R\omega$, we substitute in $\omega = \dfrac{v_G}{R}$ to obtain

$$KE_{\text{final}} = \frac{1}{2}(10)v_G^2 + \frac{1}{2}(0.8)\left(\frac{v_G}{0.4}\right)^2 == 5v_G^2 + 1v_G^2 = 6v_G^2$$

By the work-energy principle, we obtain

$$W_{\text{net}} = KE_{\text{final}} - KE_{\text{initial}}$$

$$245.25 = 6v_G^2$$

$$v_G = \sqrt{\frac{245.25}{6}} = 6.39 \text{ m/s}$$

ChatGPT Integration

- Use ChatGPT to analyze the work-energy interactions in rigid body dynamics.
- Ask ChatGPT to generate graphical representations of energy transformations in rolling motion.
- Request ChatGPT simulate the work-energy principle applied to industrial applications.

PROMPT 163

Solve a problem involving the work done by external forces on a rigid body.

Objective

To apply the work-energy principle by calculating the work done by external forces acting on a rigid body and analyzing how this work influences the body's motion.

Activity

Work is done on a rigid body when external forces cause displacement. The total work done by these forces leads to a change in kinetic energy, making the work-energy principle a useful alternative to solving force-acceleration equations. This concept is widely used in automobile crash analysis, machinery motion analysis, and lifting mechanisms. The following example is about work done on a rotating beam.

Numerical Example

A uniform beam of mass 20 kg and length 3 meters is hinged at one end and initially at rest. A force of 80 N is applied horizontally at the free end. The beam rotates about its hinge. Determine

1. the work done by the applied force when the beam rotates through 45°
2. the final angular velocity of the beam using the work-energy principle

Solution

Step 1: Compute the work done by the external force.

The work done by a force F applied over a displacement d is

$$W = \int F \, ds$$

Since the force is applied at the free end, the displacement of this point is along a circular path of radius $L = 3L = 3L = 3$ m. The arc length traveled by the free end is

$$s = L\theta$$

By substituting $\theta = 45° = \pi/4$ rad, we obtain

$$s = (3)\left(\frac{\pi}{4}\right) = \frac{3\pi}{4} \text{ m}$$

The force is always tangential to the arc, so the work done is

$$W = F \cdot s = (80)\left(\frac{3\pi}{4}\right) = 188.4 \text{ J}$$

Step 2: Compute the final angular velocity using the work-energy principle.

The work-energy equation states that

$$W_{net} = \Delta KE$$

For a rotating body, we obtain

$$W_{net} = \frac{1}{2}I_O\omega^2 - \frac{1}{2}I_O\omega_0^2$$

Since the beam starts from rest, $\omega_0 = 0$, so

$$W_{net} = \frac{1}{2}I_O\omega^2$$

For a uniform beam rotating about its hinge, the moment of inertia is

$$I_O = \frac{1}{3}mL^2$$

By substituting in $m = 20$ kg and $L = 3$ m, we obtain

$$I_O = \frac{1}{3}(20)(3)^2 = 60 \text{ kg} \cdot \text{m}^2$$

Solve for ω.

$$188.4 = \frac{1}{2}(60)\omega^2$$

$$\omega = \sqrt{\frac{188.4}{30}} = \sqrt{6.28} = 2.51 \text{ rad/s}$$

ChatGPT Integration

- Use ChatGPT to analyze the work-energy principle in real-world rotating systems.
- Ask ChatGPT to generate graphical representations of energy transformations in mechanical motion.
- Request ChatGPT compare the work-energy approach with Newton's second law in analyzing motion.

PROMPT 164

Explain the concept of kinetic energy for a rigid body and derive its mathematical expression.

Objective

To explain kinetic energy in the context of rigid body motion, derive its mathematical formulation, and describe its significance in planar motion analysis.

Activity

Kinetic energy represents the energy possessed by a rigid body due to its motion. For a rigid body in planar motion, the kinetic energy includes both translational and rotational components. Understanding this principle is essential for mechanical systems, impact analysis, and energy-based problem-solving techniques in engineering applications such as vehicle dynamics, robotic mechanisms, and industrial machinery.

Mathematical Formulation of Kinetic Energy for a Rigid Body

A rigid body moving in a plane can have both translational and rotational kinetic energy.

1. The formula for the translational kinetic energy (due to motion of the center of mass) is

$$KE_{trans} = \frac{1}{2}mv_G^2$$

where

- m is the mass of the rigid body
- v_G is the velocity of the center of mass

2. The formula for the rotational kinetic energy (due to rotation about the center of mass) is

$$KE_{rot} = \frac{1}{2}I_G\omega^2$$

where:

- I_G is the mass moment of inertia about the center of mass,
- ω is the angular velocity of the rigid body.

Thus, the total kinetic energy of a rigid body in planar motion is the sum of translational and rotational components:

$$KE_{total} = \frac{1}{2}mv_G^2 + \frac{1}{2}I_G\omega^2$$

This equation expresses the total energy a rigid body has due to its motion.

Numerical Example:

A solid cylinder of mass 15 kg and radius 0.3 meters rolls without slipping on a horizontal surface at a linear speed of 2 m/s. Determine:

The translational kinetic energy

The rotational kinetic energy

The total kinetic energy of the cylinder

Solution

Step 1: Compute the translational kinetic energy.

$$KE_{trans} = \frac{1}{2}mv_G^2$$

By substituting $m = 15$ kg and $v_G = 2$ m/s, we obtain

$$KE_{trans} = \frac{1}{2}(15)(4) = 30 \text{ J}$$

Step 2: Compute the rotational kinetic energy.

For a solid cylinder rotating about its center, the moment of inertia is

$$I_G = \frac{1}{2}mR^2$$

By substituting $m = 15$ kg and $R = 0.3$ m, we obtain

$$I_G = \frac{1}{2}(15)(0.3)^2 = 0.675 \text{ kg} \cdot \text{m}^2$$

Since the cylinder is rolling without slipping, the relationship between linear and angular velocity is

$$v_G = R\omega \quad \Rightarrow \quad \omega = \frac{v_G}{R} = \frac{2}{0.3} = 6.67 \text{ rad/s}$$

The rotational kinetic energy is

$$KE_{rot} = \frac{1}{2}I_G\omega^2$$

$$KE_{rot} = \frac{1}{2}(0.675)(6.67)^2 = 15.0 \text{ J}$$

Step 3: Compute the total kinetic energy.

$$KE_{total} = KE_{trans} + KE_{rot} = 30 + 15 = 45 \text{ J}$$

ChatGPT Integration

- Use ChatGPT to analyze kinetic energy in various mechanical systems.
- Ask ChatGPT to simulate energy distributions in rolling motion.
- Request ChatGPT compare kinetic energy expressions for different types of rigid bodies.

PROMPT 165

Apply the work-energy theorem to solve a problem involving rolling motion with and without slipping.

Objective

To apply the work-energy theorem to analyze the motion of a rolling rigid body, comparing cases of rolling with slipping and rolling without slipping, and determining how energy is distributed between translational and rotational motion.

Activity

Rolling motion is a combination of translation and rotation. The work-energy theorem provides an efficient way to analyze rolling motion without directly solving force and acceleration equations. Understanding the differences between pure rolling (no slipping) and rolling with slipping is essential in vehicle dynamics, conveyor systems, and robotic mechanisms.

Work-Energy Theorem for a Rolling Rigid Body

The work-energy theorem states that the net work done by external forces is equal to the change in the total kinetic energy:

$$W_{net} = \Delta KE_{total}$$

For a rigid body undergoing rolling motion, the total kinetic energy is

$$KE_{total} = KE_{trans} + KE_{rot}$$

$$KE_{total} = \frac{1}{2}mv_G^2 + \frac{1}{2}I_G\omega^2$$

where

- v_G is the velocity of the center of mass
- I_G is the moment of inertia about the center of mass
- ω is the angular velocity of the rolling body

For rolling without slipping, the condition

$$v_G = R\omega$$

holds, while for rolling with slipping, this condition is not satisfied.

Problem Statement:

A solid cylinder of mass 8 kg and radius 0.4 m is released from rest at the top of a frictional incline of height 3 m. Determine the final velocity of the cylinder at the bottom of the incline in the following two cases:

When the cylinder rolls without slipping

When the cylinder slides without rolling (i.e., pure translation)

Use the work-energy theorem to solve both cases.

Solution Using the Work-Energy Theorem

Step 1: Compute the initial energy.

The initial potential energy of the cylinder is

$$PE_{\text{initial}} = mgh$$

By substituting $m = 8$ kg, $g = 9.81$ m/s², and $h = 3$ m, we obtain

$$PE_{\text{initial}} = (8)(9.81)(3) = 235.44 \text{ J}$$

Since the cylinder starts from rest, the initial kinetic energy is zero:

$$KE_{\text{initial}} = 0$$

The total initial energy is

$$E_{\text{initial}} = PE_{\text{initial}} = 235.44 \text{ J}$$

Step 2: Case 1 – Rolling Without Slipping

For rolling without slipping, the energy at the bottom consists of both translational and rotational kinetic energy:

$$E_{\text{final}} = KE_{\text{trans}} + KE_{\text{rot}}$$

$$235.44 = \frac{1}{2}mv_G^2 + \frac{1}{2}I_G\omega^2$$

For a solid cylinder rotating about its center, the moment of inertia is

$$I_G = \frac{1}{2}mR^2$$

$$I_G = \frac{1}{2}(8)(0.4)^2 = 0.64 \text{ kg}\cdot\text{m}^2$$

Using the rolling condition $v_G = R\omega$, we substitute $\omega = v_G / R$ to obtain

$$235.44 = \frac{1}{2}(8)v_G^2 + \frac{1}{2}(0.64)\left(\frac{v_G}{0.4}\right)^2 = 4v_G^2 + 2v_G^2 = 6v_G^2$$

Solve for v_G:

$$v_G = \sqrt{\frac{235.44}{6}} = 6.26 \text{ m/s}$$

Step 3: Case 2 – Sliding Without Rolling

If the cylinder slides without rolling, all energy is converted into translational kinetic energy only:

$$E_{final} = KE_{trans}$$

$$235.44 = \frac{1}{2}mv_G^2$$

$$v_G = \sqrt{\frac{2(235.44)}{8}} = 7.67 \text{ m/s}$$

ChatGPT Integration

- Use ChatGPT to analyze rolling motion in mechanical systems.
- Ask ChatGPT to simulate the energy distribution between translational and rotational motion.
- Request ChatGPT compare cases of rolling with and without slipping in engineering applications.

PROMPT 166

Discuss the concept of potential energy in the context of rigid body motion and its practical applications.

Objective

To explain potential energy in the context of rigid body motion, derive its mathematical formulation, and describe its significance in engineering applications such as mechanical linkages, pendulums, lifting mechanisms, and energy storage systems.

Activity

Potential energy represents the stored energy of a rigid body due to its position in a force field. In planar motion, potential energy is mostly associated with the gravitational field and, in some cases, elastic forces (spring energy). The work-energy principle utilizes potential energy to analyze motion without explicitly solving force and acceleration equations.

Definition and Mathematical Formulation of the Potential Energy

1. Identify the definition of the gravitational potential energy.

For a rigid body in a gravitational field, the gravitational potential energy PE_g is given by

$$PE_g = mgh_G$$

where

- m is the mass of the rigid body
- g is the acceleration due to gravity
- h_G is the height of the center of mass above a reference level

2. Identify the definition of the elastic potential energy (spring energy).

For a rigid body connected to a linear elastic spring, the elastic potential energy stored in the spring is

$$PE_s = \frac{1}{2}kx^2$$

where

- k is the spring constant
- x is the spring deformation from its natural length

Work-Energy Principle Including Potential Energy

The total mechanical energy of a rigid body is the sum of its kinetic and potential energy:

$$E_{total} = KE_{total} + PE_g + PE_s$$

The work-energy principle states that the total mechanical energy remains constant in a conservative system:

$$KE_{initial} + PE_{initial} = KE_{final} + PE_{final}$$

Numerical Example

This example involves the energy conservation in a falling rod. A uniform rod of mass 6 kg and length 1.5 meters is hinged at one end and released from a horizontal position. Determine its angular velocity when it reaches the vertical position, assuming no external resistances.

Solution Using the Work-Energy Principle

Step 1: Compute the initial potential energy.

The center of mass of a uniform rod is located at its midpoint $h_G = L/2$. The initial height of the center of mass is

$$h_G = \frac{1.5}{2} = 0.75 \text{ m}$$

The initial potential energy is

$$PE_{\text{initial}} = mgh_G$$

$$PE_{\text{initial}} = (6)(9.81)(0.75) = 44.15 \text{ J}$$

Since the rod starts from rest, its initial kinetic energy is

$$KE_{\text{initial}} = 0$$

Step 2: Compute the final kinetic energy.

At the vertical position, the center of mass is at its lowest point hG=0h_G= 0hG=0, so the final potential energy is zero:

$$PE_{\text{final}} = 0$$

All initial potential energy is converted into rotational kinetic energy:

$$KE_{\text{final}} = \frac{1}{2}I_O\omega^2$$

For a uniform rod rotating about one end, the moment of inertia is

$$I_O = \frac{1}{3}mL^2$$

By substituting $m = 6$ kg and $L = 1.5$ m, we obtain

$$I_O = \frac{1}{3}(6)(1.5)^2 = 4.5 \text{ kg} \cdot \text{m}^2$$

Use the work-energy principle:

$$KE_{\text{initial}} + PE_{\text{initial}} = KE_{\text{final}} + PE_{\text{final}}$$

$$0 + 44.15 = \frac{1}{2}(4.5)\omega^2 + 0$$

Solve for ω:

$$\omega = \sqrt{\frac{44.15}{2.25}} = \sqrt{19.62} = 4.43 \text{ rad/s}$$

Practical Applications of the Potential Energy in Rigid Body Motion

1. Pendulum motion: Potential energy converts into kinetic energy as a pendulum swings.

2. Spring-mass systems: Elastic potential energy is stored in springs and released as kinetic energy.

3. Vehicle suspension systems: Potential energy in springs and dampers helps manage ride comfort.

4. Cranes and hoists: Lifting mechanisms rely on potential energy storage and controlled release.

ChatGPT Integration

- Use ChatGPT to analyze energy transformations in mechanical linkages and suspension systems.

- Ask ChatGPT to generate graphical simulations of potential and kinetic energy exchanges in dynamic systems.

- Request ChatGPT solve real-world problems involving the work-energy principle in lifting mechanisms.

PROMPT 167

Solve a problem involving the conservation of mechanical energy for a rigid body in motion.

Objective

To apply the principle of conservation of mechanical energy to a rigid body in motion, demonstrating how potential energy and kinetic energy transform in dynamic systems.

Activity

The conservation of mechanical energy states that if no non-conservative forces (such as friction or air resistance) do work on a system, the total mechanical energy remains constant. This principle is widely used in pendulum motion, free-falling bodies, impact analysis, and rolling motion in engineering applications such as robotic arms, structural dynamics, and vehicle suspensions.

Energy Conservation Principle for a Rigid Body

The total mechanical energy of a rigid body is the sum of kinetic energy and potential energy:

$$E_{total} = KE_{total} + PE_g + PE_s$$

For conservative systems, the initial mechanical energy equals the final mechanical energy:

$$KE_{initial} + PE_{initial} = KE_{final} + PE_{final}$$

where

- KE_{total} includes both translational and rotational kinetic energy
- PE_g is gravitational potential energy
- PE_s is elastic potential energy, if applicable

Numerical Example

This example involves the energy conservation in a rolling sphere. A solid sphere of mass 4 kg and radius 0.25 meters starts from rest at a height of 2 meters and rolls without slipping down an inclined plane. Determine its velocity at the bottom of the incline using the conservation of mechanical energy.

Solution Using the Energy Conservation Principle

Step 1: Compute the initial mechanical energy.

The initial potential energy is

$$PE_{initial} = mgh$$

By substituting $m = 4$ kg, $g = 9.81$ m/s², and $h = 2$ m, we obtain

$$PE_{initial} = (4)(9.81)(2) = 78.48 \text{ J}$$

Since the sphere starts from rest, the initial kinetic energy is zero:

$$KE_{initial} = 0$$

Thus, the total initial energy is

$$E_{initial} = PE_{initial} = 78.48 \text{ J}$$

Step 2: Compute the final mechanical energy.

At the bottom, the sphere has only kinetic energy, divided into translational and rotational components:

$$KE_{\text{final}} = KE_{\text{trans}} + KE_{\text{rot}}$$

For a solid sphere rotating about its center, the moment of inertia is

$$I_G = \frac{2}{5}mR^2$$

By substituting $m = 4$ kg and $R = 0.25$ m, we obtain

$$I_G = \frac{2}{5}(4)(0.25)^2 = 0.2 \text{ kg} \cdot \text{m}^2$$

Using the rolling condition $v_G = R\omega$, we substitute $\omega = v_G / R$ to obtain

$$KE_{\text{final}} = \frac{1}{2}mv_G^2 + \frac{1}{2}I_G\left(\frac{v_G}{R}\right)^2$$

$$KE_{\text{final}} = \frac{1}{2}(4)v_G^2 + \frac{1}{2}(0.2)\left(\frac{v_G}{0.25}\right)^2$$

$$KE_{\text{final}} = 2v_G^2 + 1.6v_G^2 = 3.6v_G^2$$

By the conservation of mechanical energy, we obtain

$$PE_{\text{initial}} = KE_{\text{final}}$$

$$78.48 = 3.6v_G^2$$

Solve for v_G:

$$v_G = \sqrt{\frac{78.48}{3.6}} = \sqrt{21.8} = 4.67 \text{ m/s}$$

ChatGPT Integration

- Use ChatGPT to analyze energy conservation in rolling motion.
- Ask ChatGPT to simulate the energy exchange between translational and rotational components.
- Request ChatGPT compare energy conservation results for different types of rolling objects (solid sphere, disc, and hoop).

PROMPT 168

Explain the role of power in planar kinetics and derive its mathematical formulation.

Objective

To define power in the context of planar rigid body motion, derive its mathematical formulation, and explain its significance in engineering applications such as automobiles, rotating machinery, and robotic actuators.

Activity

Power represents the rate at which work is done on a rigid body. In mechanical systems, power plays a crucial role in determining efficiency and energy transfer. It helps engineers evaluate how forces and torques influence motion over time in motors, gear systems, and energy-efficient designs.

Definition of Power in Planar Kinetics

Power is defined as the rate of work done on a body by external forces and moments:

$$P = \frac{dW}{dt}$$

where

- P is power (Watt, W)
- W is work (Joule, J)
- t is time (seconds, s)

Power Due to a Force in Translation

For a rigid body undergoing pure translation, the power associated with a force F is

$$P = \vec{F} \cdot \vec{v}$$

where

- $\vec{F}$ is the force vector
- $\vec{v}$ is the velocity of the point of application of the force

If the force is collinear with the velocity, then

$$P = F v$$

Power Due to a Torque in Rotation

For a rigid body undergoing pure rotation, the power delivered by a moment (torque) M is

$$P = M \omega$$

where

- M is the external moment (torque) applied to the body
- ω is the angular velocity of the body

This equation shows that the greater the torque and angular velocity, the higher the power output of a rotating system.

Total Power for a Rigid Body in Planar Motion

For a rigid body undergoing both translation and rotation, the total power supplied by external forces and torques is

$$P_{total} = \sum \left(\vec{F_i} \cdot \vec{v_i} \right) + \sum \left(M_i \omega \right)$$

Numerical Example

This is an example of power in a rotating disc. A solid disc of mass 12 kg and radius 0.5 meters rotates about a fixed center. A torque of 40 N·m is applied, and the disc rotates at 8 rad/s. Determine the instantaneous power delivered to the disc.

Solution

Use the power equation for rotational motion:

$$P = M \omega$$

By substituting M = 40 N·m and ω = 8 rad/s, we obtain

$$P = (40)(8) = 320 \text{ W}$$

Thus, the instantaneous power delivered to the rotating disc is 320 W.

Practical Applications of Power in Planar Kinetics

1. Automobile engines: Power determines acceleration and fuel efficiency.
2. Wind turbines: Torque and angular velocity influence power generation.
3. Robotic actuators: Power calculations help in optimizing motor selection.
4. Industrial machinery: Power efficiency affects performance and operational costs.

ChatGPT Integration

- Use ChatGPT to analyze power distribution in mechanical systems.
- Ask ChatGPT to simulate torque and power relationships in real-world applications.
- Request ChatGPT compare power outputs for different force and torque conditions.

PROMPT 169

Solve a problem involving instantaneous power delivered to a rigid body by external forces.

Objective

To apply the power equations to a rigid body experiencing external forces and torques, calculating the instantaneous power delivered and analyzing how energy transfer occurs in dynamic systems.

Activity

Instantaneous power determines the rate of energy transfer in mechanical systems. It helps engineers evaluate efficiency and performance in automobiles, turbines, robotic actuators, and industrial machines. By calculating instantaneous power, we can understand how force and torque influence real-time motion.

Mathematical Formulation of Instantaneous Power

The instantaneous power delivered to a rigid body consists of contributions from both translational and rotational motion:

$$P_{total} = P_{trans} + P_{rot}$$

For translational motion, power is given by

$$P_{trans} = \vec{F} \cdot \vec{v}$$

For rotational motion, power is given by

$$P_{rot} = M\omega$$

Thus, the total power delivered to a rigid body is

$$P_{total} = \left(\vec{F} \cdot \vec{v}\right) + \left(M\omega\right)$$

where

- $\vec{F}$ is the external force acting on the body
- $\vec{v}$ is the velocity of the point of application of the force
- M is the applied torque (moment)
- ω is the angular velocity of the rigid body

Numerical Example

This is an example involving instantaneous power in a rotating arm. A rigid robotic arm of length 1.2 meters and mass 15 kg rotates about a fixed pivot. A force of 50 N is applied tangentially at the free end, and the arm has an angular velocity of 6 rad/s. Determine

1. the instantaneous power delivered by the force
2. the instantaneous power delivered by the applied torque
3. the total instantaneous power acting on the arm

Solution

Step 1: Compute the translational power.

Since the force is applied tangentially, its direction is perpendicular to the arm, and the velocity at the point of application is

$$v = R\omega$$

By substituting $R = 1.2$ m and $\omega = 6$ rad/s, we obtain

$$v = (1.2)(6) = 7.2 \text{ m/s}$$

The instantaneous power due to force is

$$P_{\text{trans}} = Fv$$

$$P_{\text{trans}} = (50)(7.2) = 360 \text{ W}$$

Step 2: Compute the rotational power.

The torque applied by the force at the pivot is

$$\boldsymbol{M} = FR$$

$$M = (50)(1.2) = 60 \text{ N} \cdot \text{m}$$

The instantaneous power due to torque is

$$P_{\text{rot}} = M\omega$$

$$P_{\text{rot}} = (60)(6) = 360 \text{ W}$$

Step 3: Compute the total instantaneous power.

The total power is

$$P_{\text{total}} = P_{\text{trans}} + P_{\text{rot}}$$

$$P_{\text{total}} = 360 = 360 \text{ W}$$

This power accounts for both translational and rotational energy input because the force causes both.

ChatGPT Integration

- Use ChatGPT to analyze power transfer in dynamic mechanical systems.
- Ask ChatGPT to generate power distribution diagrams for different motion conditions.
- Request ChatGPT compare instantaneous and average power calculations in rotating systems.

PROMPT 170

Compare the force-based (Newtonian) and energy-based (work-energy) methods for solving a rigid body motion problem, analyzing their advantages and limitations.

Objective

To compare the force-based (Newtonian) approach and the energy-based (work-energy) approach for analyzing rigid body motion, highlighting their advantages, limitations, and best-use scenarios in mechanical engineering applications.

Activity

Rigid body motion can be analyzed using two primary methods:

1. Newton's second law (force-based approach) uses force and moment equations to determine acceleration and velocity.

2. The work-energy principle (energy-based approach) relates work done to changes in the kinetic and potential energy to find the velocity without solving force equations.

Both methods are widely applied in automotive crash testing, robotic motion, aircraft dynamics, and industrial machines, where selecting the most efficient approach can save computation time and improve system design.

TABLE 17.1 A comparison of the two approaches

Aspect	Force-Based (Newtonian) Method	Energy-Based (Work-Energy) Method
Fundamental Principle	Newton's second law: $F = ma$	Work-Energy Theorem: $W = \Delta KE$
Equations Used	$$\sum F = ma, \sum M = I\alpha$$	$$KE_{initial} + PE_{initial} = KE_{final} + PE_{final}$$
Unknowns Solved	Acceleration, forces, and reactions	Velocities and energy changes
Best for	Problems requiring force/moment analysis (contact forces and constraints)	Problems where only initial and final states are needed
Example Applications	Contact forces in robotic arms, vehicle suspensions	Projectile motion, pendulums, roller coasters
Limitations	Requires solving force equations at every step	Cannot directly find time-dependent quantities (e.g., acceleration and forces)

Numerical Example

The following is an example of determining the motion of a rotating bar using both approaches. A rigid bar of mass 10 kg and length 2 meters rotates about a hinge at one end. A force of 50 N is applied at the free end, causing the bar to rotate from rest. Determine its angular velocity after rotating 45° using

1. Newton's second law (force-based approach)
2. work-energy principle (energy-based approach)

Solution Using Newton's Second Law

Step 1: Compute the moment of inertia.

For a uniform bar rotating about one end, the moment of inertia is

$$I_O = \frac{1}{3}mL^2$$

By substituting m = 10 kg and L = 2 m, we obtain

$$I_O = \frac{1}{3}(10)(2)^2 = \frac{40}{3} \text{ kg} \cdot \text{m}^2$$

Step 2: Compute the angular acceleration.

Use Newton's second law for rotation:

$$\sum M_O = I_O$$

The applied force generates a moment:

$$M_O = (50)(2) = 100 \, N \cdot m$$

Solve for α:

$$100 = \left(\frac{40}{3}\right)\alpha$$

$$\alpha = \frac{100}{\frac{40}{3}} = 7.5 \text{ rad/s}^2$$

Use the kinematic equation:

$$\omega^2 = \omega_0^2 + 2\alpha\theta$$

$$\omega^2 = 0 + 2(7.5)\left(\frac{\pi}{4}\right)$$

$$\omega = \sqrt{\frac{15\pi}{4}} = 3.43 \text{ rad/s}$$

Solution Using the Work-Energy Principle

Apply the formula for energy conservation:

$$KE_{final} = W_{net}$$

The work done by the applied force is

$$W = F d$$

Since the arc length traveled is $d = L\theta$, we find

$$W = (50)(2)\left(\frac{\pi}{4}\right) = 78.54 \text{ J}$$

The rotational kinetic energy at the final state is

$$\frac{1}{2}I_0\omega^2 = W\frac{1}{2}\left(\frac{40}{3}\right)$$

$$\omega^2 = 78.54$$

$$\omega = \sqrt{\frac{2(78.54)}{40/3}} = 3.43 \text{ rad/s}$$

Final Comparison

- Both methods yield the same result, but the work-energy principle avoids solving for force/acceleration explicitly.
- The Newtonian method is useful when force distributions or constraints must be analyzed.

CHAPTER 18

PLANAR KINETICS OF RIGID BODIES: IMPULSE AND MOMENTUM

Prompt 171 Explain the principle of linear impulse and momentum for a rigid body in planar motion.

Prompt 172 Derive the angular impulse-momentum equation for a rigid body and discuss its significance.

Prompt 173 Solve a problem involving linear impulse and momentum for a rigid body subjected to an external force.

Prompt 174 Explain the conservation of angular momentum for a rigid body and its engineering applications.

Prompt 175 Solve a problem involving impact and angular momentum conservation in a rigid body.

Prompt 176 Discuss the concept of coefficient of restitution and its role in impact analysis.

Prompt 177 Analyze the collision of two billiard balls where their centers of mass do not remain on the same line after impact and prove that they move at right angles when e=1 (perfectly elastic collision).

Prompt 178 Analyze the oblique impact of a rigid body with a surface and determine the post-impact motion.

Prompt 179 Explain the relationship between impulse-momentum and energy methods in rigid body motion.

Prompt 180 Discuss real-world applications of impulse-momentum principles in mechanical and structural systems.

PROMPT 171

Explain the principle of linear impulse and momentum for a rigid body in planar motion.

Objective

To explain the principle of linear impulse and momentum as applied to rigid bodies in planar motion, derive its mathematical formulation, and describe its significance in mechanical systems such as vehicle crashes, robotic arm motion, and sports mechanics.

Activity

Linear impulse and momentum describe how an external force acting over time changes the momentum of a rigid body. This principle is widely used in impact analysis, collision mechanics, and propulsion systems, where forces act over short time intervals to produce significant velocity changes.

Mathematical Formulation of Linear Impulse and Momentum

For a rigid body in planar motion, the linear impulse-momentum equation is derived from Newton's second law:

$$\sum \vec{F} = m\vec{a}$$

By rewriting acceleration as the time derivative of velocity, we obtain

$$\sum \vec{F} = m\frac{d\vec{v}_G}{dt}$$

By integrating both sides from an initial time t_1 to a final time t_2, we obtain

$$\int_{t_1}^{t_2} \sum \vec{F}\, dt = m\left(\vec{v}_{G2} - \vec{v}_{G1}\right)$$

This results in the linear impulse-momentum equation:

$$\sum \vec{I} = m\left(\vec{v}_{G2} - \vec{v}_{G1}\right)$$

where

- $\sum \vec{I} = \int_{t_1}^{t_2} \sum \vec{F}\, dt$ is the linear impulse

- $m\vec{v}_{G1}$ and $m\vec{v}_{G2}$ are the initial and final linear momentum of the rigid body, respectively

- $\vec{v}_G$ is the velocity of the center of mass

Numerical Example

This is an example involving the collision of a spacecraft module. A 1500 kg spacecraft module is moving at 2 m/s in space when a thruster fires, exerting a constant force of 3000 N in the same direction as the initial velocity for 5 seconds. Determine the final velocity of the spacecraft module using the impulse-momentum principle.

Solution Using Impulse-Momentum

Step 1: Compute the linear impulse.
The impulse is given by

$$I = F \Delta t$$

By substituting F = 3000 N and Δt = 5 s, we obtain

$$I = (3000)(5) = 15000 \text{ N} \cdot \text{s}$$

Step 2: Apply the linear impulse-momentum equation.

Use the following:

$$\sum I = m v_{G2} - m v_{G1}$$

By substituting the given values, we obtain

$$15000 = (1500) v_{G2} - (1500)(2)$$

$$15000 = (1500) v_{G2} - 3000$$

$$v_{G2} = \frac{15000 + 3000}{1500} = \frac{18000}{1500} = 12 \text{ m/s}$$

Practical Applications of Impulse-Momentum in Engineering

1. Spacecraft propulsion: Thrusters exert impulse to change velocity.
2. Automotive safety: Airbags and crumple zones reduce impulse forces in collisions.
3. Athletic performance: Impulse analysis helps optimize running, jumping, and striking forces.
4. Industrial robotics: Controlling impulse in robotic arms prevents excessive force application.

PROMPT 172

Derive the angular impulse-momentum equation for a rigid body and discuss its significance.

Objective

To derive the angular impulse-momentum equation for a rigid body in planar motion, explain its significance, and describe how it applies to real-world mechanical systems such as robotic arms, automotive dynamics, and impact analysis.

Activity

Angular impulse and momentum describe how an external moment acting over time changes the angular momentum of a rigid body. This principle is widely used in crash safety analysis, torque transmission in machines, and sports biomechanics, where angular motion is influenced by applied torques.

Mathematical Formulation of Angular Impulse and Momentum

For a rigid body in planar motion, Newton's second law for rotation about the center of mass is given by

$$\sum M_G = I_G \alpha$$

By rewriting angular acceleration as the time derivative of angular velocity, we obtain

$$\sum M_G = I_G \frac{d\omega}{dt}$$

By integrating both sides from an initial time t_1 to a final time t_2, we obtain

$$\int_{t_1}^{t_2} \sum M_G dt = I_G \left(\omega_2 - \omega_1 \right)$$

This results in the angular impulse-momentum equation:

$$\sum I_\theta = I_G \omega_2 - I_G \omega_1$$

where

■ $\sum I_\theta = \int_{t_1}^{t_2} \sum M_G dt$ is the angular impulse

■ $I_G \omega_1$ and $I_G \omega_2$ are the initial and final angular momentum of the rigid body

■ ω is the angular velocity

■ I_C is the mass moment of inertia about the center of mass

This equation states that the total angular impulse applied to a rigid body over a time interval results in a change in its angular momentum.

Numerical Example

This is an example involving the impact on a rotating door. A solid door of mass 25 kg and width 1.2 meters is initially at rest and is hit at the edge by a force of 80 N applied perpendicular to the door for 0.3 seconds. Determine the angular velocity of the door immediately after impact using the angular impulse-momentum principle.

Solution Using Angular Impulse-Momentum

Step 1: Compute the angular impulse.

The impulse is given by

$$I_\theta = \int \sum M_C dt$$

Since the force is perpendicular to the door, the moment about the hinge is

$$M_C = FR$$

$$M_C = (80)(1.2) = 96 \text{ N} \cdot \text{m}$$

The impulse is

$$I_\theta = M_C \Delta t$$

$$I_\theta = (96)(0.3) = 28.8 \text{ N} \cdot \text{m} \cdot \text{s}$$

Step 2: Apply the angular impulse-momentum equation.
Use the following:

$$\sum I_\theta = I_C \omega_2 - I_C \omega_1$$

Since the door starts from rest, $\omega_1 = 0$, so we find

$$I_\theta = I_C \omega_2$$

For a solid rectangular plate rotating about one edge, the moment of inertia is

$$I_C = \frac{1}{3} mL^2$$

By substituting $m = 25$ kg and $L = 1.2$ m, we obtain

$$I_G = \frac{1}{3}(25)(1.2)^2 = 12 \text{ kg} \cdot \text{m}^2$$

Solve for ω_2:

$$\omega_2 = \frac{I_{\grave{e}}}{I_G} = \frac{28.8}{12} = 2.4 \text{ rad/s}$$

Practical Applications of Angular Impulse-Momentum in Engineering

1. vehicle crash testing: evaluating rotational motion of parts upon impact
2. robotic manipulation: analyzing how sudden forces change robotic arm velocities
3. sports mechanics: studying impulse effects in striking sports (tennis, baseball, golf)
4. machine safety mechanisms: designing impact-resistant rotating components in industrial machines

PROMPT 173

Solve a problem involving linear impulse and momentum for a rigid body subjected to an external force.

Objective

To apply the linear impulse-momentum principle to solve a problem where a rigid body experiences an external force over a time interval, demonstrating how momentum changes in mechanical systems.

Activity

Impulse-momentum analysis helps engineers understand how forces acting over time affect motion. This principle is widely used in automotive crash safety, robotic motion control, aerospace propulsion, and impact analysis in sports and machinery.

Mathematical Formulation of Linear Impulse-Momentum

The linear impulse-momentum equation for a rigid body in planar motion is given by

$$\sum I = m\vec{v}_{G2} - m\vec{v}_{G1}$$

This equation states that the total impulse applied to a body results in a change in its linear momentum.

Numerical Example

This example involves the cargo container deceleration on a ship deck. A cargo container of mass 5000 kg is moving at 6 m/s on a ship deck when a retarding force of 10000 N is applied for 4 seconds using a braking mechanism. Determine the final velocity of the container using the impulse-momentum principle.

Solution Using Impulse-Momentum

Step 1: Compute the linear impulse.
The impulse is given by

$$I = F\Delta t$$

By substituting $F = -10000$ N (negative since it acts against motion) and $\Delta t = 4\,s$, we obtain

$$I = (-10000)(4) = -40000 \text{ N} \cdot \text{s}$$

Step 2: Apply the linear impulse-momentum equation.
Use the following:

$$\sum I = mv_{G2} - mv_{G1}$$

By substituting the given values, we obtain

$$-40000 = (5000)v_{G2} - (5000)(6)$$

$$-40000 = (5000)v_{G2} - 30000$$

$$v_{G2} = \frac{-40000 + 30000}{5000} = \frac{-10000}{5000} = -2 \text{ m/s}$$

Since the final velocity is negative, the container has reversed direction and is moving at 2 m/s in the opposite direction.

This example demonstrates how impulse affects velocity changes, which is crucial in deceleration systems, safety engineering, and impact mitigation.

Practical Applications of Linear Impulse-Momentum in Engineering

1. Ship cargo handling: Braking forces slow down moving containers to prevent damage.
2. Automotive braking systems: Deceleration forces reduce vehicle momentum in controlled stops.
3. Sports mechanics: Impulse determines changes in velocity during collisions (e.g., soccer kicks).
4. Landing gear systems: Aircraft landing gears absorb impulse forces upon touchdown.

ChatGPT Integration

- Use ChatGPT to analyze impulse-momentum interactions in different braking systems.
- Ask ChatGPT to simulate deceleration effects in transport and safety engineering.
- Request ChatGPT generate impulse diagrams for different impact conditions.

PROMPT 174

Explain the conservation of angular momentum for a rigid body and its engineering applications.

Objective

To explain the principle of angular momentum conservation for a rigid body in planar motion, derive its mathematical formulation, and describe its applications in engineering systems such as gyroscopes, satellite dynamics, and vehicle stability.

Activity

Angular momentum is a fundamental concept in rotational dynamics, describing how an object's rotation remains constant in the absence of an external moment. This principle is widely used in spacecraft stabilization, mechanical balancing, and control systems in robotics.

Mathematical Formulation of Angular Momentum Conservation

The angular momentum H_G of a rigid body about its center of mass is

$$H_G = I_G \omega$$

where

- I_G is the mass moment of inertia about the center of mass
- ω is the angular velocity of the rigid body

Use Newton's second law for rotation:

$$\sum M_G = \frac{dH_G}{dt}$$

If no external moments act on the system, then

$$\sum M_G = 0 \quad \Rightarrow \quad \frac{dH_G}{dt} = 0$$

This means that angular momentum remains constant:

$$H_{G1} = H_{G2}$$

or,

$$I_G \omega_1 = I_G \omega_2$$

This is the principle of the conservation of angular momentum. This equation states that when no external moments act on a system, its angular momentum remains constant.

Numerical Example

This example involves the rotation of a satellite with deployable solar panels. A satellite with retractable solar panels is spinning at 5 rad/s with its moment of inertia $I_1 = 500\ kg \cdot m^2$. The panels extend, increasing the moment of inertia to $I_2 = 1200\ kg \cdot m^2$. Determine the new angular velocity of the satellite using the conservation of angular momentum.

Solution Using Angular Momentum Conservation

Apply the formula:

$$I_1 \omega_1 = I_2 \omega_2$$

By substituting the given values, we obtain

$$(500)(5) = (1200)\omega_2$$

$$2500 = 1200\ \omega_2$$

Solve for ω_2:

$$\omega_2 = \frac{2500}{1200} = 2.08 \text{ rad/s}$$

This example illustrates how changing the moment of inertia affects rotational speed, which is crucial in spacecraft control, engineering stabilization, and robotics.

Practical Applications of Angular Momentum Conservation

1. Gyroscopic stabilization is used in satellites, ships, and aircraft for automatic orientation control.

2. Figure skating and gymnastics: Athletes control their spin speed by adjusting their moment of inertia.

3. Automotive stability systems (ABS) and traction control use angular momentum principles to prevent skidding.

4. Robotic manipulation: Robotic arms use conservation principles to manage rotational movement efficiently.

ChatGPT Integration

- Use ChatGPT to analyze angular momentum conservation in space missions.

- Ask ChatGPT to simulate momentum transfer in dynamic systems (e.g., drones, rotating machinery).

- Request ChatGPT generate momentum balance equations for specific engineering applications.

PROMPT 175

Solve a problem involving impact and angular momentum conservation in a rigid body.

Objective

To apply the principle of angular momentum conservation to solve a problem involving impact and rotational motion, demonstrating how collisions affect angular velocity in mechanical systems.

Activity

When a rigid body undergoes an impact, angular momentum is conserved if there are no external moments acting during the collision. This principle is widely used in robotic motion control, vehicle crash analysis, and sports mechanics (e.g., bat swings, diving rotations).

Mathematical Formulation of Angular Momentum Conservation in Impact

The total angular momentum of a rigid body before and after impact is given by

$$H_{G1} = H_{G2}$$

Since angular momentum is defined as

$$H_G = I_G \omega$$

We apply conservation to obtain

$$I_G \omega_1 = I_G \omega_2$$

For a system where an object strikes a rigid body, momentum can also be considered about the impact point:

$$mvR = I_O \omega_2$$

where

- m is the mass of the impacting object
- v is its velocity before impact
- R is the distance from the impact point to the rotation axis
- I_O is the moment of inertia of the rigid body about the impact axis
- ω_2 is the angular velocity after impact

Numerical Example

This is an example using the impact on a pivoted rod. A thin rod of mass 8 kg and length 2 meters is hinged at one end and initially at rest. A 2 kg ball moving at 5 m/s strikes the free end and sticks to it. Determine the angular velocity of the rod immediately after impact using angular momentum conservation.

Solution Using Angular Momentum Conservation

Step 1: Compute the initial angular momentum.
Before impact, only the ball has linear momentum. The angular momentum of the ball about the hinge is

$$H_{O1} = mvR$$

By substituting $m = 2\,kg$, $v = 5\,m/s$, and $R = 2\,m$, we obtain

$$H_{O1} = (2)(5)(2) = 20 \text{ kg} \cdot \text{m}^2 / \text{s}$$

Step 2: Compute the moment of inertia of the system after impact.
For a thin rod pivoted at one end, the moment of inertia is

$$I_O = \frac{1}{3} mL^2$$

By substituting m = 8 kg and L = 2 m, we obtain

$$I_O = \frac{1}{3}(8)(2)^2 = \frac{32}{3} \text{ kg} \cdot \text{m}^2$$

The ball is now attached to the free end, contributing an additional moment of inertia:

$$I_{ball} = mR^2 = (2)(2)^2 = 8 \text{ kg} \cdot \text{m}^2$$

Thus, the total moment of inertia is

$$I_{total} = \frac{32}{3} + 8 = \frac{56}{3} \text{ kg} \cdot \text{m}^2$$

Step 3: Apply angular momentum conservation.
Since no external torques act during the collision, angular momentum is conserved:

$$H_{O1} = H_{O2}$$

$$mvR = I_{total}\omega_2$$

$$20 = \frac{56}{3}\omega_2$$

Solve for ω_2:

$$\omega_2 = \frac{20 \times 3}{56} = \frac{60}{56} = 1.07 \text{ rad/s}$$

ChatGPT Integration

- Use ChatGPT to analyze angular momentum changes in different impact scenarios.
- Ask ChatGPT to simulate rigid body impact responses for different materials.
- Request ChatGPT generate momentum diagrams for various engineering systems.

PROMPT 176

Discuss the concept of coefficient of restitution and its role in impact analysis.

Objective

To explain the coefficient of restitution (COR) in the context of rigid body impact analysis, derive its mathematical formulation, and describe its significance in engineering applications such as crash testing, sports biomechanics, and robotic manipulation.

Activity

The coefficient of restitution (e) is a measure of how much kinetic energy is conserved during a collision. It quantifies the elasticity of an impact, determining whether a collision is perfectly elastic, inelastic, or somewhere in between. Understanding COR is essential in automotive safety design, material testing, and mechanical system durability analysis.

Mathematical Formulation of the Coefficient of Restitution

The coefficient of restitution is defined as the ratio of relative velocity after impact to relative velocity before impact, along the line of impact:

$$e = \frac{\text{relative velocity after impact}}{\text{relative velocity before impact}}$$

For two colliding bodies (A and B), the mathematical expression is

$$e = \frac{v_{B2} - v_{A2}}{v_{A1} - v_{B1}}$$

where

- v_{A1}, v_{B1} are the initial velocities before impact
- v_{A2}, v_{B2} are the final velocities after impact
- e ranges from 0 (perfectly inelastic) to 1 (perfectly elastic)

TABLE 18.1 Types of impact based on coefficient of restitution

Impact Type	Coefficient of Restitution (e)	Description
Perfectly Elastic	$e = 1$	Kinetic energy is fully conserved. No energy loss.
Partially Elastic	$0 < e < 1$	Some kinetic energy is lost as heat, sound, or deformation.
Perfectly Inelastic	$e = 0$	Objects stick together after impact, causing maximum energy loss.

Numerical Example

This example involves crash testing a bumper with two vehicles. A 1500 kg car (A) moving at 10 m/s collides with a stationary 1000 kg car (B). After impact, car A moves at 4 m/s, while car B moves forward. Determine

1. the coefficient of restitution (e)
2. the velocity of car B after impact

Solution Using Coefficient of Restitution

Step 1: Apply the coefficient of restitution formula.
Use the following:

$$e = \frac{v_{B2} - v_{A2}}{v_{A1} - v_{B1}}$$

By substituting the given values, we obtain

$$e = \frac{v_{B2} - 4}{10}$$

Step 2: Apply the conservation of momentum.

The momentum before and after collision is conserved is as follows:

$$m_A v_{A1} + m_B v_{B1} = m_A v_{A2} + m_B v_{B2}$$

By substituting
$m_A = 1500\ kg, m_B = 1000\ kg, v_{A1} = 10\ m/s, v_{A2} = 4\ m/s,$ and $v_{B1} = 0$, we obtain

$$(1500)(10)+(1000)(0)=(1500)(4)+(1000)v_{B2}$$

$$15000 = 6000 + 1000v_{B2}$$

Solve for v_{B2}:

$$v_{B2} = \frac{15000 - 6000}{1000} = \frac{9000}{1000} = 9 \text{ m/s}$$

Step 3: Compute the coefficient of restitution.

Use the following:

$$e = \frac{9-4}{10}$$

$$e = \frac{5}{10} = 0.5$$

This example demonstrates how COR determines the elasticity of an impact, which is crucial in vehicle crash analysis, material impact testing, and sports dynamics.

ChatGPT Integration

- Use ChatGPT to analyze impact elasticity in different collision scenarios.
- Ask ChatGPT to simulate bouncing and collision responses for various materials.
- Request ChatGPT compare real-world coefficients of restitution for different impact conditions.

PROMPT 177

Analyze the collision of two billiard balls where their centers of mass do not remain on the same line after impact and prove that they move at right angles when $e = 1$ (perfectly elastic collision).

Objective

To analyze the two-dimensional collision of two billiard balls, derive the governing equations using impulse-momentum principles, and mathematically prove that when the collision is perfectly elastic ($e=1$), the two balls move at right angles after impact.

Activity

In billiards, air hockey, and particle physics, collisions often occur where the objects do not remain on the same line after impact. Unlike a direct central impact, in this case, the balls move in different directions, requiring the use of momentum conservation in two dimensions.

Mathematical Formulation of a Two-Body Oblique Collision

Definitions of the Collision Angles:

- θ = angle of Ball A's velocity after collision relative to its initial direction

- φ = angle of Ball B's velocity after collision relative to the initial direction of Ball A

Thus, after impact, the velocities v_{A2} and v_{B2} are at angles θ and ϕ from the original line of motion of Ball A, respectively.
Assumptions:

1. The two balls have equal mass m (common in billiard physics).
2. The first ball (Ball A) moves with velocity v_{A1}, and the second ball (Ball B) is initially at rest ($v_{B1} = 0$).
3. The collision occurs obliquely, meaning the balls move in different directions after impact.
4. Friction and spin effects are neglected for simplicity.
5. The collision is perfectly elastic ($e = 1$), meaning kinetic energy is conserved.

Step 1: Apply the conservation of linear momentum in two dimensions. Momentum is conserved in both the x-direction and y-direction:

$$mv_{A1} = mv_{A2}\cos\theta + mv_{B2}\cos\varphi$$

$$0 = mv_{A2}\sin\theta - mv_{B2}\sin\varphi$$

The x-axis is defined parallel to the line joining ball A and B before collision. Canceling mass mmm from both equations:

$$v_{A1} = v_{A2}\cos\theta + v_{B2}\cos\varphi$$

$$0 = v_{A2}\sin\theta - v_{B2}\sin\varphi$$

Step 2: Apply the conservation of kinetic energy.
Since the collision is perfectly elastic ($e = 1$), kinetic energy is conserved:

$$\frac{1}{2}mv_{A1}^2 = \frac{1}{2}mv_{A2}^2 + \frac{1}{2}mv_{B2}^2$$

We cancel $(m/2)$ from both sides:

$$v_{A1}^2 = v_{A2}^2 + v_{B2}^2$$

Step 3: Prove that the balls move at right angles after impact.

From conservation of momentum in the y-direction, we find

$$v_{A2}\sin\theta = v_{B2}\sin\varphi$$

Square both sides to obtain

$$\left(v_{A2}\sin\theta\right)^2 = \left(v_{B2}\sin\varphi\right)^2$$

From the conservation of momentum in the x-direction, we rearrange for v_{B2} and square both sides:

$$\left(v_{A1} - v_{A2}\cos\theta\right)^2 = \left(v_{B2}\cos\varphi\right)^2$$

Sum up the resulted equations:

$$\left(v_{A2}\sin\theta\right)^2 + \left(v_{A1} - v_{A2}\cos\theta\right)^2 = v_{B2}^2$$

Expand the equation:

$$\left(v_{A2}\sin\theta\right)^2 + \left(v_{A1}\right)^2 + \left(v_{A2}\cos\theta\right)^2 - 2v_{A1}v_{A2}\cos\theta = v_{B2}^2$$

Simplify:

$$v_{A2}^2 + v_{A1}^2 - 2v_{A1}v_{A2}\cos\theta - v_{B2}^2 = 0$$

But, from energy equation, we have

$$v_{B2}^2 = v_{A1}^2 - v_{A2}^2$$

Therefore, after plugging in for v_{B2}^2, we obtain

$$v_{A2}^2 + v_{A1}^2 - 2v_{A1}v_{A2}\cos\theta - (v_{A1}^2 - v_{A2}^2) = 0$$

Simplify:

$$\cos\theta = \frac{v_{A2}}{v_{A1}}$$

Similarly, from conservation of momentum in the x-direction, we rearrange for v_{A2} and square both sides:

$$\left(v_{A1} - v_{B2}\cos\varphi\right)^2 = \left(v_{A2}\cos\theta\right)^2$$

Sum up this result with the square of y-momentum equation:

$$\left(v_{B2}\sin\varphi\right)^2+\left(v_{A1}-v_{B2}\cos\varphi\right)^2=v_{A2}^2$$

Expand to obtain

$$\left(v_{B2}\sin\varphi\right)^2+\left(v_{A1}\right)^2+\left(v_{B2}\cos\varphi\right)^2-2v_{A1}v_{B2}\cos\varphi=v_{A2}^2$$

Simplify:

$$v_{B2}^2+v_{A1}^2-2v_{A1}v_{B2}\cos\varphi-v_{A2}^2=0$$

But, from energy equation, we have

$$v_{A2}^2=v_{A1}^2-v_{B2}^2$$

Therefore, after plugging in for v_{A2}^2, we have

$$v_{B2}^2+v_{A1}^2-2v_{A1}v_{B2}\cos\varphi-\left(v_{A1}^2-v_{B2}^2\right)=0$$

Simplify to obtain

$$\cos\varphi=\frac{v_{B2}}{v_{A1}}$$

Now, divide both sides of the energy equation by v_{A1}^2:

$$1=\left(\frac{v_{A2}}{v_{A1}}\right)^2+\left(\frac{v_{B2}}{v_{A1}}\right)^2$$

From momentum conservation, we obtain

$$\cos\theta=\frac{v_{A2}}{v_{A1}},\quad\cos\varphi=\frac{v_{B2}}{v_{A1}}$$

By substituting into the energy equation, we obtain

$$\cos^2\theta+\cos^2\varphi=1$$

But we can write $\cos\varphi=\sin\left(\dfrac{\pi}{2}-\varphi\right)$ to obtain

$$\cos^2\theta+\sin^2\left(\frac{\pi}{2}-\varphi\right)=1$$

Or, from the trigonometric identity (Pythagoras theorem), we can obtain

$$\theta=\frac{\pi}{2}-\varphi\Rightarrow\theta+\varphi=\frac{\pi}{2}=90°$$

ChatGPT Integration

- Use ChatGPT to simulate billiard ball impacts with different restitution coefficients.
- Ask ChatGPT to analyze two-body collisions in automotive crash testing.
- Request ChatGPT visualize momentum transfer in oblique impact scenarios.

PROMPT 178

Analyze the oblique impact of a rigid body with a surface and determine the post-impact motion.

Objective

To analyze an oblique impact where a rigid body collides with a surface at an angle, derive the governing equations using impulse-momentum and restitution principles, and determine the post-impact motion.

Activity

In real-world engineering systems, oblique impacts occur in bouncing objects (balls, robotic end-effectors), vehicle collisions, and manufacturing processes. Understanding how a rigid body rebounds after impact is essential for sports physics, industrial machinery, and vehicle safety analysis.

Mathematical Formulation of Oblique Impact

An oblique impact occurs when a rigid body collides with a surface at an angle, resulting in a change in both normal and tangential velocity components.

Step 1: Define the velocity components before and after impact.

Let a rigid body approach a surface with

- Velocity $\vec{v}_1$ before impact, with components
  - v_{1n} along the normal (perpendicular to the surface).
  - v_{1t} along the tangential (parallel to the surface).

- Velocity $\vec{v}_2$ after impact, with components
  - v_{2n} along the normal direction
  - v_{2t} along the tangential direction

$$\vec{v}_1 = v_{1n} + v_{1t}, \quad \vec{v}_2 = v_{2n} + v_{2t}$$

Step 2: Apply the coefficient of restitution in the normal direction. The coefficient of restitution (e) defines the ratio of relative velocity after impact to relative velocity before impact along the normal direction:

$$e = \frac{v_{2n}}{v_{1n}}$$

Since the surface is fixed, the normal velocity after impact is

$$v_{2n} = -ev_{1n}$$

The negative sign accounts for the reversal of direction upon impact.

Step 3: Apply the conservation of momentum in the tangential direction.

For an ideal case with no friction, the tangential velocity remains unchanged:

$$v_{2t} = v_{1t}$$

If friction is present, an impulse I_t acts along the tangential direction, modifying the post-impact tangential velocity:

$$mv_{2t} = mv_{1t} + I_t$$

where I_t depends on the coefficient of friction μ and the normal impulse.

Numerical Example

This is an example involving a bouncing ball with an oblique impact on a hard surface.

A rubber ball of mass 0.5 kg strikes a smooth surface at a velocity of 10 m/s at an angle of 30° with respect to the normal. The coefficient of restitution (e) is 0.8. Determine

1. the velocity components after impact
2. the new velocity direction

Solution Using Oblique Impact Equations

Step 1: Compute the initial velocity components.

Before impact:

$$v_{1n} = v_1 \cos 30°$$

$$v_{1t} = v_1 \sin 30°$$

By substituting $v_1 = 10$ m/s:

$$v_{1n} = (10) \cos 30° = (10)(0.866) = 8.66 \text{ m/s}$$

$$v_{1t} = (10) \sin 30° = (10)(0.5) = 5 \text{ m/s}$$

Step 2: Compute the post-impact normal velocity.

Use the following:

$$v_{2n} = -ev_{1n}$$

By substituting $e = 0.8$, we obtain

$$v_{2n} = -(0.8)(8.66) = -6.93 \text{ m/s}$$

Step 3: Compute the post-impact tangential velocity.
Since the surface is smooth (no friction), we obtain

$$v_{2t} = v_{1t} = 5 \text{ m/s}$$

Step 4: Compute the final velocity magnitude and direction.

The magnitude of the velocity after impact is

$$v_2 = \sqrt{v_{2n}^2 + v_{2t}^2}$$

$$v_2 = \sqrt{(-6.93)^2 + (5)^2} = \sqrt{48.06 + 25} = \sqrt{73.06} = 8.55 \text{ m/s}$$

The new velocity direction (θ_2) with respect to the normal is

$$\theta_2 = \tan^{-1}\left(\frac{v_{2t}}{v_{2n}}\right)$$

$$\theta_2 = \tan^{-1}\left(\frac{5}{6.93}\right) = \tan^{-1}(0.722) = 35.8°$$

Results

- The velocity after impact is 8.55 m/s, directed at an angle 35.8° with respect to the normal.
- The ball rebounds at a shallower angle due to energy loss in the normal direction.

ChatGPT Integration

- Use ChatGPT to simulate bouncing object trajectories with different restitution coefficients.
- Ask ChatGPT to analyze impact force distributions in angled collisions.
- Request ChatGPT compare rebound angles for different surface conditions.

PROMPT 179

Explain the relationship between impulse-momentum and energy methods in rigid body motion.

Objective

To explore the relationship between impulse-momentum and energy methods in analyzing rigid body motion, highlighting their similarities, differences, and best-use cases in engineering applications such as crash testing, mechanical linkages, and robotics.

Activity

Impulse-momentum and energy principles are two fundamental approaches for solving rigid body dynamics problems. While both describe how forces and torques influence motion, they emphasize different aspects of motion analysis. This distinction is crucial in impact mechanics, machine design, and dynamic system optimization.

Mathematical Formulation of Impulse-Momentum and Energy Methods

Impulse-Momentum Method (Time-Dependent Approach)

The impulse-momentum principle states that the change in momentum of a rigid body equals the impulse applied over a time interval:

$$\sum I = m\vec{v}_{G2} - m\vec{v}_{G1}$$

For angular motion, this is

$$\sum I_\theta = I_G \omega_2 - I_G \omega_1$$

where

- $m\vec{v}_{G1}, m\vec{v}_{G2}$ (initial and final linear momenta)
- $I_G \omega_1, I_G \omega_2$ (initial and final angular momenta)

TABLE 18.2 Comparison of impulse-momentum and energy methods

Aspect	Impulse-Momentum Method	Energy Method
Principle Used	Newton's second law	Work-Energy Theorem
Equation Type	Time-dependent	Work-based
Focus	Velocity changes due to force over time	Energy changes due to work done
Best for	Impact, collision, and short-time forces	Continuous forces and system energy balance
Example Applications	Crash testing, bouncing objects, projectiles	Lifting mechanisms, gears, and rotating machinery

Numerical Example

This is an example involving comparing both methods for a rolling drum. A 40 kg solid drum (radius = 0.3 m) is rolling without slipping at 2 m/s. A horizontal force of 150 N is applied for 3 seconds. Determine the final velocity using the following:

1. Impulse-Momentum Method
2. Work-Energy Method

Solution Using Impulse-Momentum

Step 1: Compute the impulse.

The impulse is given by

$$I = F \Delta t$$

By substituting F = 150 N and Δt = 3 s, we obtain

$$I = (150)(3) = 450 \text{ Ns}$$

Use the following:

$$\sum I = mv_{G2} - mv_{G1}$$

$$(40)v_{G2} = (40)(2) + 450$$

$$v_{G2} = \frac{80 + 450}{40} = 13.25 \text{ m/s}$$

Solution Using Work-Energy

Step 1: Compute the work done.

$$W = F d$$

Since $d = v_{avg} \cdot t$, and $v_{avg} = \frac{v_{G1} + v_{G2}}{2}$, we approximate v_{G2} using the impulse result:

$$d = \frac{2 + 13.25}{2}(3) = (7.625)(3) = 22.875 \text{ m}$$

$$W = (150)(22.875) = 3431.25 \text{ J}$$

Use the work-energy equation:

$$\frac{1}{2} m v_{G2}^2 = \frac{1}{2} m v_{G1}^2 + W$$

$$\frac{1}{2}(40) v_{G2}^2 = \frac{1}{2}(40)(2)^2 + 3431.25$$

$$20 v_{G2}^2 = 80 + 3431.25$$

$$v_{G2}^2 = \frac{3511.25}{20} = 175.56$$

$$v_{G2} = \sqrt{175.56} = 13.25 \text{ m/s}$$

ChatGPT Integration

- Compare problem-solving approaches: Use ChatGPT to analyze when impulse-momentum is preferred over energy methods in real-world scenarios.
- Simulate real-world systems: Ask ChatGPT to apply both methods to cases like automotive crash testing, sports impacts, and industrial machinery.
- Verify solutions: Request ChatGPT solve the same problem using both approaches and compare the results for accuracy and efficiency.

PROMPT 180

Discuss real-world applications of impulse-momentum principles in mechanical and structural systems.

Objective

To explore the real-world applications of impulse-momentum principles in mechanical and structural systems, demonstrating their role in impact analysis, energy absorption, and force distribution in various engineering fields.

Activity

Impulse-momentum principles are widely applied in automotive safety, aerospace engineering, robotics, and industrial machinery. Engineers use these concepts to design collision-resistant structures, optimize mechanical performance, and control dynamic forces in motion systems.

Real-World Applications of Impulse-Momentum

1. Automotive Crash Testing and Safety Systems
   - Airbags and crumple zones reduce force by extending impact time, lowering the acceleration on passengers.
   - Seatbelt design uses impulse distribution to minimize sudden deceleration injuries.
   - Crash dummy analysis measures impulse and momentum changes to improve vehicle safety ratings.

2. Aerospace Engineering and Spacecraft Dynamics
   - Satellite docking and landing use controlled impulse thrusters to match velocities before docking.
   - Asteroid deflection missions could involve adjusting asteroid trajectories using impulse-based force calculations.
   - Re-entry capsules include heat shields and parachutes that reduce impulse upon atmospheric entry.

3. Sports Science and Biomechanics
   - Tennis, baseball, and soccer ball impacts involve analyzing impulse transfer for optimizing performance.
   - Running shoe design involve studies on impulse absorption to reduce stress on joints.

- Martial arts and boxing gloves are designed to distribute impulse and protect hands from excessive impact.

4. Robotics and Industrial Automation
   - Robotic arms in assembly lines use controlled impulse for precise gripping and movement.
   - Soft robotics and collision avoidance use impulse sensors to prevent damage in collaborative workspaces.
   - Shock-absorbing manufacturing systems minimize impact forces on delicate materials during transport.

5. Structural Engineering and Earthquake Analysis
   - Shock absorbers in bridges and skyscrapers involve using impulse distribution to help buildings withstand seismic forces.
   - Protective barriers in construction sites reduce impulse forces on workers during falling object incidents.
   - Crash barriers on highways are designed to absorb momentum and reduce vehicle impact forces.

Numerical Example

The following example involves impulse in automotive crash testing. A 1500 kg car moving at 25 m/s crashes into a safety barrier, stopping in 0.8 seconds. Determine

1. the impulse exerted on the car
2. the average force experienced during the collision

Solution Using Impulse-Momentum

Step 1: Compute the impulse.

The impulse is given by

$$I = m\left(v_2 - v_1\right)$$

Since the car comes to rest ($v_2 = 0$), we obtain

$$I = (1500)(0 - 25) \quad I = -37500 \text{ Ns}$$

The negative sign indicates that the force is opposite to the car's motion.

Step 2: Compute the average force.

Use the following:

$$F = \frac{I}{\Delta t}$$

Solve for F:

$$F = \frac{-37500}{0.8} = -46875 \text{ Ns}$$

The negative sign indicates that the force acts opposite to the motion, slowing the car down.

ChatGPT Integration

- Use ChatGPT to simulate crash test impulse distributions for different vehicle models.
- Ask ChatGPT to analyze momentum conservation in real-world impact scenarios.
- Request ChatGPT suggest impact reduction strategies in structural and mechanical systems.

CHAPTER 19

THREE-DIMENSIONAL KINEMATICS OF RIGID BODIES

Prompt 181 Define three-dimensional kinematics of rigid bodies and explain its significance in mechanical systems.

Prompt 182 Describe the concept of rotational motion in three dimensions and introduce rotation matrices.

Prompt 183 Explain the concept of angular velocity and angular acceleration in three-dimensional motion.

Prompt 184 Derive the velocity of a point on a rigid body undergoing three-dimensional rotation.

Prompt 185 Explain Euler angles and their role in describing three-dimensional rotations.

Prompt 186 Solve a problem involving the angular velocity of a rotating rigid body using Euler angles.

Prompt 187 Discuss the relationship between angular velocity components and the body-fixed reference frame.

Prompt 188 Analyze instantaneous axis of rotation and its applications in three-dimensional kinematics.

Prompt 189 Explain gyroscopic motion and its significance in aerospace and mechanical systems.

Prompt 190 Discuss real-world applications of three-dimensional kinematics in robotics, aerospace, and industrial machinery.

PROMPT 181

Define three-dimensional kinematics of rigid bodies and explain its significance in mechanical systems.

Objective

To introduce the kinematics of rigid bodies in three dimensions, define key concepts, and explain its importance in mechanical systems such as robotics, aerospace, and vehicle dynamics.

Activity

Three-dimensional kinematics extends planar motion by considering rotation and translation in all three axes. Unlike planar motion, where rotation occurs about a fixed axis, three-dimensional motion involves generalized angular velocity and the use of rotation matrices to describe orientation changes.

This topic is essential for understanding complex motion in robotic arms, gyroscopic systems, and spacecraft dynamics, where objects rotate and translate simultaneously in three-dimensional space.

Mathematical Formulation of Three-Dimensional Kinematics

1. Identify the position and velocity in 3D motion.

   For a rigid body in three-dimensional motion, the position vector of any point P on the body relative to a fixed reference frame is

   $$\vec{r}_P = \vec{r}_G + \vec{r}_{P/G}$$

   where

   - $\vec{r}_G$ is the position vector of the center of mass.
   - $\vec{r}_{P/G}$ is the position of P relative to the center of mass.

   The velocity of P is obtained by differentiating:

   $$\vec{v}_P = \vec{v}_G + \vec{\omega} \times \vec{r}_{P/G}$$

   where

   - $\vec{v}_G$ is the velocity of the center of mass
   - $\vec{\omega}$ is the angular velocity vector of the rigid body

2. Determine the acceleration in 3D motion.

Differentiating the velocity gives the acceleration of point P:

$$\vec{a}_P = \vec{a}_G + \vec{\alpha} \times \vec{r}_{P/G} + \vec{\omega} \times \left(\vec{\omega} \times \vec{r}_{P/G} \right)$$

where
- $\vec{a}_G$ is the acceleration of the center of mass
- $\vec{\alpha}$ is the angular acceleration vector

Numerical Example

This example involves the velocity of a point on a rotating spacecraft.

A spacecraft is rotating with an angular velocity:

$$\vec{\omega} = \left(2\hat{i} + 3\hat{j} + 4\hat{k} \right) \text{ rad/s.}$$

A sensor is mounted at a position

$$\vec{r}_{P/G} = \left(1\hat{i} + 2\hat{j} + 3\hat{k} \right) \text{ m}$$

relative to the center of mass. If the spacecraft's center of mass moves at

$$\vec{v}_G = \left(5\hat{i} + 6\hat{j} + 2\hat{k} \right) \text{ m/s,}$$

determine the velocity of the sensor.

Solution Using the Velocity Equation

Use the following:

$$\vec{v}_P = \vec{v}_G + \vec{\omega} \times \vec{r}_{P/G}$$

Step 1: Compute the cross product $\vec{\omega} \times \vec{r}_{P/G}$.

Expand the determinant:

$$\vec{\omega} \times \vec{r}_{P/G} = \begin{vmatrix} \hat{i} & \hat{j} & \hat{k} \\ 2 & 3 & 4 \\ 1 & 2 & 3 \end{vmatrix} = \hat{i}(9-8) - \hat{j}(6-4) + \hat{k}(4-3) = \left(1\hat{i} - 2\hat{j} + 1\hat{k} \right) \text{ m/s}$$

Step 2: Compute $\vec{v}_P$.

$$\vec{v}_P = \left(5\hat{i} + 6\hat{j} + 2\hat{k} \right) + \left(1\hat{i} - 2\hat{j} + 1\hat{k} \right) = \left(6\hat{i} + 4\hat{j} + 3\hat{k} \right) \text{ m/s}$$

ChatGPT Integration

- Verify three-dimensional motion equations: Ask ChatGPT to compute velocity and acceleration for different angular velocity vectors.

- Simulate spacecraft motion: Use ChatGPT to simulate the trajectory of objects in microgravity using rotation matrices.

- Ask ChatGPT to compute the acceleration of a point on a rotating spacecraft, given in the numerical example.

PROMPT 182

Describe the concept of rotational motion in three dimensions and introduce rotation matrices.

Objective

To describe rotational motion in three dimensions, introduce rotation matrices, and clarify that we are discussing active rotations, where the rigid body rotates while the coordinate system remains fixed.

Activity

In three-dimensional kinematics, rigid bodies can rotate simultaneously about multiple axes, unlike planar motion, which involves rotation about a single fixed axis. To mathematically describe these rotations, we use rotation matrices, which allow us to track the movement of points within the rotating body.

This discussion focuses on active rotations, meaning the object itself rotates, while the coordinate system remains unchanged. This is in contrast to passive rotations, where the coordinate system rotates while the object remains stationary. Active rotations are commonly used in robotics, aerospace navigation, vehicle dynamics, and multi-body systems analysis, where the motion of mechanical components must be tracked in a fixed reference frame.

Mathematical Formulation of Rotational Motion in 3D

Step 1: Define the angular velocity in three dimensions.

The angular velocity vector $\vec{\omega}$ represents rotation about an arbitrary axis

$$\vec{\omega} = \omega_x \hat{i} + \omega_y \hat{j} + \omega_z \hat{k}$$

where

- $\omega_x, \omega_y, \omega_z$ are the angular velocity components along the x, y, and z axes, respectively
- $\hat{i}, \hat{j}, \hat{k}$ are the unit vectors along the coordinate axes

Since we are considering active rotations, this vector represents how the rigid body is rotating relative to the fixed coordinate system.

Step 2: Create the active rotation matrices.

For an active rotation, the rigid body rotates while the coordinate system remains fixed. The transformation equations describe how a point on the rotating body moves due to the applied rotation.
Active Rotation About the X-Axis

$$R_x(\theta) = \begin{bmatrix} 1 & 0 & 0 \\ 0 & \cos\theta & \sin\theta \\ 0 & -\sin\theta & \cos\theta \end{bmatrix}$$

Active Rotation About the Y-Axis

$$R_y(\theta) = \begin{bmatrix} \cos\theta & 0 & -\sin\theta \\ 0 & 1 & 0 \\ \sin\theta & 0 & \cos\theta \end{bmatrix}$$

Active Rotation About the Z-Axis

$$R_z(\theta) = \begin{bmatrix} \cos\theta & \sin\theta & 0 \\ -\sin\theta & \cos\theta & 0 \\ 0 & 0 & 1 \end{bmatrix}$$

These matrices define how the coordinates of points attached to the rigid body change as it rotates within a fixed reference frame. Please note that each rotation angle is defined accordingly and independently.

Step 3: Transform a point in a rotating body.

If a point at vector location $\vec{r}$ is attached to a rotating rigid body, its new position after an active rotation is given by the new vector $\vec{r}'$:

$$\vec{r}' = R\vec{r}$$

For a sequence of active rotations (e.g., first about x, then y, and then z), we have

$$R_{\text{total}} = R_z R_y R_x$$

Thus, the transformed point is

$$\vec{r}' = R_{total}\vec{r}$$

Numerical Example

This example involves the rotation of a point using an active rotation matrix. A rigid body undergoes a 90° active rotation about the z-axis. A point originally at (2,3,1) moves due to the rotation. Determine the new coordinates using the active rotation matrix for the z-axis.

Solution Using the Active Rotation Matrix

Step 1: Define the initial position vector.

$$\vec{r} = \begin{bmatrix} 2 \\ 3 \\ 1 \end{bmatrix}$$

Step 2: Use the active Z-axis rotation matrix for the 90° rotation.

$$R_z(90°) = \begin{bmatrix} \cos 90° & \sin 90° & 0 \\ -\sin 90° & \cos 90° & 0 \\ 0 & 0 & 1 \end{bmatrix} = \begin{bmatrix} 0 & 1 & 0 \\ -1 & 0 & 0 \\ 0 & 0 & 1 \end{bmatrix}$$

Step 3: Compute the new coordinates.

$$\vec{r}' = R_z(90°)\vec{r} = \begin{bmatrix} 0 & 1 & 0 \\ -1 & 0 & 0 \\ 0 & 0 & 1 \end{bmatrix}\begin{bmatrix} 2 \\ 3 \\ 1 \end{bmatrix} = \begin{bmatrix} 3 \\ -2 \\ 1 \end{bmatrix}$$

Thus, the new coordinates of the point after rotation are (3,−2,1).

ChatGPT Integration

- Use ChatGPT to generate rotation matrices for different angles and axes.
- Ask ChatGPT to compute the coordinate transformations for a passive rotation in the given numerical example and compare the results with the active rotation case.
- Request ChatGPT simulate multi-axis rotations in aerospace applications.

PROMPT 183

Explain the concept of angular velocity and angular acceleration in three-dimensional motion.

Objective

To explain angular velocity and angular acceleration in three-dimensional rigid body motion, discuss their vector representation, and describe how they are used to analyze rotational kinematics in engineering applications.

Activity

In three-dimensional motion, a rigid body can rotate about multiple axes simultaneously, making angular velocity and angular acceleration crucial for understanding rotational behavior. Unlike in planar motion, where these quantities are scalars, in 3D motion, they must be treated as vectors. This vector-based approach is widely used in robotic systems, aerospace dynamics, and vehicle stability control.

In this discussion, we continue using the active rotation convention, where the rigid body rotates while the coordinate system remains fixed.

Mathematical Formulation of Angular Velocity and Angular Acceleration

Step 1: Define the angular velocity in three dimensions.

The angular velocity vector $\vec{\omega}$ represents the instantaneous rotational speed and direction of a rigid body,

$$\vec{\omega} = \omega_x \hat{i} + \omega_y \hat{j} + \omega_z \hat{k}$$

where

- $\omega_x, \omega_y, \omega_z$ are the angular velocity components along the x, y, and z axes, respectively
- $\hat{i}, \hat{j}, \hat{k}$ are the unit vectors along the coordinate axes

Since angular velocity is a vector, its direction is determined by the right-hand rule (R.H.R.), meaning it is aligned with the instantaneous axis of rotation, which may be inclined relative to the coordinate axes.

Step 2: Define the angular acceleration in three dimensions.

The angular acceleration vector $\vec{\alpha}$ describes how the angular velocity $\vec{\omega}$ changes over time:

$$\vec{\alpha} = \frac{d\vec{\omega}}{dt}$$

Expand this into component form:

$$\vec{\alpha} = \alpha_x \hat{i} + \alpha_y \hat{j} + \alpha_z \hat{k}$$

where

- $\alpha_x, \alpha_y, \alpha_z$ are the angular acceleration components along the x, y, and z axes, respectively

Step 3: Determine the relationship between the angular velocity and position vector.

For a point P moving in a rotating rigid body, its instantaneous velocity is given by

$$\vec{v}_P = \vec{\omega} \times \vec{r}_{P/G}$$

where

- $\vec{v}_P$ is the velocity of the point relative to a fixed coordinate system.
- $\vec{r}_{P/G}$ is the position vector of point P relative to the center of rotation G.
- The $\times$ symbol represents the cross product, which ensures the velocity direction is perpendicular to both $\vec{\omega}$ and $\vec{r}_{P/G}$.

Similarly, the acceleration of point P is given by

$$\vec{a}_P = \vec{\alpha} \times \vec{r}_{P/G} + \vec{\omega} \times \left(\vec{\omega} \times \vec{r}_{P/G} \right)$$

where

- the first term $\vec{\alpha} \times \vec{r}_{P/G}$ represents the tangential acceleration
- the second term $\vec{\omega} \times \left(\vec{\omega} \times \vec{r}_{P/G} \right)$ represents the centripetal acceleration

Numerical Example

The following is an example involving the angular velocity and acceleration in a rotating arm.

A rigid robotic arm rotates about the z-axis with an angular velocity of $\omega_z = 4$ rad/s and an angular acceleration of $\alpha_z = 2$ rad/s². A point P on the arm is located 1.5 m from the rotation axis. Determine

1. the velocity of point P
2. the acceleration of point P

Solution Using Angular Kinematics Equations

Step 1: Compute the velocity of point P.

Use the following:

$$\vec{v}_P = \vec{\omega} \times \vec{r}_{P/G}$$

Since rotation is about the z-axis, we write

$$\vec{\omega} = 4\hat{k}, \quad \vec{r}_{P/G} = 1.5\hat{i}$$

Compute the cross product:

$$\vec{v}_P = \left(4\hat{k}\right) \times \left(1.5\hat{i}\right) = 6\hat{j} \text{ m/s}$$

Thus, the velocity of point P is 6 m/s in the +y-direction.

Step 2: Compute the acceleration of point P.

Use the following:

$$\vec{a}_P = \vec{\alpha} \times \vec{r}_{P/G} + \vec{\omega} \times \left(\vec{\omega} \times \vec{r}_{P/G}\right)$$

Compute each term:

- Tangential acceleration:

$$\vec{\alpha} \times \vec{r}_{P/G} = \left(2\hat{k}\right) \times \left(1.5\hat{i}\right) = 3\hat{j}$$

- Centripetal acceleration:

$$\vec{\omega} \times \vec{v}_P = \left(4\hat{k}\right) \times \left(6\hat{j}\right) = -24\hat{i}$$

Thus, we obtain

$$\vec{a}_P = \left(-24\hat{i} + 3\hat{j}\right) \text{ m/s}^2$$

ChatGPT Integration

- Use ChatGPT to derive angular velocity and angular acceleration equations for different types of three-dimensional motion, such as pure rotation about an inclined axis or a combination of rotational and translational motion.

- Ask ChatGPT to analyze the relationship between angular velocity components in different reference frames, comparing results in the inertial frame versus the body-fixed frame.

- Request ChatGPT simulate the motion of a rotating rigid body with changing angular acceleration, visualizing how angular velocity evolves over time for complex motion scenarios.

PROMPT 184

Derive the velocity of a point on a rigid body undergoing three-dimensional rotation.

Objective

To derive the velocity equation for a point on a rigid body undergoing three-dimensional rotational motion, using angular velocity and position vectors.

Activity

When a rigid body rotates in three dimensions, every point on the body moves along a curved path dictated by the angular velocity of the body. Unlike in two-dimensional motion, where velocity is determined by a single rotational axis, three-dimensional motion requires vector analysis. We use active rotations, meaning the rigid body rotates while the coordinate system remains fixed. This ensures that the velocity of any point on the rigid body is computed relative to a fixed reference frame.

Mathematical Derivation of Velocity for a Rotating Rigid Body

Step 1: Define the position vector of the point.
Consider a rigid body rotating in three-dimensional space. Let

- G be a reference point (e.g., the center of mass or a fixed point on the body)
- P be any point on the rigid body, with position relative to G given by

$$\vec{r}_{P/G} = x\hat{i} + y\hat{j} + z\hat{k}$$

where x, y, z are the coordinates of point P relative to G.

Step 2: Express the angular velocity as a vector.

The angular velocity of the rigid body is represented as

$$\vec{\omega} = \omega_x \hat{i} + \omega_y \hat{j} + \omega_z \hat{k}$$

where $\omega_x, \omega_y, \omega_z$ are the components of angular velocity along the coordinate axes.

Step 3: Compute the velocity of point P.

The velocity of point P relative to a fixed coordinate system is given by

$$\vec{v}_P = \vec{v}_G + \vec{\omega} \times \vec{r}_{P/G}$$

where

▪ $\vec{v}_P$ is the velocity of point P in the fixed coordinate system

▪ $\vec{v}_G$ is the velocity of the reference point G

▪ $\times$ represents the cross product, which ensures that the velocity direction is perpendicular to both $\vec{\omega}$ and $\vec{r}_{P/G}$

Expand the cross product:

$$\vec{\omega} \times \vec{r}_{P/G} = \begin{vmatrix} \hat{i} & \hat{j} & \hat{k} \\ \omega_x & \omega_y & \omega_z \\ x & y & z \end{vmatrix} = \begin{bmatrix} \omega_y z - \omega_z y \\ \omega_z x - \omega_x z \\ \omega_x y - \omega_y x \end{bmatrix}$$

Thus, the velocity equation becomes

$$\vec{v}_P = \vec{v}_G + \begin{bmatrix} \omega_y z - \omega_z y \\ \omega_z x - \omega_x z \\ \omega_x y - \omega_y x \end{bmatrix}$$

Numerical Example

This example involves using the velocity of a point on a rotating arm. A rigid robotic arm rotates about the y-axis with an angular velocity $\vec{\omega} = 5\hat{j}$ rad/s. A point P on the arm is located at $(2,0,4)$ meters relative to the base. The base of the arm has zero translational velocity. Determine the velocity of point P.

Solution Using the Velocity Equation

Step 1: Define the given values.

▪ $\vec{\omega} = 5\hat{j}$ rad/s

▪ $\vec{r}_{P/G} = \left(2\hat{i}, 0\hat{j}, 4\hat{k}\right)$ m

▪ $\vec{v}_G = 0$ (since the base is fixed)

Step 2: Compute $\vec{\omega} \times \vec{r}_{P/G}$.

$$\vec{\omega} \times \vec{r}_{P/G} = \begin{vmatrix} \hat{i} & \hat{j} & \hat{k} \\ 0 & 5 & 0 \\ 2 & 0 & 4 \end{vmatrix} = \left(20\hat{i} - 10\hat{k} \right) \text{ m/s}$$

Thus, the velocity of point P is

$$\vec{v}_P = 20\hat{i} - 10\hat{k} \text{ m/s}$$

ChatGPT Integration

- Use ChatGPT to verify velocity calculations for different rigid body configurations, such as multi-link robotic arms or rotating spacecraft components.

- Ask ChatGPT to generate simulations for three-dimensional velocity fields in rotating systems, visualizing how velocity vectors change over time.

- Request ChatGPT analyze the effects of varying angular velocities on velocity distributions in engineering systems like turbines, robotic joints, and gyroscopes.

PROMPT 185

Explain Euler angles and their role in describing three-dimensional rotations.

Objective

To explain Euler angles, how they are used to describe three-dimensional rotations, and their significance in engineering applications such as aerospace dynamics, robotic control, and vehicle motion analysis.

Activity

When a rigid body undergoes three-dimensional rotation, tracking its orientation is crucial for kinematic analysis. One of the most common methods is using Euler angles, which define rotation using three sequential angles.

Euler angles describe how a rigid body rotates from one coordinate system to another using a sequence of three rotations about different axes. This method is widely used in robotics, flight dynamics, and vehicle navigation systems.

This discussion follows the passive rotation convention.

Mathematical Definition of Euler Angles

Euler angles represent the orientation of a rigid body using three sequential rotations:

1. φ (Precession Angle) – rotation about the Z-axis of the original frame

2. θ (Nutation Angle) – rotation about the new X-axis after the first rotation

3. ψ (Spin Angle) – rotation about the new Z-axis after the second rotation

These rotations are applied in a specific order to fully describe the body's orientation. A total of twelve rotation combinations is possible.

Step 1: Define the Euler angle rotations.

Each rotation is represented by a rotation matrix:

1. Rotation About the Z-Axis by φ (Precession Angle or yaw)

$$R_z(\varphi) = \begin{bmatrix} \cos\varphi & \sin\varphi & 0 \\ -\sin\varphi & \cos\varphi & 0 \\ 0 & 0 & 1 \end{bmatrix}$$

2. Rotation About the New X-Axis by θ (Nutation Angle or roll)

$$R_x(\theta) = \begin{bmatrix} 1 & 0 & 0 \\ 0 & \cos\theta & \sin\theta \\ 0 & -\sin\theta & \cos\theta \end{bmatrix}$$

3. Rotation About the New Z-Axis by ψ (Spin Angle)

$$R'_z(\psi) = \begin{bmatrix} \cos\psi & \sin\psi & 0 \\ -\sin\psi & \cos\psi & 0 \\ 0 & 0 & 1 \end{bmatrix}$$

Step 2: Compute the overall rotation matrix.

The total rotation matrix is obtained by multiplying the three matrices in the correct sequence:

$$R_{total} = R_z'(\psi)R_x(\theta)R_z(\varphi)$$

Expand the multiplication to obtain the following:

$$R_{total} = \begin{bmatrix} \cos\psi & \sin\psi & 0 \\ -\sin\psi & \cos\psi & 0 \\ 0 & 0 & 1 \end{bmatrix} \begin{bmatrix} 1 & 0 & 0 \\ 0 & \cos\theta & \sin\theta \\ 0 & -\sin\theta & \cos\theta \end{bmatrix} \begin{bmatrix} \cos\varphi & \sin\varphi & 0 \\ -\sin\varphi & \cos\varphi & 0 \\ 0 & 0 & 1 \end{bmatrix}$$

Numerical Example

The following example involves computing the orientation using Euler angles. A rigid body undergoes three successive rotations:

- 30° about the Z-axis (φ)
- 45° about the new X-axis (θ)
- 60° about the new Z-axis (ψ)

Determine the total rotation matrix.

Solution Using Rotation Matrices

Step 1: Compute each rotation matrix.

By substituting the given angles into the individual rotation matrices, we obtain

$$R_z(30°) = \begin{bmatrix} 0.866 & 0.5 & 0 \\ -0.5 & 0.866 & 0 \\ 0 & 0 & 1 \end{bmatrix}$$

$$R_x(45°) = \begin{bmatrix} 1 & 0 & 0 \\ 0 & 0.707 & 0.707 \\ 0 & -0.707 & 0.707 \end{bmatrix}$$

$$R_z'(60°) = \begin{bmatrix} 0.5 & 0.866 & 0 \\ -0.866 & 0.5 & 0 \\ 0 & 0 & 1 \end{bmatrix}$$

Step 2: Compute the total rotation matrix.

$$R_{\text{total}} = R_z'(60°)R_x(45°)R_z(30°)$$

By multiplying the matrices in the correct order,
$\left[R_z'(60°)\right] \times \left[R_x(45°) \times R_z(30°)\right]$, we obtain

$$R_{\text{total}} = \begin{bmatrix} 0.1268 & 0.7803 & 0.6124 \\ -0.9268 & -0.1268 & 0.3536 \\ 0.3536 & -0.6124 & 0.7071 \end{bmatrix}$$

Thus, the final orientation of the body is represented by this matrix.

ChatGPT Integration

- Use ChatGPT to analyze Euler angle sequences for different rotation conventions, such as $z - x - z$ or $z - y - x$ sequences.
- Ask ChatGPT to simulate the effect of sequential Euler rotations, visualizing how orientation changes over time.
- Request ChatGPT compute alternative rotation matrices using quaternion representation, comparing results with Euler angles.

PROMPT 186

Solve a problem involving the angular velocity of a rotating rigid body using Euler angles.

Objective

To apply Euler angles to determine the angular velocity of a rotating rigid body, derive the mathematical expression for angular velocity, and solve a numerical example.

Activity

Euler angles describe the orientation of a rigid body in space using three sequential rotations. However, when a rigid body undergoes continuous rotation, its angular velocity is expressed in terms of the time derivatives of the Euler angles. This is critical for flight dynamics, robotic motion planning, and spacecraft navigation, where orientation and rotational speed must be precisely controlled.

Mathematical Formulation of Angular Velocity in Terms of Euler Angles

Step 1: Define the Euler angles and rotation sequence.

The three Euler angles are

- φ (Precession Angle) – rotation about the Z-axis of the fixed coordinate system
- θ (Nutation Angle) – rotation about the new X-axis after the first rotation
- ψ (Spin Angle) – rotation about the new Z-axis after the second rotation

Step 2: Express the angular velocity in terms of Euler angles.
The total angular velocity of the rigid body is the sum of the contributions from the three rotations (i.e., the first rotation contributes $\dot{\varphi}$, second rotation $\dot{\theta}$, and third rotation $\dot{\psi}$):

$$\vec{\omega} = \dot{\varphi}\hat{k} + \dot{\theta}\hat{e}_1 + \dot{\psi}\hat{e}_3$$

where

- $\dot{\varphi}$ is the rate of precession (change in φ)
- $\dot{\theta}$ is the rate of nutation (change in θ)
- $\dot{\psi}$ is the rate of spin (change in ψ)
- $\hat{e}_1$ and $\hat{e}_3$ are unit vectors in the rotated coordinate system

Rewrite this in terms of the fixed coordinate system:

$$\vec{\omega} = \begin{pmatrix} \dot{\varphi}\sin\theta\sin\psi + \dot{\theta}\cos\psi \\ \dot{\varphi}\sin\theta\cos\psi - \dot{\theta}\sin\psi \\ \dot{\varphi}\cos\theta + \dot{\psi} \end{pmatrix}$$

Numerical Example

This example involves computing the angular velocity using Euler angles. A gyroscope undergoes rotation characterized by the following Euler angle rates:

- Precession rate: $\dot{\phi} = 2\ \text{rad}/\text{s}$
- Nutation rate: $\dot{\theta} = 1.5\ \text{rad}/\text{s}$
- Spin rate: $\dot{\psi} = 3\ \text{rad}/\text{s}$

▪ Instantaneous nutation angle: $\theta = 45°$

▪ Instantaneous spin angle: $\psi = 60°$

Determine the angular velocity vector $\vec{\omega}$.

Solution Using Angular Velocity Equations

Step 1: Compute the components of $\vec{\omega}$.

Use the following:

$$\omega_x = (2)\left(\sin 45° \sin 60°\right) + \left(1.5\cos 60°\right) = 1.974 \text{ rad/s}$$

$$\omega_y = (2)\left(\sin 45° \cos 60°\right) - \left(1.5\sin 60°\right) = -0.592 \text{ rad/s}$$

$$\omega_z = \left(2\cos 45°\right) + (3) = 4.414 \text{ rad/s}$$

Thus, the angular velocity vector is

$$\vec{\omega} = \left(1.974\hat{i} - 0.592\hat{j} + 4.414\hat{k}\right) \text{ rad/s}$$

This vector is defined in the body-fixed coordinates resulted from three rotations.

ChatGPT Integration

▪ Use ChatGPT to compute angular velocity vectors for different Euler angle sequences, such as $Z - X - Z$ or $Y - Z - Y$ configurations.

▪ Ask ChatGPT to visualize how angular velocity evolves over time in a multi-axis rotating system.

▪ Request ChatGPT compare the Euler angle approach with quaternion representation, evaluating their advantages and limitations in practical applications.

PROMPT 187

Discuss the relationship between angular velocity components and the body-fixed reference frame.

Objective

To describe how the angular velocity vector is expressed in body-fixed and inertial reference frames, explain the decomposition of angular velocity components, and discuss their significance in aerospace, robotics, and vehicle dynamics.

Activity

In three-dimensional motion, a rigid body can rotate about multiple axes simultaneously. The angular velocity vector $\vec{\omega}$ can be described in two ways:

1. in an inertial (fixed) reference frame, where rotation is measured relative to a global coordinate system

2. in a body-fixed reference frame, where rotation is measured relative to the moving coordinate system attached to the rigid body

In many engineering applications (e.g., flight dynamics, satellite motion, and robotic arms), expressing angular velocity in the body-fixed frame is advantageous because the equations of motion simplify when formulated in this moving frame.

Mathematical Formulation of Angular Velocity in a Body-Fixed Frame

Step 1: Express the angular velocity as a vector.

The angular velocity vector $\vec{\omega}$ in the inertial frame is

$$\vec{\omega} = \omega_x \hat{i} + \omega_y \hat{j} + \omega_z \hat{k}$$

where

- $\omega_x, \omega_y, \omega_z$ are the angular velocity components along the inertial x, y, z axes

- $\hat{i}$, $\hat{j}$, $\hat{k}$ are the unit vectors of the inertial coordinate system

However, when expressed in the body-fixed frame, the angular velocity components change based on the object's orientation.

Step 2: Decompose the angular velocity in the body-fixed frame.

In the body-fixed frame, the angular velocity is often expressed in terms of Euler angles (φ, θ, ψ), where, for example

- φ (Yaw/Precession) defines rotation about the Z-axis of the original frame

- θ (Pitch/Nutation) defines rotation about the new Y-axis after the first rotation

- ψ (Yaw/Spin) defines rotation about the new Z-axis after the second rotation

Instead of using the fixed frame components $\omega_x, \omega_y, \omega_z$, we express angular velocity in terms of these Euler angles. Here, we assume that the body undergoes a sequence of three rotations as yaw-pitch-yaw, for example.

Step 3: Determine how the angular velocity is expressed in the body-fixed frame.

The total angular velocity vector $\vec{\omega}$ is given by (in body-fixed resulted coordinates):

$$\begin{Bmatrix} \omega_x \\ \omega_y \\ \omega_z \end{Bmatrix} = \begin{bmatrix} R'_z(\psi) \end{bmatrix} \begin{pmatrix} 0 \\ 0 \\ \dot{\psi} \end{pmatrix} + \begin{bmatrix} R'_z(\psi) \end{bmatrix}\begin{bmatrix} R_y(\theta) \end{bmatrix} \begin{pmatrix} 0 \\ \dot{\theta} \\ 0 \end{pmatrix} + \begin{bmatrix} R'_z(\psi) \end{bmatrix}\begin{bmatrix} R_y(\theta) \end{bmatrix}\begin{bmatrix} R_z(\varphi) \end{bmatrix} \begin{pmatrix} 0 \\ 0 \\ \dot{\varphi} \end{pmatrix}$$

where

■ Rotation About the Z-Axis by φ (Precession Angle or yaw)

$$R_z(\varphi) = \begin{bmatrix} \cos\varphi & \sin\varphi & 0 \\ -\sin\varphi & \cos\varphi & 0 \\ 0 & 0 & 1 \end{bmatrix}$$

■ Rotation About the New Y-Axis by θ (Nutation Angle or pitch)

$$R_x(\theta) = \begin{bmatrix} \cos\theta & 0 & -\sin\theta \\ 0 & 1 & 0 \\ \sin\theta & 0 & \cos\theta \end{bmatrix}$$

■ Rotation About the New Z-Axis by ψ (Spin Angle, or final yaw)

$$R'_z(\psi) = \begin{bmatrix} \cos\psi & \sin\psi & 0 \\ -\sin\psi & \cos\psi & 0 \\ 0 & 0 & 1 \end{bmatrix}.$$

This equation decomposes total angular velocity into three rotational components, each occurring about a different axis in sequence.

Step 4: Transform to body-fixed coordinates.

To express $\vec{\omega}$ in terms of the body-fixed coordinate system, we use the transformation for yaw-pitch-yaw combination:

$$\vec{\omega} = \begin{Bmatrix} \dot{\theta}\cos\psi - \dot{\varphi}\sin\theta\sin\psi \\ \dot{\theta}\sin\psi + \dot{\varphi}\sin\theta\cos\psi \\ \dot{\varphi}\cos\theta + \dot{\psi} \end{Bmatrix}$$

where

- the first term ω_x, corresponds to the x-component of angular velocity in the body-fixed frame
- the second term ω_y, corresponds to the y-component
- the third term ω_z, corresponds to the z-component

Numerical Example

The following example involves computing the angular velocity components in a body-fixed frame.

Given Data

- Yaw rate $(\dot{\varphi})$ = 3 rad/s
- Pitch rate $(\dot{\theta})$ = 2 rad/s
- Yaw rate $(\dot{\psi})$ = 1 rad/s
- Current pitch angle $(\theta = 45°)$
- Current yaw angle $(\psi = 30°)$

Formulas for Angular Velocity Components (Yaw-Pitch-Yaw Sequence)

$$\omega_x = (3)\left(\sin 45° \sin 30°\right) + \left(2\cos 30°\right) = 2.793 \text{ rad/s}$$

$$\omega_y = (3)\left(\sin 45° \cos 30°\right) - \left(2\sin 30°\right) = 0.837 \text{ rad/s}$$

$$\omega_z = \left(3\cos 45°\right) + 1 = 3.121 \text{ rad/s}$$

or

$$\vec{\omega} = \left(2.793\hat{i} + 0.837\hat{j} + 3.121\hat{k}\right) \text{ rad/s}$$

ChatGPT Integration

- Use ChatGPT to compute angular velocity transformations between body-fixed and inertial frames for different Euler angle sequences (a total of 12).
- Ask ChatGPT to visualize angular velocity components in rotating objects, such as satellites or robotic joints.
- Request ChatGPT compare body-fixed and inertial frame angular velocity representations and discuss their advantages in engineering applications.

PROMPT 188

Analyze instantaneous axis of rotation and its applications in three-dimensional kinematics.

Objective

To explain the instantaneous axis of rotation (IAR) in three-dimensional motion, derive its mathematical formulation, and discuss its significance in robotics, vehicle dynamics, and aerospace applications.

Activity

A rigid body undergoing general three-dimensional motion can have a complex combination of translational and rotational movement. However, at any instant, there exists a unique axis called the instantaneous axis of rotation (IAR), about which the body appears to rotate without translation.

Understanding the IAR is crucial in

- vehicle rollover dynamics (e.g., in crash testing)
- robot arm motion analysis (e.g., minimizing vibration effects)
- gyroscopic and satellite control systems (e.g., stabilizing spacecraft)

This discussion follows the active rotation convention, meaning the rigid body rotates while the coordinate system remains fixed.

Mathematical Formulation of the Instantaneous Axis of Rotation (IAR)

Step 1: Define the angular velocity and velocity relations.

The velocity of any point P on a rigid body is given by

$$\vec{v}_P = \vec{v}_G + \vec{\omega} \times \vec{r}_{P/G}$$

where

- $\vec{v}_P$ = velocity of point P
- $\vec{v}_G$ = velocity of the reference point G (e.g., center of mass)
- $\vec{\omega}$ = angular velocity vector
- $\vec{r}_{P/G}$ = position vector of P relative to G

For a pure rotation about the IAR, the velocity must satisfy

$$\vec{v}_P = \vec{\omega} \times \vec{r}_I$$

where $\vec{r}_I$ is the position of the instantaneous center of rotation along the IAR.

Step 2: Find the instantaneous axis of rotation.

The instantaneous axis of rotation is a unit vector $\hat{e}_I$ aligned with the angular velocity

$$\hat{e}_I = \frac{\vec{\omega}}{|\vec{\omega}|}$$

where

$$|\vec{\omega}| = \sqrt{\omega_x^2 + \omega_y^2 + \omega_z^2}$$

This unit vector gives the direction of the IAR.

Step 3: Compute the location of the instantaneous center of rotation. To find the point along the IAR where velocity is zero, we solve

$$\vec{v}_I = \vec{v}_G + \vec{\omega} \times \vec{r}_{I/G} = 0$$

We rearrange for $\vec{r}_{I/G}$:

$$\vec{r}_{I/G} = -\frac{\vec{\omega} \times \vec{v}_G}{|\vec{\omega}|^2}$$

where

- $\vec{r}_{I/G}$ = position of the instantaneous center of rotation relative to G
- $\vec{\omega} \times \vec{v}_G$ determines how far the center is from the body's center of mass

Numerical Example

The following example involves computing the instantaneous axis of rotation. A rigid body in three-dimensional motion has

- angular velocity

$$\vec{\omega} = 2\hat{i} + 3\hat{j} + 6\hat{k} \quad \text{rad/s}$$

- velocity of its center of mass

$$\vec{v}_G = 4\hat{i} + 2\hat{j} + 1\hat{k} \quad \text{m/s}$$

Determine

1. the unit vector along the instantaneous axis of rotation (IAR)
2. the position of the instantaneous center of rotation relative to the center of mass

Solution Using Instantaneous Rotation Formulas

Step 1: Compute the unit vector along the IAR.

Use the following:

$$\hat{e}_I = \frac{\vec{\omega}}{|\vec{\omega}|}$$

First, compute $|\vec{\omega}|$:

$$|\vec{\omega}| = \sqrt{2^2 + 3^2 + 6^2} = \sqrt{4 + 9 + 36} = \sqrt{49} = 7$$

Now, compute $\hat{e}_I$:

$$\hat{e}_I = \frac{1}{7}\left(2\hat{i} + 3\hat{j} + 6\hat{k}\right)$$

Step 2: Compute the position of the instantaneous center of rotation.

Use the following:

$$\vec{r}_{I/G} = -\frac{\vec{\omega} \times \vec{v}_G}{|\vec{\omega}|^2}$$

Expand this to obtain the following:

$$\vec{\omega} \times \vec{v}_G = \begin{vmatrix} \hat{i} & \hat{j} & \hat{k} \\ 2 & 3 & 6 \\ 4 & 2 & 1 \end{vmatrix} = \left(-9\hat{i} + 22\hat{j} - 8\hat{k}\right)$$

Now, we find

$$\vec{r}_{I/G} = \left(0.184\hat{i} - 0.449\hat{j} + 0.163\hat{k}\right)$$

ChatGPT Integration

- Use ChatGPT to compute the instantaneous axis of rotation for different cases, such as robotic arms and spacecraft motion.
- Ask ChatGPT to analyze real-world applications of IAR in mechanical systems, such as bicycle wheels or rotating machines.
- Request ChatGPT simulate the motion of a rigid body around its instantaneous axis of rotation in different reference frames.

PROMPT 189

Explain gyroscopic motion and its significance in aerospace and mechanical systems.

Objective

To explain the principles of gyroscopic motion, derive its governing equations, and discuss its applications in aerospace navigation, stability control, and mechanical systems.

Activity

Gyroscopic motion describes the behavior of a spinning rigid body when subjected to external torques. The key property of gyroscopes is their ability to maintain orientation due to angular momentum conservation. This principle is used in navigation systems, aerospace stability, and precision mechanical devices.

Applications of gyroscopic motion include

- aircraft and spacecraft attitude control (gyroscopic stabilization)
- marine and land vehicle stabilization (gyroscopic precession in steering)
- rotating machinery and robotics (gyroscopic effects in precision mechanisms)

Mathematical Formulation of Gyroscopic Motion

Step 1: Define the angular momentum of a rotating body.
For a rigid body rotating about an axis, the angular momentum is given by

$$\vec{H} = I\vec{\omega}$$

where

- $\vec{H}$ = angular momentum of the rotating body
- I = moment of inertia about the spin axis
- $\vec{\omega}$ = angular velocity vector

In the absence of external torques, the angular momentum remains constant, meaning

$$\frac{d\vec{H}}{dt} = 0$$

Step 2: Determine the effect of an applied torque (precession motion).

If an external torque $\vec{M}$ is applied, it causes a change in the angular momentum:

$$\vec{M} = \frac{d\vec{H}}{dt}$$

For a gyroscope spinning about its principal axis, the torque produces precession, a slow rotation of the axis about another direction. The precession angular velocity $\vec{\Omega}$ is related to the applied torque by

$$\vec{\Omega} = \frac{\vec{M} \times \vec{H}}{|H|^2}$$

where

- $\vec{\Omega}$ = precession angular velocity
- $\vec{M}$ = applied torque vector
- $|H|^2$ = squared magnitude of angular momentum

Numerical Example

The following example involves computing precession in a gyroscope. A gyroscope with a spinning disc has the following parameters:

- mass of disc = 5 kg
- radius of disc = 0.2 m
- angular velocity of spin (ω) = 300 rad/s
- applied torque about an axis perpendicular to spin = 4 N·m
- moment of inertia of the disc about its spin axis

$$I = \frac{1}{2}mR^2$$

Determine the precession angular velocity.

Solution Using Gyroscopic Precession Formula

Step 1: Compute the moment of inertia.

Using I:

$$I = \frac{1}{2}(5)(0.2)^2 = 0.1 \ kg.m^2$$

Step 2: Compute the angular momentum.

Use H:

$$H = (0.1)(300) = 30 \, kg.m^2 / s$$

Step 3: Compute the precession angular velocity.

Use Ω:

$$\Omega = \frac{4}{30} = 0.133 \, rad/s$$

This is the value for the precession angular velocity.

ChatGPT Integration

- Use ChatGPT to analyze the gyroscopic effect in real-world systems, such as drones, motorcycles, and spacecraft.
- Ask ChatGPT to simulate the motion of a gyroscope under different torque conditions, visualizing the precession and nutation effects.
- Request ChatGPT compare the gyroscopic stability in different mechanical systems, such as ships, satellites, and robotic arms.

PROMPT 190

Discuss real-world applications of three-dimensional kinematics in robotics, aerospace, and industrial machinery.

Objective

To explore real-world applications of three-dimensional kinematics in robotics, aerospace systems, and industrial machinery, highlighting the role of rigid body motion, Euler angles, and angular velocity decomposition in practical engineering solutions.

Activity

Three-dimensional kinematics plays a crucial role in the design, control, and analysis of mechanical systems that undergo complex motion. Engineers use kinematic equations, rotation matrices, and angular velocity formulations to model motion accurately in robotics, aerospace vehicles, and industrial automation.

Important applications include

* robotics: motion planning, inverse kinematics, and dynamic control of robotic arms and mobile robots
* aerospace systems: flight dynamics, spacecraft attitude control, and gyroscopic stabilization
* industrial machinery: automated systems, conveyor mechanisms, and multi-axis CNC machines.

Mathematical Formulation of Three-Dimensional Kinematics in Engineering Applications

Step 1: Model the rigid body motion using rotation matrices.

For a rigid body in three-dimensional motion, its orientation can be represented using a rotation matrix:

$$R_{\text{total}} = R_z'(\psi)R_y(\theta)R_z(\varphi)$$

where, for example

* $R_z(\varphi)$ represents yaw (precession)
* $R_y(\theta)$ represents pitch (nutation)
* $R_z'(\psi)$ represents roll (spin)

This transformation allows engineers to **track the orientation** in robotics, aerospace, and mechanical systems.

Step 2: Determine the angular velocity in a moving system.
The angular velocity of a rigid body is given by

$$\vec{\omega} = \left\{ \begin{array}{c} \dot{\psi} - \dot{\varphi}\sin\theta \\ \dot{\theta}\cos\psi + \dot{\varphi}\cos\theta\sin\psi \\ \dot{\varphi}\cos\theta\cos\psi - \dot{\theta}\sin\psi \end{array} \right\}$$

This equation is critical for robotic arm movement, drone stabilization, and satellite attitude control.

Numerical Example

The following example involves the application of three-dimensional kinematics in a robotic arm. A robotic arm moves with the following kinematic parameters:

* Yaw rate $(\dot{\varphi})$ = 1.5 rad/s
* Pitch rate $(\dot{\theta})$ = 1.2 rad/s

- Roll rate $(\dot{\psi})$ = 0.8 rad/s
- Pitch angle (θ) = 30°
- Roll angle (ψ) = 45°

Determine the angular velocity components in the body-fixed reference frame.

Solution Using Body-Fixed Angular Velocity Formulation

By substituting, we obtain

$$\omega x = (0.8) - (1.5)(sin\,30°) = 0.05\,rad/s$$

$$\omega y = (1.2)(cos\,45°) + (1.5)(sin\,30°sin\,45°) = 1.38\,rad/s$$

$$\omega_z = (1.5)(cos\,30°cos\,45°) - 1.2\,sin\,45° = 0.07\,rad/s$$

Thus, the angular velocity components in the robotic arm's body-fixed frame are

$$\vec{\omega} = \left(0.05\hat{i} + 1.38\hat{j} + 0.07\hat{k}\right)\,rad/s$$

ChatGPT Integration

- Use ChatGPT to analyze the role of three-dimensional kinematics in robot motion planning, such as inverse kinematics and path optimization.
- Ask ChatGPT to simulate the effects of different Euler angle sequences on aerospace vehicle orientation, such as spacecraft re-entry dynamics.
- Request ChatGPT compare kinematic constraints in industrial automation systems, such as robotic arms versus CNC machines.

CHAPTER 20

THREE-DIMENSIONAL KINETICS OF RIGID BODIES

Prompt 191 Derive the general equations of motion for a rigid body in three-dimensional space using Newton-Euler equations.

Prompt 192 Explain the conservation of angular momentum in three-dimensional motion and its applications in engineering systems.

Prompt 193 Solve a problem involving the effect of external torques on a rotating spacecraft.

Prompt 194 Discuss the principle of work and energy for a rigid body undergoing three-dimensional motion.

Prompt 195 Solve a problem using the work-energy principle to determine the motion of a gyroscopic rotor in a three-dimensional system.

Prompt 196 Explain the impulse-momentum principle for a rigid body and derive its mathematical formulation in three dimensions.

Prompt 197 Solve a problem involving an impact on a rotating rigid body, using the impulse-momentum equations.

Prompt 198 Analyze rotational stability in aerospace and mechanical systems, focusing on gyroscopic effects and control mechanisms.

Prompt 199 Analyze the motion of a robotic manipulator with multiple degrees of freedom using three-dimensional kinetics principles.

Prompt 200 Solve a real-world engineering problem involving three-dimensional kinetics, such as the motion of a drone or an industrial robotic arm.

PROMPT 191

Derive the general equations of motion for a rigid body in three-dimensional space using Newton-Euler equations.

Objective

To derive the Newton-Euler equations of motion for a rigid body in three-dimensional space, explaining their significance in analyzing forces, moments, and motion in engineering applications such as robotics, aerospace, and mechanical systems.

Activity

Rigid body motion in three dimensions requires considering both translational and rotational dynamics. The Newton-Euler equations combine Newton's second law for linear motion and Euler's Equations for rotational motion to describe the complete dynamics of a rigid body.

These equations are used in

- robotic manipulators for motion control
- spacecraft and satellite dynamics for attitude control
- vehicle and aircraft stability analysis for maneuverability

Mathematical Formulation of Newton-Euler Equations

Step 1: Use Newton's second law for translational motion.

For a rigid body of mass mm moving under an external force $\vec{F}$\vec{F},
Newton's second law states:

$$\sum \vec{F} = m\vec{a}_G$$

where
- $\sum \vec{F}$ = net external force acting on the rigid body
- m = mass of the rigid body
- $\vec{a}_G$ = acceleration of the center of mass

This equation governs the linear motion of the rigid body.

Step 2: Use Euler's equations for rotational motion.

The rotational motion of a rigid body is described using Euler's Equations, which relate the net external moment to the time rate of change of angular momentum:

$$\sum \vec{M}_G = \frac{d}{dt} \vec{H}_G$$

where

- $\sum \vec{M}_G$ = net external moment about the center of mass G

- $\vec{H}_G$ = angular momentum about G

For a rigid body with a time-dependent moment of inertia tensor I_G, the angular momentum is

$$\vec{H}_G = I_G \vec{\omega}$$

Applying the product rule to differentiate $\vec{H}_G$ {\vec{H}}_G:

$$\sum \vec{M}_G = I_G \frac{d\vec{\omega}}{dt} + \vec{\omega} \times (I_G \vec{\omega})$$

This equation governs the rotational motion of a rigid body.

Step 3: Express the equations in a body-fixed frame.

In a rotating body-fixed frame, the equations of motion take the following form:

$$\sum F_x = m\left(\dot{v}_x + \omega_y v_z - \omega_z v_y \right)$$

$$\sum F_y = m\left(\dot{v}_y + \omega_z v_x - \omega_x v_z \right)$$

$$\sum F_z = m\left(\dot{v}_z + \omega_x v_y - \omega_y v_x \right)$$

For rotational motion, the Euler equations in a rotating frame are

$$\sum M_x = I_x \dot{\omega}_x + \left(I_z - I_y \right) \omega_y \omega_z$$

$$\sum M_y = I_y \dot{\omega}_y + \left(I_x - I_z \right) \omega_z \omega_x$$

$$\sum M_z = I_z \dot{\omega}_z + \left(I_y - I_x \right) \omega_x \omega_y$$

These equations describe how forces and moments cause linear and angular accelerations in a three-dimensional rigid body system.

Numerical Example

This example involves applying the Newton-Euler equations to a rotating satellite.

A satellite has

- mass = 500 kg
- moment of inertia about principal axes:
  - $I_x = 200$ kg·m²,
  - $I_y = 300$ kg·m²,
  - $I_z = 400$ kg·m²
- initial angular velocity: $\omega_x = 0.5$ rad/s, $\omega_y = 0.3$ rad/s, $\omega_z = 0.2$ rad/s
- applied external torques:
  - $M_x = 5$ N·m,
  - $M_y = 3$ N·m,
  - $M_z = 4$ N·m

Determine the angular acceleration components.

Solution Using Euler's Equations

Step 1: Compute the angular accelerations.

Use the following:

$$\dot{\omega}_x = \frac{M_x - \left(I_z - I_y\right)\omega_y\omega_z}{I_x} = \frac{5 - (400 - 300)(0.3)(0.2)}{200} = -0.005 \text{ rad}/s^2$$

$$\dot{\omega}_y = \frac{M_y - \left(I_x - I_z\right)\omega_z\omega_x}{I_y} = \frac{3 - (200 - 400)(0.2)(0.5)}{300} = 0.0767 \text{ rad}/s^2$$

$$\dot{\omega}_z = \frac{M_z - \left(I_y - I_x\right)\omega_x\omega_y}{I_z} = \frac{4 - (300 - 200)(0.5)(0.3)}{400} = -0.0275 \text{ rad}/s^2$$

ChatGPT Integration:

- Use ChatGPT to analyze spacecraft attitude control using Newton-Euler equations.
- Ask ChatGPT to simulate the motion of a rigid body under different torque conditions.
- Request ChatGPT compare Newton-Euler dynamics in robotic arms and aerospace systems.

PROMPT 192

Explain the conservation of angular momentum in three-dimensional motion and its applications in engineering systems.

Objective

To explain the principle of angular momentum conservation in three-dimensional motion, derive its governing equation, and discuss its applications in aerospace, mechanical, and robotic systems.

Activity

The conservation of angular momentum states that in the absence of external torques, a system's angular momentum remains constant. This principle is widely applied in

- spacecraft attitude control (e.g., reaction wheels and momentum exchange)
- gyroscopic stabilization (e.g., motorcycles, ships, and aircraft)
- robotics and precision motion control (e.g., gyroscopic sensors in robotic arms)

This prompt focuses on deriving and applying angular momentum conservation, differentiating it from Newton-Euler equations in Prompt 191, which consider external torques.

Mathematical Formulation of Angular Momentum Conservation

Step 1: Define the angular momentum.

For a rigid body rotating about a fixed point G (center of mass), the angular momentum is defined as

$$\vec{H}_G = I_G \vec{\omega}$$

where

- $\vec{H}_G$ = angular momentum about G
- I_G = moment of inertia tensor about G, expressed in the body-fixed frame
- $\vec{\omega}$ = angular velocity vector in the body-fixed frame

If no external moments act on the system, then the total angular momentum remains constant:

$$\frac{d\vec{H}_G}{dt} = 0$$

This equation implies that any changes in moment of inertia I_G must be countered by changes in angular velocity $\vec{\omega}$ to maintain a constant angular momentum.

Step 2: Apply the formula to a symmetric rotating body.
For a body rotating freely in space, with no external torque, the angular momentum remains constant:

$$I_G\vec{\omega} = \text{constant}$$

If the moment of inertia changes, then the angular velocity must adjust accordingly. This is observed in figure skaters pulling in their arms to spin faster, reducing I_G and increasing ω.

For an axisymmetric body with its symmetry axis aligned with z, the angular momentum component along that axis is

$$H_z = I_z\omega_z$$

Step 3: Determine the gyroscopic motion and precession.
When an external torque $\vec{M}$ is not zero but perpendicular to $\vec{H}$, it causes precession, a slow rotation of the angular momentum vector:

$$\vec{M} = \frac{d\vec{H}}{dt} = \vec{\Omega} \times \vec{H}$$

where

- $\vec{\Omega}$ = precession angular velocity
- $\vec{H}$ = angular momentum

This principle is used in gyroscopic sensors, which measure orientation changes in aircraft, ships, and robotic systems.

Numerical Example

This example involves the spacecraft attitude control using angular momentum conservation.

A spacecraft uses a reaction wheel to adjust its attitude in orbit. Given

- initial angular velocity: $\omega_i = 0.5$ rad/s
- initial moment of inertia: $I_i = 400$ kg·m²
- final moment of inertia: $I_f = 250$ kg·m²
- no external torques are acting on the spacecraft

Determine the final angular velocity ω_f after the change in moment of inertia.

Solution Using Angular Momentum Conservation

Use the following:

$$I_i \omega_i = I_f \omega_f$$

Solve for ω_f:

$$\omega_f = \frac{I_i \omega_i}{I_f} = \frac{(400)(0.5)}{250} = 0.8 \text{ rad/s}$$

Thus, the spacecraft spins faster as its moment of inertia decreases.

ChatGPT Integration

- Use ChatGPT to analyze how figure skaters, satellites, and gyroscopic systems utilize angular momentum conservation.
- Ask ChatGPT to simulate gyroscopic motion and precession in mechanical and aerospace applications.
- Request ChatGPT compare angular momentum conservation in different engineering fields, such as robotics, aerospace, and vehicle stability.

PROMPT 193

Solve a problem involving the effect of external torques on a rotating spacecraft.

Objective

To apply Newton-Euler equations to analyze the effect of external torques on the motion of a rigid spacecraft, demonstrating how torques influence angular acceleration and orientation changes in three-dimensional space.

Activity

Spacecraft in orbit or deep space experience minimal external forces, but they are often subjected to control torques generated by thrusters, reaction wheels, or gyroscopic actuators. These torques influence

- attitude control and stabilization (correcting spacecraft orientation)
- maneuvering and trajectory corrections (adjusting spin rates)
- effects of disturbances from gravity gradients and solar radiation pressure

In this problem, we apply the Newton-Euler equations to determine how external torques influence rotational motion.

Mathematical Formulation of Torque-Induced Motion

Step 1: Use the Newton-Euler equations for rotational motion.

The Newton-Euler equations describe the effect of external torques on a rigid body's rotation:

$$\sum \vec{M}_G = \frac{d\vec{H}_G}{dt}$$

Since angular momentum is defined as

$$\vec{H}_G = I_G \vec{\omega}$$

We can apply differentiation in a rotating frame:

$$\sum \vec{M}_G = I_G \frac{d\vec{\omega}}{dt} + \vec{\omega} \times \left(I_G \vec{\omega} \right)$$

These equations determine how torques affect rotational motion in a spacecraft.

Step 2: Express the equations in a body-fixed frame.

For a rigid spacecraft, the Euler equations simplify the system into three scalar equations:

$$\sum M_x = I_x \dot{\omega}_x + \left(I_z - I_y \right) \omega_y \omega_z$$

$$\sum M_y = I_y \dot{\omega}_y + \left(I_x - I_z \right) \omega_z \omega_x$$

$$\sum M_z = I_z \dot{\omega}_z + \left(I_y - I_x \right) \omega_x \omega_y$$

These equations show how external moments influence the rotational acceleration of a spacecraft.

Numerical Example

This example involves the spacecraft attitude control under the external torque.

A rigid spacecraft is rotating with the following properties:

- Moments of inertia:
  - $I_x = 250$ kg·m²
  - $I_y = 300$ kg·m²
  - $I_z = 450$ kg·m²
- Initial angular velocity:
  - $\omega_x = 0.4$ rad/s
  - $\omega_y = 0.2$ rad/s
  - $\omega_z = 0.3$ rad/s
- External control torques applied:
  - $M_x = 8$ N·m
  - $M_y = 5$ N·m
  - $M_z = 7$ N·m

Determine the angular acceleration components.

Solution Using Euler's Equations

Step 1: Compute the angular accelerations.

Use the values:

$$\dot{\omega}_x = \frac{M_x - \left(I_z - I_y\right)\omega_y\omega_z}{I_x} = \frac{8 - \left(450 - 300\right)\left(0.2\right)\left(0.3\right)}{250} = -0.004 \text{ rad} / \text{s2}$$

$$\dot{\omega}_y = \frac{M_y - \left(I_x - I_z\right)\omega_z\omega_x}{I_y} = \frac{5 - \left(250 - 450\right)\left(0.3\right)\left(0.4\right)}{300} = 0.0967 \text{ rad} / \text{s2}$$

$$\dot{\omega}_z = \frac{M_z - \left(I_y - I_x\right)\omega_x\omega_y}{I_z} = \frac{7 - \left(300 - 250\right)\left(0.4\right)\left(0.2\right)}{450} = 0.00667 \text{ rad} / \text{s2}$$

These values describe how external torques applied to a spacecraft induce changes in its angular velocity over time.

ChatGPT Integration

- Use ChatGPT to analyze spacecraft stability under external torques using Euler's equations.
- Ask ChatGPT to simulate torque-induced attitude adjustments in space missions.
- Request ChatGPT compare torque-based motion control in satellites and robotic arms.

PROMPT 194

Discuss the principle of work and energy for a rigid body undergoing three-dimensional motion.

Objective

To explain the work-energy principle for a rigid body undergoing three-dimensional motion, derive the governing equations, and discuss its applications in mechanical, aerospace, and robotic systems.

Activity

The work-energy principle states that the net work done on a rigid body is equal to its change in kinetic energy. This principle is used in

- vehicle crash analysis (impact energy absorption)
- spacecraft trajectory planning (energy management in orbital maneuvers)
- robotic system design (energy-efficient motion and actuation)

This discussion derives the work-energy equation for a rigid body in three-dimensional motion and applies it to real-world problems.

Mathematical Formulation of the Work-Energy Principle

Step 1: Identify the work done by forces in three dimensions.

The work W done by a force $\vec{F}$ acting on a rigid body moving from position A to B is given by

$$W = \int_{A}^{B} \vec{F} \cdot d\vec{r}$$

where
- W = total work done by external forces
- $\vec{F}$ = force applied to the rigid body
- $d\vec{r}$ = differential displacement of the body's center of mass

The net work done by all external forces is

$$W_{net} = \Delta KE$$

where ΔKE is the change in kinetic energy.

Step 2: Determine the kinetic energy of a rigid body.

For a rigid body in three-dimensional motion, the total kinetic energy consists of both translational and rotational components:

$$KE = KE_{trans} + KE_{rot}$$

$$KE_{trans} = \frac{1}{2}mv_G^2$$

$$KE_{rot} = \frac{1}{2}\vec{\omega}^T I_G \vec{\omega}$$

where
- m = mass of the rigid body
- v_G = velocity of the center of mass
- I_G = moment of inertia tensor
- $\vec{\omega}$ = angular velocity vector

Thus, the work-energy theorem for a rigid body is

$$W_{net} = \frac{1}{2}mv_{G2}^2 + \frac{1}{2}\vec{\omega}_2^T I_G \vec{\omega}_2 - \left(\frac{1}{2}mv_{G1}^2 + \frac{1}{2}\vec{\omega}_1^T I_G \vec{\omega}_1 \right)$$

Numerical Example

This example involves the work done on a spinning object in free space. A rigid satellite in deep space has the following properties:

- mass: m = 500 m = 500 m = 500 kg
- moment of inertia tensor (principal axes):
  - I_x = 200 kg·m²
  - I_y = 300 kg·m²
  - I_z = 400 kg·m²

- initial conditions:
  - $v_{G1} = 2.0$ m/s
  - $\omega_{x1} = 0.5$ rad/s
  - $\omega_{y1} = 0.3$ rad/s
  - $\omega_{z1} = 0.2$ rad/s
- final conditions after applied force and torque:
  - $v_{G2} = 4.0$ m/s
  - $\omega_{x2} = 0.8$ rad/s
  - $\omega_{y2} = 0.4$ rad/s
  - $\omega_{z2} = 0.3$ rad/s

Determine the total work done on the satellite.

Solution Using the Work-Energy Theorem

Step 1: Compute the initial kinetic energy.

$$KE_{trans,1} = \frac{1}{2}(500)(2.0)^2 = 1000 \text{ J}$$

$$KE_{rot,1} = \frac{1}{2}(50 + 27 + 16) = 46.5 \text{ J}$$

$$KE_1 = 1000 + 46.5 = 1046.5 \text{ J}$$

Step 2: Compute the final kinetic energy.

$$KE_{trans,2} = \frac{1}{2}(500)(4.0)^2 = 4000 \text{ J}$$

$$KE_{rot,2} = \frac{1}{2}\left[(200)(0.8)^2 + (300)(0.4)^2 + (400)(0.3)^2\right] = \frac{1}{2}(128 + 48 + 36) = 106 \text{ J}$$

$$KE_2 = 4000 + 106 = 4106 \text{ J}$$

Step 3: Compute the work done.

$$W_{net} = KE_2 - KE_1 = 4106 - 1046.5 = 3059.5 \text{ J}$$

ChatGPT Integration

- Use ChatGPT to simulate work-energy applications in aerospace, such as satellite maneuvers.
- Ask ChatGPT to analyze energy-efficient motion planning for robotic manipulators.
- Request ChatGPT compare work-energy concepts in different mechanical systems.

PROMPT 195

Solve a problem using the work-energy principle to determine the motion of a gyroscopic rotor in a three-dimensional system.

Objective

To apply the work-energy principle to analyze the motion of a gyroscopic rotor, derive the governing equations, and solve a problem involving kinetic energy transfer in a rotating system.

Activity

Gyroscopic rotors are fundamental in mechanical, aerospace, and robotic systems, where they

- stabilize vehicles and spacecraft (e.g., reaction wheels and gyrostabilizers)
- store and transfer rotational energy (e.g., flywheels in regenerative braking)
- enhance precision motion control (e.g., robotic arms and industrial machinery)

This prompt applies the work-energy theorem to determine how forces and moments influence the motion of a spinning rotor in three-dimensional motion.

Mathematical Formulation of Work-Energy Principle for a Gyroscopic Rotor

Step 1: Determine the work done by forces and moments.
For a rigid rotor subjected to external forces and torques, the work-energy principle states

$$W_{net} = \Delta KE$$

The total kinetic energy of a gyroscopic rotor includes both translational and rotational components:

$$KE = KE_{trans} + KE_{rot}$$

$$KE_{trans} = \frac{1}{2} m v_G^2$$

$$KE_{rot} = \frac{1}{2} \vec{\omega}^T I_G \vec{\omega}$$

where

- m = mass of the rotor
- v_G = velocity of the rotor's center of mass
- I_G = moment of inertia tensor
- $\vec{\omega}$ = angular velocity vector

The work done on the rotor is given by

$$W_{net} = \int_{t_1}^{t_2} \sum \vec{M} \cdot \vec{\omega} \, dt$$

This equation describes how applied moments affect rotational energy.

Numerical Example

This example involves the work-energy analysis of a spinning flywheel. A flywheel system in a regenerative braking unit has the following properties:

- mass: $m = 50$ kg
- moment of inertia (about spin axis): $I_G = 10$ kg·m²
- initial conditions:
  - $v_{G1} = 5.0$ m/s
  - $\omega_1 = 300$ rad/s
- final conditions after braking force is applied:
  - $v_{G2} = 2.0$ m/s
  - $\omega_2 = 150$ rad/s
- Net external torque applied: $M = 5$ N·m

Determine the work done on the flywheel during braking.

Solution Using the Work-Energy Principle

Step 1: Compute the initial kinetic energy.

$$KE_{trans,1} = \frac{1}{2}(50)(5.0)^2 = 625 \text{ J}$$

$$KE_{rot,1} = \frac{1}{2}(10)(300)^2 = 450000 \text{ J}$$

$$KE_1 = 625 + 450000 = 450625 \text{ J}$$

Step 2: Compute the final kinetic energy.

$$KE_{trans,2} = \frac{1}{2}(50)(2.0)^2 = 100 \text{ J}$$

$$KE_{rot,2} = \frac{1}{2}(10)(150)^2 = 112500 \text{ J}$$

$$KE_2 = 100 + 112500 = 112600 \text{ J}$$

Step 3: Compute the work done.

$$W_{net} = KE_2 - KE_1 = 112600 - 450625 = -338025 \text{ J}$$

Thus, 338,025 joules of energy are removed from the flywheel due to braking.

ChatGPT Integration

- Use ChatGPT to analyze energy storage and recovery in flywheel systems.
- Ask ChatGPT to simulate the effects of regenerative braking on energy dissipation.
- Request ChatGPT compare gyroscopic energy storage with other energy management methods.

PROMPT 196

Explain the impulse-momentum principle for a rigid body and derive its mathematical formulation in three dimensions.

Objective

To derive the impulse-momentum principle for a rigid body in three-dimensional motion, explain its significance, and discuss its applications in robotics, vehicle collisions, and aerospace engineering.

Activity

The impulse-momentum principle states that the impulse applied to a rigid body equals its change in momentum. This principle is widely applied in

- crash testing and impact analysis (e.g., airbags and vehicle safety systems)

- spacecraft docking and orbital maneuvers (e.g., thruster-induced impulse)
- robotic manipulator dynamics (e.g., impulse-based motion planning)

This discussion derives the impulse-momentum equation and applies it to a real-world impact scenario.

Mathematical Formulation of the Impulse-Momentum Principle

Step 1: Use the linear impulse-momentum equation.
For a rigid body subjected to an external force $\vec{F}$ over a time interval t_1 to t_2, the impulse is

$$\int_{t_1}^{t_2} \sum \vec{F}\, dt = m\vec{v}_G(t_2) - m\vec{v}_G(t_1)$$

where
- $\sum \vec{F}$ = net external force acting on the rigid body
- m = mass of the body
- $\vec{v}_G$ = velocity of the center of mass at different times

This equation describes how an applied impulse changes a body's linear momentum.

Step 2: Use the angular impulse-momentum equation.
For rotational motion, the impulse-momentum equation relates the applied torque to the change in angular momentum:

$$\int_{t_1}^{t_2} \sum \vec{M}_G\, dt = \vec{H}_G(t_2) - \vec{H}_G(t_1)$$

Since the angular momentum is

$$\vec{H}_G = I_G \vec{\omega}$$

the impulse-momentum equation in matrix form becomes

$$I_G \vec{\omega}(t_2) - I_G \vec{\omega}(t_1) = \int_{t_1}^{t_2} \sum \vec{M}_G\, dt$$

This equation governs how a torque impulse changes a body's angular momentum.

Numerical Example

This example involves the spacecraft docking maneuver using impulse-momentum. A small spacecraft with

- mass: $m = 1000$ kg
- moment of inertia:
  - $I_x = 800$ kg·m²
  - $I_y = 900$ kg·m²
  - $I_z = 1100$ kg·m²
- initial velocity: $v_{G1} = 0.5$ m/s
- initial angular velocity:
  - $\omega_{x1} = 0.2$ rad/s
  - $\omega_{y1} = 0.1$ rad/s
  - $\omega_{z1} = 0.05$ rad/s
- impulse applied by thrusters:
  - $J_x = 400$ Ns
  - $J_y = 300$ Ns
  - $J_z = 200$ Ns

Determine the final linear and angular velocities after the impulse is applied.

Solution Using the Impulse-Momentum Principle

Step 1: Compute the final linear velocity.
Use the following:

$$mv_{G2} = mv_{G1} + J$$

Solve for v_{G2}:

$$v_{G2} = v_{G1} + \frac{J}{m}$$

By substituting values, we obtain

$$v_{G2} = 0.5 + \frac{400}{1000} = 0.9 \text{ m/s}$$

Step 2: Compute the final angular velocity.

Use the following:

$$I_G \vec{\omega}_2 = I_G \vec{\omega}_1 + \sum \vec{M} \cdot \Delta t$$

Solve for ω_{x2}:

$$\omega_{x2} = \omega_{x1} + \frac{J_x}{I_x}$$

By substituting values, we obtain

$$\omega_{x2} = 0.2 + \frac{400}{800} = 0.7 \text{ rad/s}$$

We repeat the calculation for ω_{y2} to obtain

$$\omega_{y2} = 0.1 + \frac{300}{900} = 0.433 \text{ rad/s}$$

We repeat the calculation for ω_{z2} to obtain

$$\omega_{z2} = 0.05 + \frac{200}{1100} = 0.232 \text{ rad/s}$$

These results demonstrate how small thruster impulses can modify spacecraft trajectory and attitude.

ChatGPT Integration

- Use ChatGPT to simulate spacecraft docking maneuvers based on impulse-momentum principles.
- Ask ChatGPT to analyze impulse-based collision responses in robotic systems.
- Request ChatGPT compare impulse applications in spacecraft, vehicles, and robotics.

PROMPT 197

Solve a problem involving an impact on a rotating rigid body, using the impulse-momentum equations.

Objective

To apply the impulse-momentum principle to analyze the impact on a rotating rigid body, derive the governing equations, and solve a problem involving linear and angular momentum changes due to impact.

Activity

Rigid bodies undergoing collisions or impulsive forces experience sudden changes in linear and angular momentum. This principle is widely applied in

- automotive crash analysis (e.g., energy transfer during a vehicle collision)
- sports biomechanics (e.g., impact between a bat and a ball)
- industrial machine impacts (e.g., robotic arm striking an object)

This discussion derives the impulse-momentum equations and applies them to a real-world impact scenario.

Mathematical Formulation of Impulse-Momentum Equations for Impact

Step 1: Use the linear impulse-momentum equation.

For a rigid body subjected to an impulsive force $\vec{J}$, the change in linear momentum is

$$\vec{J} = m\vec{v}_G(t_2) - m\vec{v}_G(t_1)$$

where

- $\vec{J}$ = impulse applied during impact
- m = mass of the rigid body
- $\vec{v}_G$ = velocity of the center of mass before and after impact

This equation describes how impulse forces affect the body's linear motion.

Step 2: Use the angular impulse-momentum equation.

For rotational motion, the change in angular momentum due to impact is given by

$$\int_{t_1}^{t_2} \sum \vec{M}_G \, dt = \vec{H}_G(t_2) - \vec{H}_G(t_1)$$

Since angular momentum is

$$\vec{H}_G = I_G\vec{\omega}$$

the impulse-momentum equation in matrix form becomes

$$I_G\vec{\omega}(t_2) - I_G\vec{\omega}(t_1) = \int_{t_1}^{t_2} \sum \vec{M}_G \, dt$$

This equation governs how an impacting force changes the rotational motion of a rigid body.

Numerical Example

The following example involves the impact between a baseball bat and a ball. A baseball bat strikes a ball with the following attributes:

- mass of ball: $m = 0.15$ kg
- moment of inertia of the bat: $I_G = 1.5$ kg·m²
- initial velocity of ball: $v_{G1} = -30$ m/s (toward the bat)
- initial angular velocity of the bat: $\omega_1 = 0$ rad/s
- impulse applied by the bat: $J = 6.0$ Ns
- impact distance from bat's center of mass: $r = 0.8$ m

Determine the final velocity of the ball and the bat's angular velocity after impact.

Solution Using the Impulse-Momentum Principle

Step 1: Compute the final velocity of the ball.

Use the following:

$$mv_{G2} = mv_{G1} + J$$

Solve for v_{G2}:

$$v_{G2} = v_{G1} + \frac{J}{m}$$

By substituting in values, we obtain

$$v_{G2} = -30 + \frac{6.0}{0.15} = 10 \text{ m/s}$$

Step 2: Compute the final angular velocity of the bat.

Use the following:

$$I_G\omega_2 = I_G\omega_1 + rJ$$

Solve for ω_2:

$$\omega_2 = \omega_1 + \frac{rJ}{I_G}$$

By substituting in values, we obtain

$$\omega_2 = \frac{4.8}{1.5} = 3.2 \text{ rad/s}$$

Thus, the bat begins to rotate at 3.2 rad/s after impact.

This demonstrates how impulse-momentum principles apply to impact events in sports and engineering systems.

ChatGPT Integration

- Use ChatGPT to simulate impact events in sports, such as a baseball bat striking a ball.
- Ask ChatGPT to analyze impact forces in crash testing and industrial machinery.
- Request ChatGPT compare impulse-momentum applications in different engineering systems.

PROMPT 198

Analyze rotational stability in aerospace and mechanical systems, focusing on gyroscopic effects and control mechanisms.

Objective

To explain rotational stability in aerospace and mechanical systems, derive the governing equations, and discuss how gyroscopic effects influence stability and control mechanisms in real-world applications.

Activity

Rotational stability is crucial in many engineering applications, including

- aerospace systems (e.g., satellite attitude control, and aircraft stability)
- vehicle dynamics (e.g., motorcycles, ships, and high-speed trains)
- mechanical precision systems (e.g., robotic arms and industrial machines)

This discussion focuses on the role of gyroscopic effects in stabilizing motion and enabling precise control in rotating systems.

Mathematical Formulation of Rotational Stability

Step 1: Use the gyroscopic motion and torque-induced stability.

A gyroscope resists changes in its orientation due to angular momentum conservation. The torque acting on a gyroscope is

$$\sum \vec{M} = \frac{d\vec{H}}{dt}$$

Since angular momentum is

$$\vec{H} = I\vec{\omega}$$

We apply differentiation to obtain

$$\sum \vec{M} = I\frac{d\vec{\omega}}{dt} + \vec{\omega} \times \left(I\vec{\omega}\right)$$

This equation describes how torques affect rotational stability.

Step 2: Use the precession motion and gyroscopic stability.

For a spinning system under an external moment, precession occurs, defined as

$$\vec{M} = \vec{\Omega} \times \vec{H}$$

where

- $\vec{\Omega}$ = precession angular velocity
- $\vec{M}$ = applied moment
- $\vec{H}$ = angular momentum

This principle is used in stabilization mechanisms such as gyroscopes and control moment gyroscopes (CMGs).

Numerical Example

The following example involves the gyroscopic stabilization in a satellite. A satellite stabilization system has the following features:

- moment of inertia: I = 150 kg·m²
- initial angular velocity: ω = 6 rad/s
- applied control moment: M = 3 N·m

Determine the precession angular velocity of the satellite.

Solution Using Gyroscopic Equations

Use the following:

$$\Omega = \frac{M}{H}$$

By substituting $H = I\omega$, we obtain

$$\Omega = \frac{M}{I\omega}$$

By substituting values, we obtain

$$\Omega = \frac{3}{(150)(6)} = 0.0033 \text{ rad/s}$$

Thus, the satellite experiences a precession motion at 0.0033 rad/s.

ChatGPT Integration

- Use ChatGPT to analyze how gyroscopic effects stabilize aerospace and marine systems.
- Ask ChatGPT to simulate precession motion under different torque conditions.
- Request ChatGPT compare active vs. passive stabilization methods in engineering.

PROMPT 199

Analyze the motion of a robotic manipulator with multiple degrees of freedom using three-dimensional kinetics principles.

Objective

To analyze the three-dimensional motion of a robotic manipulator with multiple degrees of freedom (DOF), derive the governing equations, and discuss how kinetics principles are applied to robotic control and automation.

Activity

Robotic manipulators are fundamental in industrial automation, medical robotics, and aerospace systems. The analysis of their motion involves

- kinematics and kinetics modeling (position, velocity, acceleration, forces, and torques)
- joint dynamics and actuator forces (controlling motion with motors and actuators)
- energy-efficient trajectory planning (optimizing force and torque distribution)

This discussion applies Newton-Euler and Lagrangian dynamics to study the motion of a multi-DOF robotic arm.

Mathematical Formulation of Robotic Manipulator Dynamics

Step 1: Use the Newton-Euler equations for a multi-DOF robot.

For each link of a robotic arm, the Newton-Euler equations describe the relationship between forces, torques, and motion:

$$\sum \vec{F} = m\vec{a}_C$$

$$\sum \vec{M} = I_C \frac{d\vec{\omega}}{dt} + \vec{\omega} \times \left(I_C \vec{\omega} \right)$$

where
- $\vec{F}$ = net force acting on the robotic link
- m = mass of the link
- $\vec{a}_C$ = acceleration of the center of mass
- $\vec{M}$ = torque applied to the link
- I_C = moment of inertia tensor
- $\vec{\omega}$ = angular velocity of the link

These equations describe how forces and torques control robotic motion.

Step 2: Use the Lagrangian formulation for motion planning.

An alternative approach uses Lagrange's equation:

$$\frac{d}{dt}\left(\frac{\partial L}{\partial \dot{q}_i} \right) - \frac{\partial L}{\partial q_i} = \tau_i$$

where
- L = KE−PE is the Lagrangian function (kinetic energy minus potential energy)
- q_i = generalized coordinate (joint position)

- $\dot{q}_i$ = joint velocity
- τ_i = joint torque/force

This method is widely used in robotics control systems to optimize motion planning.

Numerical Example

This example involves the motion analysis of a 2-DOF robotic arm. A two-link robotic arm has

- Link 1: m_1 = 5 kg, I_1 = 2 kg·m², length L_1 = 1.0 m
- Link 2: m_2 = 3 kg, I_2 = 1.2 kg·m², length L_2 = 0.8 m
- Initial angular velocities:
  - ω_1 = 2.0 rad/s
  - ω_2 = 1.5 rad/s
- External torques applied:
  - τ_1 = 8 N·m
  - τ_2 = 5 N·m

Determine the angular accelerations of both links.

Solution Using Newton-Euler Equations

Step 1: Write Newton-Euler equations for each link, including coupling terms

$$I_1\dot{\omega}_1 + h_{12} = \tau_1$$
$$I_2\dot{\omega}_2 + h_{21} = \tau_2$$

Where h_{12} and h_{21} are nonlinear coupling terms due to Coriolis and centripetal effects. For a 2-DOF planar arm, commonly:

$$h_{12} = m_2 L_1 L_2 \omega_1 \omega_2$$
$$h_{21} = -m_2 L_1 L_2 \omega_1^2$$

Step 2: Calculate coupling terms

$$h_{12} = 3 \times 1.0 \times 0.8 \times 2.0 \times 1.5 = 7.2 \text{N} \cdot \text{m}$$
$$h_{21} = -3 \times 1.0 \times 0.8 \times (2.0)^2 = -9.6 \text{N} \cdot \text{m}$$

Step 3: Solve for angular accelerations $\dot{\omega}_1$ and $\dot{\omega}_2$

$$I_1\dot{\omega}_1 = \tau_1 - h_{12} = 8 - 7.2 = 0.8 \Rightarrow \dot{\omega}_1 = \frac{0.8}{2} = 0.4 \text{rad/s}^2$$

$$I_2\dot{\omega}_2 = \tau_2 - h_{21} = 5 - (-9.6) = +14.6 \Rightarrow \dot{\omega}_2 = \frac{14.6}{1.2} \approx 12.17 \text{rad/s}^2$$

ChatGPT Integration

- Use ChatGPT to simulate the motion of a robotic arm under different torque inputs.
- Ask ChatGPT to analyze inverse kinematics and motion planning in robotic systems.
- Request ChatGPT compare Newton-Euler and Lagrangian approaches in robotic control.

PROMPT 200

Solve a real-world engineering problem involving three-dimensional kinetics, such as the motion of a drone or an industrial robotic arm.

Objective

To apply three-dimensional kinetics principles to a real-world engineering problem, derive the governing equations, and solve a problem involving the motion of a drone or an industrial robotic arm under external forces and torques.

Activity

Many engineering applications involve three-dimensional motion, including

- drones and UAVs (unmanned aerial vehicles) navigating under external forces and torques
- industrial robotic arms executing precision tasks with controlled motion
- autonomous vehicles maintaining stability and maneuvering efficiently

This discussion applies Newton-Euler equations and impulse-momentum principles to analyze a practical engineering problem.

Mathematical Formulation of Drone/Robotic Arm Motion

Step 1: Use the Newton-Euler equations for a drone or robotic arm.

The equations governing the motion of a rigid body in three-dimensional space are

$$\sum \vec{F} = m\vec{a}_G$$

$$\sum \vec{M} = I_G \frac{d\vec{\omega}}{dt} + \vec{\omega} \times \left(I_G \vec{\omega}\right)$$

where

- $\vec{F}$ = external forces acting on the system
- $\vec{M}$ = external torques affecting orientation
- $\vec{\omega}$ = angular velocity of the system

These equations describe how drones and robotic arms respond to applied forces and moments.

Numerical Example

This example involves the motion analysis of a drone in flight. A quadrotor drone experiences external forces and moments:

- mass: m = 2.5 kg
- moments of inertia:
  - I_x = 0.02 kg·m²
  - I_y = 0.025 kg·m²
  - I_z = 0.03 kg·m²
- initial angular velocity:
  - ω_{x1} = 1.0 rad/s
  - ω_{y1} = 0.5 rad/s
  - ω_{z1} = 0.2 rad/s
- external forces applied:
  - F_x =5 N
  - F_y = 3 N
  - F_z = 10 N
- external torques applied:
  - M_x=1.5 N·m
  - M_y=1.2 N·m
  - M_z=0.8 N·m

Determine the linear acceleration and angular acceleration of the drone.

Solution Using Newton-Euler Equations

Step 1: Compute the linear acceleration.

Use the following:

$$a_G = \frac{\sum \vec{F}}{m}$$

By substituting values, we obtain

$$a_{Gx} = \frac{5}{2.5} = 2.0 \text{ m/s}^2$$

$$a_{Gy} = \frac{3}{2.5} = 1.2 \text{ m/s}^2$$

$$a_{Gz} = \frac{10}{2.5} = 4.0 \text{ m/s}^2$$

Thus, the drone experiences an acceleration of (2.0, 1.2, 4.0) m/s².

Step 2: Compute the Angular Acceleration.

Use the following:

$$I_C \dot{\omega} = \sum \vec{M} - \vec{\omega} \times (I_C \vec{\omega})$$

By approximating for small angular velocity changes, we obtain

$$\dot{\omega}_x = \frac{M_x}{I_x}$$

$$\dot{\omega}_y = \frac{M_y}{I_y}$$

$$\dot{\omega}_z = \frac{M_z}{I_z}$$

By substituting values, we obtain

$$\dot{\omega}_x = \frac{1.5}{0.02} = 75.0 \text{ rad/s}^2$$

$$\dot{\omega}_y = \frac{1.2}{0.025} = 48.0 \text{ rad/s}^2$$

$$\dot{\omega}_z = \frac{0.8}{0.03} = 26.67 \text{ rad/s}^2$$

Thus, the drone's angular acceleration is (75.0, 48.0, 26.67) rad/s².

ChatGPT Integration

▪ Use ChatGPT to simulate the effect of external forces on drone flight stability.

▪ Ask ChatGPT to analyze robotic arm motion under different torque inputs.

▪ Request ChatGPT compare UAV dynamics and robotic control systems.

Dynamics Wizard Tutor App

This GPT App is designed to complement Part 2 of this book (Dynamics) by offering an interactive and problem-solving environment. Users can explore their preferred chapters and prompts from the book's structured list, allowing them to gain a deeper understanding of topics such as kinematics, kinetics, work-energy principles, impulse-momentum methods, and rigid body dynamics. The app features step-by-step walkthroughs for complex dynamics problems, motion simulations for visualizing particle and rigid body motion, real-time feedback and conceptual explanations to reinforce learning, custom problem-solving tools for force-acceleration, work-energy, and impulse-momentum methods, and personalized progress tracking to help students strengthen weak areas.

Whether you are analyzing particle motion, solving rigid body equations, or applying energy methods to real-world dynamics problems, this app serves as a powerful extension of the book to support hands-on learning in engineering dynamics.

Users can obtain access and run the Dynamics Wizard Tutor App directly through OpenAI GPTs. To get started, when available, navigate to the GPT Apps section within OpenAI's platform and search for "Dynamics Wizard Tutor App." Once launched, the app's intuitive interface allows users to select chapters, prompts, or specific topics from this book. Simply input your questions or problems, and the app will provide step-by-step solutions, interactive explanations, and engineering insights to enhance your understanding of dynamics.

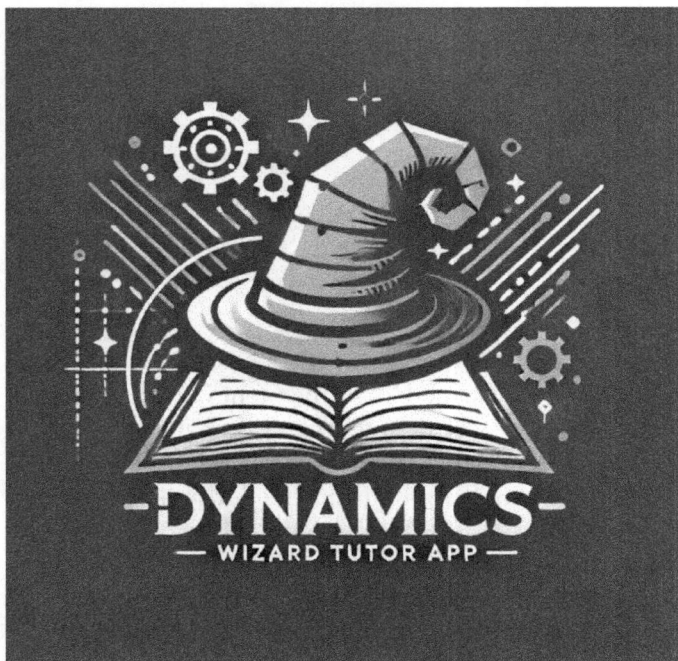

FIGURE 20.1 The logo for the Wizard Tutor

Transformation of Angular Velocities in Euler Angle Rotations

In Part 2, several prompts include numerical examples using Euler angles that were introduced to describe the motion of a rigid body by transforming the coordinate system attached to it. Each of the three successive rotations contributes to a corresponding local angular velocity. For example, a rotation about the x-axis is associated with a specific angular velocity, and similar contributions arise from rotations about the y- and z-axes.

To provide background, this section introduces a method for expressing these angular velocities in terms of their components in the final coordinate system, denoted as (x',y',z'). We begin by deriving the transformation equations for a specific motion and then extend these results to establish a generalized formulation for all 12 possible sequences of Euler angle rotations (including proper and Tait-Bryan categories), as summarized below:

TABLE 20.1 All 12 possible sequences of Euler angle rotations

Proper Euler Angles	Tait-Bryan Angles
$x \to y \to x$	$x \to y \to z$
$x \to z \to x$	$x \to z \to y$
$y \to x \to y$	$y \to x \to z$
$y \to z \to y$	$y \to z \to x$
$z \to x \to z$	$z \to x \to y$
$z \to y \to z$	$z \to y \to x$

Generally, rotation about the z-axis is called *yaw*, about the y-axis is referred as *pitch*, and about the x-axis is known as *roll*. Consider a rigid body undergoing a sequence of yaw-roll-yaw rotations. This motion consists of three successive rotations: θ_1 about the initial z-axis, θ_2 about the newly established x-axis, and θ_3 about the updated z-axis. The resulting coordinate system after these transformations is denoted as (x', y', z'). The instantaneous angular velocities associated with each of these rotations are defined as follows:

$$\omega_{yaw} = \begin{pmatrix} 0 \\ 0 \\ \dot{\theta}_1 \end{pmatrix}, \text{ first rotation-yaw angular velocity.}$$

$$\omega_{roll} = \begin{pmatrix} \dot{\theta}_2 \\ 0 \\ 0 \end{pmatrix}, \text{ second rotation-roll angular velocity.}$$

$$\omega_{yaw} = \begin{pmatrix} 0 \\ 0 \\ \dot{\theta}_3 \end{pmatrix}, \text{ third rotation-yaw angular velocity.}$$

where, $\omega_i = \dot{\theta}_i = \dfrac{d\theta_i}{dt}$. It is important to note that these vectors do not constitute an orthogonal set.

The vector $\begin{pmatrix} 0 \\ 0 \\ \dot{\theta}_1 \end{pmatrix}$ is transform through the first yaw rotation as

$$[R_{\theta_1}]\begin{pmatrix} 0 \\ 0 \\ \dot{\theta}_1 \end{pmatrix} = \begin{bmatrix} \cos\theta_1 & \sin\theta_1 & 0 \\ -\sin\theta_1 & \cos\theta_1 & 0 \\ 0 & 0 & 1 \end{bmatrix}\begin{pmatrix} 0 \\ 0 \\ \dot{\theta}_1 \end{pmatrix} = \begin{pmatrix} 0 \\ 0 \\ \dot{\theta}_1 \end{pmatrix}$$

This vector must be added to the $\begin{pmatrix} \dot{\theta}_2 \\ 0 \\ 0 \end{pmatrix}$ to obtain $\begin{pmatrix} \dot{\theta}_2 \\ 0 \\ \dot{\theta}_1 \end{pmatrix}$, which repre-

sents the angular velocity expressed in the coordinate system that goes through the roll rotation. Therefore,

$$[R_{\theta_1}]\begin{pmatrix} \dot{\theta}_2 \\ 0 \\ \dot{\theta}_1 \end{pmatrix} = \begin{bmatrix} 1 & 0 & 0 \\ 1 & \sin\theta_2 & \sin\theta_2 \\ 0 & -\sin\theta_2 & \cos\theta_2 \end{bmatrix}\begin{pmatrix} \dot{\theta}_2 \\ 0 \\ \dot{\theta}_3 \end{pmatrix} = \begin{pmatrix} \dot{\theta}_2 \\ \dot{\theta}_1\sin\theta_2 \\ \dot{\theta}_1\cos\theta_2 + \dot{\theta}_3 \end{pmatrix}.$$

This vector must be added to the $\begin{pmatrix} 0 \\ 0 \\ \dot{\theta}_3 \end{pmatrix}$ to obtain $\begin{pmatrix} \dot{\theta}_2 \\ \dot{\theta}_1\sin\theta_2 \\ \dot{\theta}_1\cos\theta_2 + \dot{\theta}_3 \end{pmatrix}$, which

represents the angular velocity defined in the coordinate system that goes through the last yaw rotation. Therefore,

$$[R_{\theta_3}]\begin{pmatrix} \dot{\theta}_2 \\ \dot{\theta}_1\sin\theta_2 \\ \dot{\theta}_1\cos\theta_2 + \dot{\theta}_3 \end{pmatrix} = \begin{bmatrix} \cos\theta_3 & \sin\theta_3 & 0 \\ -\sin\theta_2 & \cos\theta_3 & 0 \\ 0 & 0 & 1 \end{bmatrix}\begin{pmatrix} \dot{\theta}_2 \\ \dot{\theta}_1\sin\theta_2 \\ \dot{\theta}_1\cos\theta_2 + \dot{\theta}_3 \end{pmatrix}$$

$$= \begin{pmatrix} \dot{\theta}_1\cos\theta_3 + \dot{\theta}_1\cos\theta_2\sin\theta_3 \\ -\dot{\theta}_2\sin\theta_3 + \dot{\theta}_1\sin\theta_2\cos\theta_3 \\ \dot{\theta}_1\cos\theta_2 + \dot{\theta}_3 \end{pmatrix}$$

The resulted vector shows the components of the angular velocities in the final coordinate system (x', y', z'):

$$\begin{cases} \omega_{x'} = \dot{\theta}_2\cos\theta_3 + \dot{\theta}_1\sin\theta_2\sin\theta_3 \\ \omega_{y'} = -\dot{\theta}_2\sin\theta_3 + \dot{\theta}_1\sin\theta_2\cos\theta_3 \\ \omega_{z'} = \dot{\theta}_1\cos\theta_2 + \dot{\theta}_3 \end{cases}$$

Combining all the steps, we obtain

$$
\begin{Bmatrix} \omega_{x'} \\ \omega_{y'} \\ \omega_{z'} \end{Bmatrix} = \begin{bmatrix} R_{\theta_3} \end{bmatrix} \begin{pmatrix} 0 \\ 0 \\ \dot{\theta}_3 \end{pmatrix} + \begin{bmatrix} R_{\theta_3} \end{bmatrix} \begin{bmatrix} R_{\theta_2} \end{bmatrix} \begin{pmatrix} \dot{\theta}_2 \\ 0 \\ 0 \end{pmatrix} + \begin{bmatrix} R_{\theta_3} \end{bmatrix} \begin{bmatrix} R_{\theta_2} \end{bmatrix} \begin{bmatrix} R_{\theta_1} \end{bmatrix} \begin{pmatrix} 0 \\ 0 \\ \dot{\theta}_1 \end{pmatrix}
$$

The detail expressions for $\begin{bmatrix} R_{\theta_1} \end{bmatrix}$, $\begin{bmatrix} R_{\theta_2} \end{bmatrix}$, and $\begin{bmatrix} R_{\theta_3} \end{bmatrix}$ can be found in Tables 15.2 and 15.3 of *Tensor Analysis for Engineers: Transformations – Mathematics – Applications* (3rd ed.) by Tabatabaian, M. (2023), Mercury Learning and Information.

Similarly, the remaining possible Euler angle rotation sequences and their corresponding angular velocity expressions can be derived in the same manner. The complete list is provided below:

Roll-Pitch-Yaw: $(\theta_1, \theta_2, \theta_3)$

$$
\begin{Bmatrix} \omega_{x'} \\ \omega_{y'} \\ \omega_{z'} \end{Bmatrix} = [R\theta_3] \begin{pmatrix} 0 \\ 0 \\ \dot{\theta}_3 \end{pmatrix} + [R\theta_3][R\theta_2] = \begin{pmatrix} 0 \\ \dot{\theta}_2 \\ 0 \end{pmatrix} = [R\theta_3][R\theta_2][R\theta_1] \begin{pmatrix} \dot{\theta}_1 \\ 0 \\ 0 \end{pmatrix} = \begin{Bmatrix} \dot{\theta}_1 \cos\theta_2 \cos\theta_3 + \dot{\theta}_2 \sin\theta_3 \\ \dot{\theta}_2 \cos\theta_3 - \dot{\theta}_1 \cos\theta_2 \sin\theta_3 \\ \dot{\theta}_1 \sin\theta_2 + \dot{\theta}_3 \end{Bmatrix}
$$

Roll-Pitch-Roll: $(\theta_1, \theta_2, \theta_3)$

$$
\begin{Bmatrix} \omega_{x'} \\ \omega_{y'} \\ \omega_{z'} \end{Bmatrix} = [R\theta_3] \begin{pmatrix} \dot{\theta}_3 \\ 0 \\ 0 \end{pmatrix} + [R\theta_3][R\theta_2] = \begin{pmatrix} 0 \\ \dot{\theta}_2 \\ 0 \end{pmatrix} = [R\theta_3][R\theta_2][R\theta_1] \begin{pmatrix} \dot{\theta}_1 \\ 0 \\ 0 \end{pmatrix} = \begin{Bmatrix} \dot{\theta}_1 \cos\theta_2 + \dot{\theta}_3 \\ \dot{\theta}_2 \cos\theta_3 - \dot{\theta}_1 \sin\theta_2 \sin\theta_3 \\ \dot{\theta}_2 \sin\theta_3 - \dot{\theta}_1 \sin\theta_2 \cos\theta_3 \end{Bmatrix}
$$

Roll-Yaw-Pitch: $(\theta_1, \theta_2, \theta_3)$

$$
\begin{Bmatrix} \omega_{x'} \\ \omega_{y'} \\ \omega_{z'} \end{Bmatrix} = [R\theta_3] \begin{pmatrix} 0 \\ \dot{\theta}_3 \\ 0 \end{pmatrix} + [R\theta_3][R\theta_2] = \begin{pmatrix} 0 \\ 0 \\ \dot{\theta}_2 \end{pmatrix} = [R\theta_3][R\theta_2][R\theta_1] \begin{pmatrix} \dot{\theta}_1 \\ 0 \\ 0 \end{pmatrix} = \begin{Bmatrix} \dot{\theta}_1 \cos\theta_2 \cos\theta_3 - \dot{\theta}_2 \sin\theta_3 \\ \dot{\theta}_3 - \dot{\theta}_1 \sin\theta_2 \\ \dot{\theta}_2 \cos\theta_2 \sin\theta_3 - \dot{\theta}_2 \cos\theta_3 \end{Bmatrix}
$$

Roll-Yaw-Roll: $(\theta_1, \theta_2, \theta_3)$

$$\begin{Bmatrix} \omega_{x'} \\ \omega_{y'} \\ \omega_{z'} \end{Bmatrix} = [R\theta_3]\begin{pmatrix} \dot\theta_3 \\ 0 \\ 0 \end{pmatrix} + [R\theta_3][R\theta_2]\begin{pmatrix} 0 \\ 0 \\ \dot\theta_2 \end{pmatrix} = [R\theta_3][R\theta_2][R\theta_1]\begin{pmatrix} \dot\theta_1 \\ 0 \\ 0 \end{pmatrix} = \begin{Bmatrix} \dot\theta_1\cos\theta_2 + \dot\theta_3 \\ \dot\theta_2\sin\theta_3 - \dot\theta_1\sin\theta_2\cos\theta_3 \\ \dot\theta_1\cos\theta_2\sin\theta_3 - \dot\theta_2\cos\theta_3 \end{Bmatrix}$$

Pitch-Roll-Yaw: $(\theta_1, \theta_2, \theta_3)$

$$\begin{Bmatrix} \omega_{x'} \\ \omega_{y'} \\ \omega_{z'} \end{Bmatrix} = [R\theta_3]\begin{pmatrix} 0 \\ 0 \\ \dot\theta_3 \end{pmatrix} + [R\theta_3][R\theta_2]\begin{pmatrix} \dot\theta_2 \\ 0 \\ 0 \end{pmatrix} = [R\theta_3][R\theta_2][R\theta_1]\begin{pmatrix} 0 \\ \dot\theta_1 \\ 0 \end{pmatrix} = \begin{Bmatrix} \dot\theta_1\cos\theta_2\sin\theta_3 + \dot\theta_2\cos\theta_3 \\ \dot\theta_1\cos\theta_2\cos\theta_3 - \dot\theta_2\cos\theta_3 \\ \dot\theta_1\sin\theta_2 + \dot\theta_3 \end{Bmatrix}$$

Pitch-Roll-Pitch: $(\theta_1, \theta_2, \theta_3)$

$$\begin{Bmatrix} \omega_{x'} \\ \omega_{y'} \\ \omega_{z'} \end{Bmatrix} = [R\theta_3]\begin{pmatrix} 0 \\ \dot\theta_3 \\ 0 \end{pmatrix} + [R\theta_3][R\theta_2]\begin{pmatrix} \dot\theta_2 \\ 0 \\ 0 \end{pmatrix} = [R\theta_3][R\theta_2][R\theta_1]\begin{pmatrix} 0 \\ \dot\theta_1 \\ 0 \end{pmatrix} = \begin{Bmatrix} \dot\theta_1\sin\theta_2\sin\theta_3 + \dot\theta_2\cos\theta_3 \\ \dot\theta_1\cos\theta_2 + \dot\theta_3 \\ \dot\theta_2\cos\theta_3 - \dot\theta_1\sin\theta_2\cos\theta_3 \end{Bmatrix}$$

Pitch-Yaw-Roll: $(\theta_1, \theta_2, \theta_3)$

$$\begin{Bmatrix} \omega_{x'} \\ \omega_{y'} \\ \omega_{z'} \end{Bmatrix} = [R\theta_3]\begin{pmatrix} \dot\theta_3 \\ 0 \\ 0 \end{pmatrix} + [R\theta_3][R\theta_2]\begin{pmatrix} 0 \\ 0 \\ \dot\theta_2 \end{pmatrix} = [R\theta_3][R\theta_2][R\theta_1]\begin{pmatrix} 0 \\ \dot\theta_1 \\ 0 \end{pmatrix} = \begin{Bmatrix} \dot\theta_1\sin\theta_2 + \dot\theta_3 \\ \dot\theta_1\cos\theta_2\sin\theta_3 + \dot\theta_2\sin\theta_3 \\ \dot\theta_2\cos\theta_3 - \dot\theta_1\cos\theta_2\sin\theta_3 \end{Bmatrix}$$

Pitch-Yaw-Pitch: $(\theta_1, \theta_2, \theta_3)$

$$\begin{Bmatrix} \omega_{x'} \\ \omega_{y'} \\ \omega_{z'} \end{Bmatrix} = [R\theta_3]\begin{pmatrix} 0 \\ \dot\theta_3 \\ 0 \end{pmatrix} + [R\theta_3][R\theta_2]\begin{pmatrix} 0 \\ 0 \\ \dot\theta_2 \end{pmatrix} = [R\theta_3][R\theta_2][R\theta_1]\begin{pmatrix} 0 \\ \dot\theta_1 \\ 0 \end{pmatrix} = \begin{Bmatrix} \dot\theta_1\sin\theta_2\cos\theta_3 + \dot\theta_2\sin\theta_3 \\ \dot\theta_1\cos\theta_2 + \dot\theta_3 \\ \dot\theta_2\sin\theta_2\sin\theta_3 - \dot\theta_2\cos\theta_3 \end{Bmatrix}$$

Yaw-Roll-Pitch: $(\theta_1, \theta_2, \theta_3)$

$$\begin{Bmatrix} \omega_{x'} \\ \omega_{y'} \\ \omega_{z'} \end{Bmatrix} = [R\theta_3]\begin{pmatrix} 0 \\ \dot{\theta}_3 \\ 0 \end{pmatrix} + [R\theta_3][R\theta_2] = \begin{pmatrix} \dot{\theta}_2 \\ 0 \\ 0 \end{pmatrix} = [R\theta_3][R\theta_2][R\theta_1]\begin{pmatrix} 0 \\ 0 \\ \dot{\theta}_1 \end{pmatrix} = \begin{Bmatrix} \dot{\theta}_2 \cos\theta_3 - \cos\theta_3 + \dot{\theta}_2 \sin\theta_3 \\ \dot{\theta}_1 \cos\theta_2 + \dot{\theta}_3 \\ \dot{\theta}_2 \sin\theta_2 \sin\theta_3 + \dot{\theta}_2 \sin\theta_3 \end{Bmatrix}$$

Yaw-Roll-Yaw: $(\theta_1, \theta_2, \theta_3)$

$$\begin{Bmatrix} \omega_{x'} \\ \omega_{y'} \\ \omega_{z'} \end{Bmatrix} = [R\theta_3]\begin{pmatrix} 0 \\ 0 \\ \dot{\theta}_3 \end{pmatrix} + [R\theta_3][R\theta_2] = \begin{pmatrix} \dot{\theta}_2 \\ 0 \\ 0 \end{pmatrix} = [R\theta_3][R\theta_2][R\theta_1]\begin{pmatrix} 0 \\ 0 \\ \dot{\theta}_1 \end{pmatrix} = \begin{Bmatrix} \dot{\theta}_2 \cos\theta_3 - \dot{\theta}_1 \sin\theta_2 \sin\theta_3 \\ -\dot{\theta}_2 \sin\theta_3 + \dot{\theta}_1 \sin\theta_2 \cos\theta_3 \\ -\dot{\theta}_1 \sin\theta_2 + \dot{\theta}_3 \end{Bmatrix}$$

Yaw-Pitch-Roll: $(\theta_1, \theta_2, \theta_3)$

$$\begin{Bmatrix} \omega_{x'} \\ \omega_{y'} \\ \omega_{z'} \end{Bmatrix} = [R\theta_3]\begin{pmatrix} \dot{\theta}_3 \\ 0 \\ 0 \end{pmatrix} + [R\theta_3][R\theta_2] = \begin{pmatrix} 0 \\ \dot{\theta}_2 \\ 0 \end{pmatrix} = [R\theta_3][R\theta_2][R\theta_1]\begin{pmatrix} 0 \\ 0 \\ \dot{\theta}_1 \end{pmatrix} = \begin{Bmatrix} \dot{\theta}_3 - \dot{\theta}_1 \sin\theta_2 \\ \dot{\theta}_1 \cos\theta_2 \sin\theta_3 + \dot{\theta}_2 \cos\theta_3 \\ \dot{\theta}_1 \cos\theta_2 \cos\theta_3 - \dot{\theta}_2 \sin\theta_3 \end{Bmatrix}$$

Yaw-Pitch-Yaw: $(\theta_1, \theta_2, \theta_3)$

$$\begin{Bmatrix} \omega_{x'} \\ \omega_{y'} \\ \omega_{z'} \end{Bmatrix} = [R\theta_3]\begin{pmatrix} 0 \\ 0 \\ \dot{\theta}_3 \end{pmatrix} + [R\theta_3][R\theta_2] = \begin{pmatrix} 0 \\ \dot{\theta}_2 \\ 0 \end{pmatrix} = [R\theta_3][R\theta_2][R\theta_1]\begin{pmatrix} 0 \\ 0 \\ \dot{\theta}_1 \end{pmatrix} = \begin{Bmatrix} \dot{\theta}_1 \sin\theta_3 + \dot{\theta}_1 \sin\theta_2 \cos\theta_3 \\ \dot{\theta}_2 \cos\theta_3 - \dot{\theta}_1 \sin\theta_2 \sin\theta_3 \\ \dot{\theta}_1 \cos\theta_2 - \dot{\theta}_3 \end{Bmatrix}$$

Further analysis enables similar calculations for the moment of inertia, momentum, and their variations in relation to the motion of a rigid body.

PART 3: MECHANICAL VIBRATIONS

INTRODUCTION TO MECHANICAL VIBRATIONS

Mechanical vibrations are an essential aspect of engineering mechanics, dealing with oscillatory motion of mechanical systems. The study of vibrations is critical for understanding how systems respond to dynamic forces, and it plays a vital role in the design, analysis, and maintenance of machines, vehicles, structures, and even electronic devices. Mechanical vibrations occur in various forms, such as free vibrations, forced vibrations, damped vibrations, and undamped vibrations. The ability to predict and control these vibrations ensures that systems can operate safely and efficiently without excessive wear or failure.

In the context of engineering, vibrations are often categorized based on their causes and effects. Understanding key topics such as natural frequency, damping, resonance, and mode shapes is crucial for engineers aiming to design vibration-resistant structures and systems. From the vibration of a guitar string to the oscillations of bridges or the behavior of mechanical components in engines, vibrations are omnipresent and must be carefully controlled to prevent damage, fatigue, or undesirable noise. In this section, we explore the foundational principles of mechanical vibrations, their types, and their applications in various engineering fields, providing insights into how vibrations are analyzed and mitigated in real-world systems.

The following chapters are included in Part 3:

Each chapter includes approximately ten interconnected prompts, designed to guide readers in exploring the subject matter in greater depth, incorporating numerical examples and additional exploratory questions.

CHAPTER 21

BASIC CONCEPTS OF VIBRATION

Prompt 201 Define mechanical vibration and describe its significance in engineering applications.

Prompt 202 Explain the difference between free and forced vibrations. Provide real-world examples for each.

Prompt 203 Discuss the concept of natural frequency and how it relates to resonance.

Prompt 204 Explain the difference between harmonic and non-harmonic vibrations. Provide examples of each type.

Prompt 205 Describe the concept of degrees of freedom in vibrating systems.

Prompt 206 Discuss the role of stiffness and mass in determining the natural frequency of a system.

Prompt 207 Define damping in the context of vibrations and explain its effect on system behavior. Include the definition of critical damping.

Prompt 208 Explain the importance of the damping ratio and how it classifies underdamped, overdamped, and critically damped systems.

Prompt 209 Discuss the types of external forces that can cause vibrations in mechanical systems.

Prompt 210 Describe the importance of initial conditions (displacement and velocity) in the analysis of vibrating systems.

PROMPT 201

Define mechanical vibration and describe its significance in engineering applications.

Objective

Understand the basic definition of mechanical vibration and its importance in engineering.

Activity

- Define the mechanical vibration as the oscillatory motion of a physical system about an equilibrium point.
- Discuss the role of vibrations in various engineering systems such as bridges, aircraft, and automobiles. Highlight both beneficial (e.g., vibratory conveyors) and harmful effects (e.g., fatigue failure due to excessive vibration).

Numerical Example

A cantilever beam is subjected to a harmonic force causing it to vibrate. The beam's displacement follows a sinusoidal pattern over time, where the maximum displacement is 2 cm, and the frequency of vibration is 5 Hz. Calculate the time it takes for the beam to complete one full cycle of vibration.

Solution

The period T is the reciprocal of the frequency: $T = 1/f = 1/5 = 0.2$ seconds. So, the beam takes 0.2 seconds to complete one cycle.

ChatGPT Integration

- Ask ChatGPT to define mechanical vibration and compare it with your own definition.
- Use ChatGPT to explore additional real-world examples where vibration plays a critical role in both design and failure modes of engineering systems.

PROMPT 202

Explain the difference between free and forced vibrations. Provide real-world examples for each.

Objective

Understand the distinction between free and forced vibrations and identify their real-world applications.

Activity

- Define free vibration as the natural response of a system when displaced from its equilibrium position and released, without any external force acting after the initial disturbance.
- Define forced vibration as the response of a system to a continuous external force.
- Provide examples:
  - Free Vibration: A tuning fork vibrating after being struck.
  - Forced Vibration: A building swaying due to wind forces.

Numerical Example

A simple pendulum with a length of 1.5 meters is displaced and released, undergoing free vibration. Calculate the natural frequency of the pendulum.

Solution

The formula for the natural frequency of a simple pendulum is

$$f = \frac{1}{2\pi}\sqrt{\frac{g}{L}}$$

where $g = 9.81$ m/s^2 and $L = 1.5$ m.

$$f = \frac{1}{2\pi}\sqrt{\frac{9.81}{1.5}} = 0.407\,Hz$$

The natural frequency of the pendulum is approximately 0.407 Hz.

ChatGPT Integration

- Ask ChatGPT to clarify the differences between free and forced vibrations.
- Explore with ChatGPT additional examples from various fields, such as how buildings or bridges respond to seismic forces (forced vibrations) versus natural vibrations.

PROMPT 203

Discuss the concept of natural frequency and how it relates to resonance.

Objective

Understand the definition of natural frequency and the phenomenon of resonance in vibrating systems.

Activity

- Define the natural frequency as the frequency at which a system naturally vibrates when disturbed from equilibrium.
- Explain resonance as the condition that occurs when the frequency of an external force matches the natural frequency of the system, often leading to large amplitude vibrations.
- Provide real-world examples:
  - natural frequency: the oscillation of a car's suspension system after hitting a bump
  - resonance: the collapse of the Tacoma Narrows Bridge due to wind-induced vibrations at the bridge's natural frequency

Numerical Example

A spring-mass system has a mass of 2 kg and a spring stiffness of 200 N/m. Calculate the natural frequency of the system.

Solution

The natural frequency for a spring-mass system is given by

$$f_n = \frac{1}{2\pi}\sqrt{\frac{k}{m}}$$

where $k = 200$ N/m and $m = 2$ kg.

$$f_n = \frac{1}{2\pi}\sqrt{\frac{200}{2}} = 1.59\,Hz$$

The natural frequency of the system is approximately 1.59 Hz.

ChatGPT Integration

- Ask ChatGPT to define the natural frequency and resonance and provide more examples of systems where resonance can be harmful or beneficial.
- Use ChatGPT to explore famous engineering failures caused by resonance, such as bridges and machinery.

PROMPT 204

Explain the difference between harmonic and non-harmonic vibrations. Provide examples of each type.

Objective

Understand the distinction between harmonic and non-harmonic vibrations and recognize examples in practical scenarios.

Activity

- Define harmonic vibrations as vibrations that follow a 'sinusoidal' waveform over time, typically caused by periodic forces.
- Define non-harmonic vibrations as vibrations that do not follow a "sinusoidal" pattern and may be caused by irregular or transient forces.
- Provide examples:
  - harmonic vibrations: a rotating machine causing a sinusoidal force on its support structure
  - non-harmonic vibrations: the sudden impact of a hammer on a structure causing irregular vibrations

Numerical Example

A mechanical system is subjected to two forces:

1. harmonic force: $F_h(t) = 50\sin(10t)\,N$
2. non-harmonic force: $F_{nh}(t) = 50t^2 N$

The system has a mass $m = 5$ kg and stiffness $k = 1000$ N/m. The damping is negligible. Calculate the displacement response for the harmonic and non-harmonic forces and compare the results.

Solution

Response to the Harmonic Force:

The natural frequency is $\omega_n = \sqrt{\dfrac{k}{m}} = \sqrt{\dfrac{1000}{5}} = 14.14\,\text{rad/s}.$

Since the excitation frequency is $\omega = 10$ rad/s, resonance does not occur.

The displacement response is $x(t) = \dfrac{F_0/m}{\sqrt{\left(\omega_n^2 - \omega^2\right)^2}}\sin(\omega t),$ where

$F_0 = 50\,\text{N}.$ Substituting: $x(t) = \dfrac{50/5}{\sqrt{\left(14.14^2 - 10^2\right)^2}}\sin(10t)$ and simplify:

$x(t) = 0.1\sin(10t).$ The response is a sinusoidal oscillation with a frequency of $\omega = 10$ rad/s.

Response to the Non-Harmonic Force:

For the non-harmonic force $F_{nh}(t) = 50t^2$, the displacement satisfies:
$m\ddot{x} + kx = F_{nh}(t)$

Substituting: $5\ddot{x} + 1000x = 50t^2$. Solving this differential equation gives:

$x(t) = \dfrac{t^2}{20} - 0.0005.$

The response grows quadratically with time and does not oscillate.

Thus, the general solution is:

$$x(t) = C_1\cos(14.14t) + C_2\sin(14.14t) + 0.005t^2 - 0.0005$$

If initial conditions are zero, the constants can be solved. But for long-term response, the dominant term is:

$$x(t) \approx 0.005t^2$$

Comparison:

- The harmonic response is sinusoidal with constant amplitude, determined by the excitation force.
- The non-harmonic response increases quadratically with time and does not involve oscillation.

ChatGPT Integration

- Ask ChatGPT to define harmonic and non-harmonic vibrations and verify your understanding.
- Use ChatGPT to find real-world applications where harmonic vibrations are deliberately used (e.g., tuning forks) and situations where non-harmonic vibrations occur unexpectedly.

PROMPT 205

Describe the concept of degrees of freedom in vibrating systems.

Objective

Understand the meaning of degrees of freedom (DOF) and its relevance to vibration analysis.

Activity

- Define degrees of freedom as the number of independent coordinates required to describe the motion of a system.
- Explain that a system's DOF determines the complexity of its vibratory motion. For example, a simple pendulum has one degree of freedom, while a multi-body system can have several.
- Provide examples:
  - one DOF: a mass-spring system that can move only vertically
  - two DOF: a system of two masses connected by springs, each capable of moving independently

Numerical Example

A two-mass system consists of masses $m_1 = 2\ kg$ and $m_2 = 3\ kg$, connected by a spring with stiffness $k = 100\ N/m$. Each mass can move independently along the same line. How many degrees of freedom does the system have?

Solution

The system has two degrees of freedom because both masses can move independently in one direction.

ChatGPT Integration

- Ask ChatGPT to define degrees of freedom and provide additional examples of systems with varying DOF.
- Use ChatGPT to explore the implications of DOF in more complex systems, such as robotic arms or multi-body mechanical structures.

PROMPT 206

Discuss the role of stiffness and mass in determining the natural frequency of a system.

Objective

Understand how stiffness and mass affect the natural frequency of a vibrating system.

Activity

- Define stiffness (k) as a measure of a system's resistance to deformation and mass (m) as the quantity of matter in the system.
- Explain that the natural frequency of a system increases with greater stiffness and decreases with greater mass. Use the equation $f_n = \dfrac{1}{2\pi}\sqrt{\dfrac{k}{m}}$ to describe this relationship.
- Provide examples:
  - A stiffer spring will result in a higher natural frequency.
  - A heavier mass will result in a lower natural frequency.

Numerical Example

A mass-spring system consists of a mass $m = 4$ kg and a spring with stiffness $k = 400$ N/m.

Calculate the natural frequency of the system.

Solution

The natural frequency is given by

$$f_n = \frac{1}{2\pi}\sqrt{\frac{400}{4}} = 1.59\,Hz$$

The natural frequency of the system is approximately 1.59 Hz.

ChatGPT Integration

- Ask ChatGPT to explain the relationship between stiffness, mass, and natural frequency in more detail.
- Use ChatGPT to explore the impact of these parameters in complex engineering applications such as buildings, machinery, and vehicle suspensions.

PROMPT 207

Define damping in the context of vibrations and explain its effect on system behavior. Include the definition of critical damping.

Objective

Understand the concept of damping, including critical damping, and its role in modifying the behavior of vibrating systems.

Activity

- Define *damping* as the mechanism by which energy is dissipated from a vibrating system, typically through friction, air resistance, or material deformation.
- Explain that damping reduces the amplitude of oscillations over time and prevents indefinite oscillation in practical systems. Discuss different types of damping (e.g., viscous and Coulomb).
- Define critical damping as the condition where the damping is just sufficient to prevent oscillation, allowing the system to return to equilibrium as quickly as possible without overshooting. Systems with damping below this level are underdamped, while those above are overdamped.

- Provide examples:
  - viscous damping: shock absorbers in a car
  - Coulomb damping: sliding friction between mechanical components
  - critical damping: automotive suspension designed to prevent oscillation after hitting a bump

Numerical Example

A mass-spring-damper system consists of a mass $m = 5$ kg, a spring with stiffness $k = 200$ N/m, and a damping coefficient $c = 10$ Ns/m. Calculate the damping ratio ζ of the system. Determine if the system is under-damped, critically damped, or overdamped.

Solution

The damping ratio is given by

$$\xi = \frac{c}{2\sqrt{k.m}} = \frac{10}{2\sqrt{200 \times 5}} = 0.158$$

Since $\zeta < 1$, the system is underdamped with a damping ratio of approximately 0.158. For critical damping, $\zeta = 1$, and for overdamping, $\zeta > 1$.

ChatGPT Integration

- Ask ChatGPT to explain the different types of damping, including critical damping, and their effects on system behavior.
- Use ChatGPT to explore real-world applications where critical damping is crucial, such as in automotive or civil engineering systems designed to minimize oscillations.

PROMPT 208

Explain the importance of the damping ratio and how it classifies under-damped, overdamped, and critically damped systems.

Objective

Understand the significance of the damping ratio and how it is used to classify the response of vibrating systems.

Activity

- Define the damping ratio (ζ) as a dimensionless measure of damping in a system. The damping ratio compares the actual damping to the critical damping needed to prevent oscillations.
- Explain the classification of systems based on the damping ratio:
  - Underdamped ($\zeta < 1$): The system oscillates with gradually decreasing amplitude.
  - Critically damped ($\zeta = 1$): The system returns to equilibrium as quickly as possible without oscillating.
  - Overdamped ($\zeta > 1$): The system returns to equilibrium without oscillating but more slowly than in the critically damped case.
- Provide examples:
  - underdamped: a car's suspension system when hitting a bump
  - overdamped: a door closer that closes a door slowly without slamming
  - critically damped: high-performance shock absorbers in racing cars

Numerical Example

A mass-spring-damper system has a damping ratio of $\zeta = 0.5$ Is this system underdamped, overdamped, or critically damped?

Solution

Since $\zeta = 0.5$ is less than 1, the system is underdamped, meaning it will oscillate with decreasing amplitude.

ChatGPT Integration

- Ask ChatGPT to explain the significance of the damping ratio and how it classifies systems.
- Use ChatGPT to explore additional examples of underdamped, over-damped, and critically damped systems in engineering.

PROMPT 209

Discuss the types of external forces that can cause vibrations in mechanical systems.

Objective

Understand the various types of external forces that can induce vibrations in mechanical systems.

Activity

- Identify and describe different types of external forces that can cause vibrations:
  - periodic forces: repeating forces over time, such as those caused by rotating machinery or unbalanced loads
  - impulse forces: sudden forces, such as a hammer striking a surface or an impact in a collision
  - random forces: unpredictable forces, such as wind gusts or seismic activity
- Provide examples:
  - periodic forces: the constant rotation of a fan causing vibrations in the support structure
  - impulse forces: a pile driver impacting the ground
  - random forces: vibrations caused by an earthquake

Numerical Example

A rotating machine exerts a periodic force on its foundation with an amplitude of 100 N at a frequency of 10 Hz. How many cycles does the machine complete in one minute?

Solution

The machine operates at 10 cycles per second (Hz), so in one minute (60 seconds), it completes the following:

$$10 \text{ Hz} \times 60 \text{ seconds} = 600 \text{ cycles}$$

The machine completes 600 cycles in one minute.

ChatGPT Integration

- Ask ChatGPT to provide additional examples of external forces that cause vibrations in various mechanical systems.
- Use ChatGPT to explore how engineers design systems to withstand or mitigate vibrations caused by these forces.

PROMPT 210

Describe the importance of initial conditions (displacement and velocity) in the analysis of vibrating systems.

Objective

Understand how initial conditions, such as displacement and velocity, affect the behavior of a vibrating system.

Activity

- Define initial displacement as the position of the system when vibration begins, and initial velocity as the rate of change of displacement at the start of vibration.
- Explain that these initial conditions determine the system's response, such as amplitude and phase of the vibration. In free vibration, the system's motion depends on both initial displacement and initial velocity.
- Provide examples:
  - A pendulum displaced from its equilibrium position and released with no initial velocity.
  - A spring-mass system with initial velocity but no initial displacement.

Numerical Example

A spring-mass system is released from an initial displacement of 0.1 meters with an initial velocity of 0 m/s. The system has a natural frequency of 2 Hz. What is the system's displacement as a function of time?

Solution

The equation for displacement in free vibration is

$$x(t) = X_0 \cos(2\pi f_n t) = X_0 \cos(\omega_n t)$$

where $X_0 = 0.1$ m, $f_n = 2$ Hz, and t is time.

By substituting values, we obtain

$$x(t) = 0.1 \cos(4\pi t)$$

The displacement varies sinusoidally with time, oscillating between ± 0.1 m.

ChatGPT Integration

- Ask ChatGPT to explain the role of initial conditions in vibrating systems and how they affect the system's long-term response.
- Use ChatGPT to explore additional scenarios where initial displacement and velocity play critical roles, such as in earthquake simulations or machinery startup behavior.

CHAPTER 22

FREE VIBRATION OF SINGLE-DEGREE-OF-FREEDOM (SDOF) SYSTEMS

Prompt 211 Define the concept of free vibration in the context of single-degree-of-freedom systems.

Prompt 212 Derive the equation of motion for a free vibrating single-degree-of-freedom system and explain its physical meaning.

Prompt 213 Explain the role of inertia and restoring force in free vibrations of an SDOF system.

Prompt 214 Discuss the significance of natural frequency in the context of free vibration of SDOF systems.

Prompt 215 Explain how the initial conditions (initial displacement and velocity) affect the response of a free vibrating SDOF system.

Prompt 216 Derive the solution for the undamped free vibration of a mass-spring system and explain the significance of each term.

Prompt 217 Solve a problem involving the natural frequency and period of a spring-mass system.

Prompt 218 Discuss the energy balance in a free vibrating SDOF system and explain how potential and kinetic energy exchange during vibration.

Prompt 219 Explain how damping can be neglected in free vibration and under what conditions this assumption is valid.

Prompt 220 Solve a numerical example involving the free vibration of an SDOF system with given mass and stiffness.

PROMPT 211

Define the concept of free vibration in the context of single-degree-of-freedom systems.

Objective

Understand the concept of free vibration and its application to single-degree-of-freedom systems.

Activity

- Define free vibration as the motion of a system after it has been displaced from its equilibrium position and then released, with no external forces acting on it after the release.
- Explain that for single-degree-of-freedom (SDOF) systems, free vibration refers to the system oscillating freely with one independent mode of motion.
- Provide examples:
  - a mass-spring system oscillating up and down after being stretched and released
  - a pendulum swinging back and forth after being displaced and released

Numerical Example

A mass-spring system with a mass of 3 kg and a spring stiffness of 150 N/m is displaced and released, causing free vibration. Calculate the natural frequency of the system.

Solution

The natural frequency is given by

$$f_n = \frac{1}{2\pi}\sqrt{\frac{k}{m}} = \frac{1}{2\pi}\sqrt{\frac{150}{3}} = 1.125 \text{ Hz}$$

The natural frequency of the system is approximately 1.125 Hz.

ChatGPT Integration

- Ask ChatGPT to define free vibration and compare the concept across different examples of SDOF systems.

▪ Use ChatGPT to explore additional real-world applications of free vibration, such as in mechanical systems, structures, and vehicle suspensions.

PROMPT 212

Derive the equation of motion for a free vibrating single-degree-of-freedom system and explain its physical meaning.

Objective

Derive and understand the equation of motion for a free vibrating SDOF system and interpret its physical meaning.

Activity

▪ Derive the equation of motion for an undamped SDOF system, starting with Newton's second law. For a mass-spring system, the equation is

$$m\ddot{x}(t) + kx(t) = 0$$

where m is the mass, k is the stiffness, and $x(t)$ is the displacement as a function of time.

▪ Explain that this is a second-order differential equation that describes the motion of the system.

▪ Discuss how the solution to this equation is a sinusoidal function of time, representing oscillations with constant amplitude and frequency.

Numerical Example

A mass of 5 kg is attached to a spring with stiffness 200 N/m and undergoes free vibration.

Write the equation of motion for the system and calculate the natural frequency.

Solution

The equation of motion is

$$5\ddot{x}(t) + 200x(t) = 0$$

The natural frequency is given by

$$f_n = \frac{1}{2\pi}\sqrt{\frac{k}{m}} = \frac{1}{2\pi}\sqrt{\frac{200}{5}} = 1 \text{ Hz}$$

The natural frequency of the system is 1 Hz. The general solution is $x(t) = C_1 \cos(6.32t) + C_2 \sin(6.32t)$.

ChatGPT Integration

- Ask ChatGPT to describe the steps of deriving the equation of motion for an SDOF system and verify your understanding.
- Use ChatGPT to explore more complex examples where additional forces or damping might affect the system.

PROMPT 213

Explain the role of inertia and restoring force in free vibrations of an SDOF system.

Objective

Understand the relationship between inertia and restoring force in the dynamics of a free vibrating SDOF system.

Activity

- Define *inertia* as the tendency of the mass to resist changes in its motion. In a vibrating system, inertia is represented by the mass m, and it causes the system to continue moving once displaced.
- Define the restoring force as the force that opposes the displacement and tries to return the system to equilibrium. In a spring-mass system, this force is proportional to the displacement and governed by Hooke's Law: $F = -kx$.
- Discuss how the balance between inertia and the restoring force leads to oscillatory motion, with inertia driving the system away from equilibrium and the restoring force pulling it back.

Numerical Example

A mass of 2 kg is attached to a spring with stiffness 100 N/m. The mass is displaced 0.1 meters from its equilibrium position and released. Calculate the maximum restoring force acting on the mass.

Solution

The restoring force is given by Hooke's Law:

$$F = -kx = -(100)(0.1) = -10 \text{ N}$$

The maximum restoring force is 10 N in the direction opposite to the displacement.

ChatGPT Integration

- Ask ChatGPT to explain how inertia and restoring forces interact during free vibration and compare different examples.
- Use ChatGPT to explore systems with varying mass and stiffness values to see how these factors influence the role of inertia and restoring force in free vibrations.

PROMPT 214

Discuss the significance of natural frequency in the context of free vibration of SDOF systems.

Objective

Understand the importance of natural frequency in determining the behavior of free vibrating SDOF systems.

Activity

- Define *natural frequency* as the frequency at which a system vibrates when disturbed from equilibrium and allowed to oscillate freely.
- Explain that the natural frequency is determined by the system's mass and stiffness, and that it is a critical parameter in predicting the system's response.
- Discuss the role of natural frequency in engineering applications, such as designing structures to avoid resonance (when the natural frequency matches an external force frequency).

Numerical Example

A mass of 4 kg is attached to a spring with stiffness 250 N/m. Calculate the natural frequency of the system.

Solution

The natural frequency is given by

$$f_n = \frac{1}{2\pi}\sqrt{\frac{k}{m}} = \frac{1}{2\pi}\sqrt{\frac{250}{4}} = 1.26 \text{ Hz}$$

The natural frequency of the system is approximately 1.26 Hz.

ChatGPT Integration

- Ask ChatGPT to define natural frequency and discuss why it is important in the design of vibrating systems.
- Use ChatGPT to explore real-world cases where avoiding resonance through proper natural frequency calculation was crucial, such as in bridge or machinery design.

PROMPT 215

Explain how the initial conditions (initial displacement and velocity) affect the response of a free vibrating SDOF system.

Objective

Understand the influence of initial displacement and velocity on the motion of a free vibrating SDOF system.

Activity

- Define *initial displacement* as the position of the system when vibration begins and *initial velocity* as the speed of the system at that moment.
- Explain that these initial conditions determine the amplitude and phase of the resulting motion.
- Discuss how different combinations of initial displacement and velocity create different types of oscillations, such as purely harmonic motion or damped oscillation, depending on the system's properties.

Numerical Example

A mass-spring system is given an initial displacement of 0.2 meters and an initial velocity of 0 m/s. The system has a natural frequency of 3 Hz. Write the equation of motion for this system.

Solution

The equation of motion for an undamped free vibration system is

$$x(t) = X_0 \cos(2\pi f_n t)$$

where $X_0 = 0.2$ m and $f_n = 3$ Hz.
By substituting values, we obtain

$$x(t) = 0.2\cos(6\pi t)$$

The displacement oscillates with a maximum amplitude of 0.2 meters.

ChatGPT Integration

- Ask ChatGPT to explain how initial conditions affect the long-term behavior of vibrating systems and verify your equation of motion.
- Use ChatGPT to explore more complex cases, such as those with non-zero initial velocities, and how they modify the oscillatory response.

PROMPT 216

Derive the solution for the undamped free vibration of a mass-spring system and explain the significance of each term.

Objective

Derive the mathematical solution for undamped free vibration and understand the meaning of each component.

Activity

- Start with the equation of motion for undamped free vibration:
$$m\ddot{x}(t) + kx(t) = 0$$
- Derive the solution: $x(t) = X_0 cos(\omega_n t) + \dfrac{\dot{X}_0}{\omega_n} sin(\omega_n t)$

- Where
  - X_0 is the initial displacement.
  - $\dot{X}_0$ is the initial velocity.
  - $\omega_n = \sqrt{\dfrac{k}{m}}$ is the natural angular frequency.
- Explain that the system oscillates with constant amplitude and frequency, and that the sine and cosine terms represent the effect of initial displacement and velocity on the motion.

Numerical Example

A mass-spring system with $m = 3$ kg and $k = 300$ N/m has an initial displacement of 0.1 m and zero initial velocity. Write the solution for the displacement as a function of time.

Solution

The natural angular frequency is

$$\omega_n = \sqrt{\frac{k}{m}} = \sqrt{\frac{300}{3}} = 10 \text{ rad/s}$$

The displacement is

$$x(t) = 0.1 \, cos(10t)$$

Since the initial velocity is zero, the sine term drops out.

ChatGPT Integration

- Ask ChatGPT to derive the general solution for undamped free vibration and confirm your understanding of each term in the equation.
- Use ChatGPT to explore different scenarios, such as non-zero initial velocities, and how they affect the displacement equation.

PROMPT 217

Solve a problem involving the natural frequency and period of a spring-mass system.

Objective

Understand how to calculate both the natural frequency and the period of oscillation for a spring-mass system.

Activity

- Define the natural frequency f_n as the number of oscillations per second, and the period T as the time it takes for one complete oscillation.
- The natural frequency is given by $f_n = \dfrac{1}{2\pi}\sqrt{\dfrac{k}{m}}$
- The period is the reciprocal of the natural frequency: $T = \dfrac{1}{f_n}$
- Provide an example of a spring-mass system and solve for both the natural frequency and the period.

Numerical Example

A 4 kg mass is attached to a spring with stiffness k = 100 N/m. Calculate the natural frequency and the period of oscillation.

Solution

The natural frequency is

$$f_n = \frac{1}{2\pi}\sqrt{\frac{100}{4}} = 0.796 \text{ Hz}$$

The period is

$$T = 1/0.796 = 1.26 \text{ seconds } T$$

So, the system oscillates with a frequency of 0.796 Hz and a period of 1.26 seconds.

ChatGPT Integration

- Ask ChatGPT to solve for the natural frequency and period in similar spring-mass systems and verify your calculations.
- Use ChatGPT to explore the effect of changing mass or stiffness on both natural frequency and period.

PROMPT 218

Discuss the energy balance in a free vibrating SDOF system and explain how potential and kinetic energy exchange during vibration.

Objective

Understand the energy balance in a free vibrating SDOF system and how energy transitions between potential and kinetic forms.

Activity

- Explain that in a free vibrating system, energy oscillates between potential energy (stored in the spring) and kinetic energy (associated with the motion of the mass).

- The total mechanical energy of the system remains constant, assuming no damping or energy loss. At the maximum displacement, potential energy is at its peak, while kinetic energy is zero. At the equilibrium point, the kinetic energy is at its peak, and potential energy is zero.

- The potential energy is given by $PE = \dfrac{1}{2}kx^2$

- The kinetic energy is given by $PE = \dfrac{1}{2}mv^2$

Numerical Example

A 2 kg mass attached to a spring with stiffness 150 N/m is displaced 0.1 meters and released. Calculate the maximum potential energy in the system.

Solution

The maximum potential energy is

$$PE = \frac{1}{2}kx^2 = \frac{1}{2}(150)(0.1)^2 = 0.75 \text{ J}$$

The maximum potential energy stored in the system is 0.75 joules.

ChatGPT Integration

▪ Ask ChatGPT to explain how potential and kinetic energy exchange during vibration and verify your understanding of energy conservation in free vibrating systems.

▪ Use ChatGPT to explore the energy behavior in systems with varying masses and stiffness.

PROMPT 219

Explain how damping can be neglected in free vibration and under what conditions this assumption is valid.

Objective

Understand the conditions under which damping can be neglected in the analysis of free vibration.

Activity

▪ Define *damping* as the process by which vibrational energy is dissipated, usually through friction or resistance. In some cases, the damping is small enough that its effects can be ignored in the analysis.

▪ Explain that damping can be neglected when the damping ratio is very small (i.e., when the system is lightly damped), meaning that the reduction in amplitude over time is minimal.

▪ Discuss real-world scenarios where ignoring damping is appropriate, such as in systems that are designed to have low friction or where the vibration duration is short enough that damping has little effect.

Numerical Example

A spring-mass system with a mass of 3 kg and stiffness of 120 N/m has a damping coefficient of 0.05 Ns/m. Calculate the damping ratio and determine if damping can be neglected.

Solution

The damping ratio is given by

$$\zeta = \frac{c}{2\sqrt{k \cdot m}} = 0.05 / \left(2\sqrt{120 \times 3}\right) = 0.0013$$

Since ζ is very small, the system is lightly damped, and damping can be neglected for most practical purposes.

ChatGPT Integration

- Ask ChatGPT to explain situations where damping can be neglected and why this assumption is commonly made in free vibration analysis.
- Use ChatGPT to explore systems with varying levels of damping and how this affects the decision to ignore or include damping in the analysis.

PROMPT 220

Solve a numerical example involving the free vibration of an SDOF system with given mass and stiffness.

Objective

Apply the principles of free vibration to solve a numerical example for an SDOF system.

Activity

- Provide a mass-spring system with given mass and stiffness and solve for the system's natural frequency, period, and displacement over time.

Numerical Example

A 5 kg mass is attached to a spring with stiffness $k = 200 \, \text{N}/\text{m}$ and is displaced by 0.15 meters from its equilibrium position.
Do the following:

1. Calculate the natural frequency of the system.
2. Determine the period of oscillation.
3. Write the equation of motion for the displacement.

Solution

The natural frequency is

$$f_n = \frac{1}{2\pi}\sqrt{\frac{k}{m}} = \frac{1}{2\pi}\sqrt{\frac{200}{5}} = 1.01 \text{ Hz}$$

The period is

$$T = \frac{1}{f_n} = 1/1.01 = 0.99 \text{ seconds}$$

The equation of motion, assuming zero initial velocity, is

$$x(t) = 0.15 \, cos(2\pi \times 1.01t)$$

ChatGPT Integration

- Ask ChatGPT to walk through similar numerical problems to verify your understanding of free vibration calculations.
- Use ChatGPT to explore more advanced examples, such as cases with non-zero initial velocities or systems with damping included.

CHAPTER 23

FORCED VIBRATION OF SINGLE-DEGREE-OF-FREEDOM SYSTEMS

Prompt 221 Define forced vibration and explain how it differs from free vibration in an SDOF system.

Prompt 222 Derive the equation of motion for a forced vibration system subjected to a harmonic external force.

Prompt 223 Explain the concept of resonance in forced vibration and describe its potential effects on mechanical systems.

Prompt 224 Discuss the role of damping in forced vibration and how it influences the system's response to external forces.

Prompt 225 Solve a problem involving the steady-state response of an SDOF system subjected to a sinusoidal external force.

Prompt 226 Explain the concept of phase difference in forced vibration and how it depends on frequency and damping.

Prompt 227 Discuss the frequency response of an SDOF system and how the amplitude of vibration changes with varying excitation frequency.

Prompt 228 Solve a problem involving the amplitude of forced vibration at a frequency close to resonance.

Prompt 229 Explain the concept of transmissibility in forced vibration and its application in vibration isolation systems.

Prompt 230 Discuss real-world examples of forced vibration in mechanical systems and how engineers mitigate harmful resonance effects.

PROMPT 221

Define forced vibration and explain how it differs from free vibration in an SDOF system.

Objective

Understand the concept of forced vibration and distinguish it from free vibration in an SDOF system.

Activity

- Define *forced vibration* as the motion of a system caused by an external force that continuously acts on it, as opposed to free vibration, which occurs when the system is disturbed and allowed to oscillate freely without external force after the initial disturbance.

- Explain that in forced vibration, the system responds to an external periodic force, often leading to different vibrational characteristics compared to free vibration, such as phase lag and amplitude changes depending on the frequency of the external force.

- Provide examples:
  - forced vibration: a washing machine vibrating due to the rotating imbalance in the drum
  - free vibration: a mass-spring system vibrating after being displaced and released

Numerical Example

A 3 kg mass is attached to a spring with stiffness $k = 150$ N/m. The system is subjected to a harmonic external force with an amplitude of 10 N at a frequency of 2 Hz. Describe the difference in response if the external force is removed (free vibration) versus the system being continuously driven by the external force (forced vibration).

Solution

In free vibration, the system will oscillate at its natural frequency, determined by $f_n = \dfrac{1}{2\pi}\sqrt{\dfrac{k}{m}} = 1.12$ Hz.

In forced vibration, the system oscillates at the frequency of the external force (2 Hz in this case) and may experience a phase difference depending on the forcing frequency and the system's natural frequency.

ChatGPT Integration

- Ask ChatGPT to explain the difference between forced and free vibration and explore more real-world examples of systems undergoing forced vibrations. (Hint: For solving a complex problem like this, you may want to use OpenAI o1 (released in Sep. 2024, by openai.com/01, subscription required.))
- Use ChatGPT to verify the conditions under which forced vibration differs significantly from free vibration in various systems.

PROMPT 222

Derive the equation of motion for a forced vibration system subjected to a harmonic external force.

Objective

Derive and understand the equation of motion for a forced vibration system with a harmonic external force.

Activity

- Start with Newton's second law for a damped mass-spring system under the influence of a harmonic external force $F(t) = F_0 cos(\omega t)$, where $x(t)$ is the displacement, F_0 is the amplitude of the external force, and ω is the forcing frequency:

$$m\ddot{x}(t) + c\dot{x}(t) + kx(t) = F_0 cos(\omega t)$$

where m is the mass, c is the damping coefficient, and k is the stiffness of the spring.

- Explain that this is a second-order non-homogeneous differential equation. The solution consists of two parts:
  - the transient solution, which decays over time due to damping
  - the steady-state solution, which is the long-term response of the system and has the same frequency as the external force

Numerical Example

A 5 kg mass is attached to a spring with stiffness k = 300 N/m, and a damping coefficient c = 20 Ns/m. The system is subjected to a harmonic external force with an amplitude of $F_0 = 50\ N$ and a frequency $\omega = 5$ rad/s. Write the equation of motion for this system.

Solution

The equation of motion is

$$5\ddot{x}(t) + 20\dot{x}(t) + 300x(t) = 50\cos(5t)$$

The solution describes the forced vibration of the system under the given conditions.

$$x(t) = e^{-2t}\left[A\cos\left(2\sqrt{14}t\right) + B\sin\left(2\sqrt{14}t\right)\right] + \frac{14}{65}\cos(5t) + \frac{56}{455}\sin(5t)$$

- A and B are constants determined by initial conditions.
- $2\sqrt{14} \approx 7.483$. So, the homogeneous part oscillates at approximately 7.483 rad/s with an exponential decay due to e^{-2t}.
- The particular solution oscillates at the driving frequency of 5 rad/s.

ChatGPT Integration

- Ask ChatGPT to verify the derivation of the equation of motion and walk through the steps of solving it for a harmonic external force.
- Use ChatGPT to explore more complex cases with varying damping levels and forcing frequencies and examine their effects on the system's response.

PROMPT 223

Explain the concept of resonance in forced vibration and describe its potential effects on mechanical systems.

Objective

Understand the phenomenon of resonance in forced vibration and its potential impact on mechanical systems.

Activity

- Define *resonance* as the condition that occurs when the frequency of the external force matches the natural frequency of the system. Under resonance, the amplitude of the vibration becomes significantly larger due to constructive interference between the external force and the system's natural oscillations.

■ Explain that resonance can lead to catastrophic failures in mechanical systems if not properly managed. Examples include bridges collapsing (e.g., Tacoma Narrows Bridge) or machinery breaking down due to excessive vibrations at resonance.

Numerical Example

A mass-spring system with a natural frequency $f_n = 3$ Hz is subjected to an external harmonic force at $f = 3$ Hz. Describe what happens to the amplitude of vibration when the system reaches resonance.

Solution

When the external force frequency matches the natural frequency, the amplitude of vibration increases dramatically. Without proper damping, this can cause the system to oscillate with large amplitudes, potentially leading to failure. For example, the amplitude X of the steady-state response is given by

$$X = \frac{F_0}{\sqrt{\left(k - m\omega^2\right)^2 + \left(c\omega\right)^2}}$$

At resonance ($\omega = \omega_n = 3$, for example), where the natural angular frequency $\omega_n = \sqrt{\frac{k}{m}}$, the term $\left(k - m\omega^2\right)$ becomes zero, and the amplitude reaches a maximum value $X_{max} = \frac{F_0}{c\omega_n}$.

ChatGPT Integration

■ Ask ChatGPT to explain the conditions under which resonance occurs and explore real-world examples where resonance was beneficial or harmful.

■ Use ChatGPT to investigate methods of preventing or controlling resonance in mechanical systems, such as adding damping or tuning the system's natural frequency.

PROMPT 224

Discuss the role of damping in forced vibration and how it influences the system's response to external forces.

Objective

Understand the effect of damping on the response of a forced vibrating system.

Activity

- Define *damping* as the mechanism by which vibrational energy is dissipated, usually due to friction, resistance, or material deformation.
- Explain that in forced vibration, damping reduces the amplitude of the system's response, especially near resonance. In systems with low damping, the amplitude can increase dramatically at resonance, while in highly damped systems, the amplitude remains lower and more controlled.
- Discuss how damping affects both the transient response (which decays faster with higher damping) and the steady-state response (which is less sensitive to resonance when damping is present).

Numerical Example

A 4 kg mass is attached to a spring with stiffness k = 200 N/m, and a damping coefficient of c = 10 Ns/m. The system is subjected to a harmonic force with an amplitude of F_0 = 20 N and frequency f = 2 Hz. Compare the system's amplitude of response at resonance with and without damping.

Solution

With damping, the amplitude at resonance is significantly reduced compared to the undamped case. The system experiences less dramatic oscillations due to the energy dissipated by damping forces. The solution reads, assuming zero initial conditions:

$$x(t) = e^{-1.25t} \left[0.04268 \; cos(6.959t) - 0.01485 \; sin(6.959t) \right]$$
$$-0.04268 \; cos(4\pi t) + 0.01243 \; sin(4\pi t)$$

ChatGPT Integration

- Ask ChatGPT to explain the influence of damping on forced vibration and verify how damping alters both the transient and steady-state responses.
- Use ChatGPT to explore scenarios with different damping ratios and how they affect the overall behavior of the system under forced vibration.

PROMPT 225

Solve a problem involving the steady-state response of an SDOF system subjected to a sinusoidal external force.

Objective

Understand how to calculate the steady-state response of an SDOF system subjected to a harmonic external force.

Activity

- The displacement amplitude in the steady-state is given by

$$X = \frac{F_0/m}{\sqrt{\left(\omega_n^2 - \omega^2\right)2 + \left(2\zeta\omega_n\omega\right)^2}}$$

where
- F_0 is the amplitude of the external force
- ω is the forcing frequency
- ω_n is the natural frequency
- ζ is the damping ratio $\zeta = \dfrac{c}{2\sqrt{k \cdot m}}$

Numerical Example

A 6 kg mass is attached to a spring with stiffness $k = 300$ N/m and damping coefficient $c = 15$ Ns/m. The system is subjected to a harmonic force with an amplitude of $F_0 = 50$ and frequency $f = 4$ Hz.

Solution

1. Calculate the natural angular frequency ω_n:

$$\omega_n = \sqrt{k/m} = \sqrt{300/6} = 7.07\,rad/s$$

2. The forcing angular frequency ω is $\omega = 2\pi f = 2\pi \times 4 = 25.13$

3. Calculate the damping ratio ζ: $\zeta = \dfrac{c}{2\sqrt{k \cdot m}} = \dfrac{15}{2\sqrt{300 \times 6}} = 0.182$

4. Now substitute the values into the displacement amplitude formula:

$$X = \frac{50/6}{\sqrt{\left(7.07^2 - 25.13^2\right)^2 + \left(2 \times 0.182 \times 7.07 \times 25.13\right)^2}} \approx 0.0143 \text{ m}$$

The steady-state displacement amplitude is approximately 0.0143 meters.

ChatGPT Integration

▪ Ask ChatGPT to verify the solution for the steady-state response and explore how changing parameters like damping or frequency would affect the result.

PROMPT 226

Explain the concept of phase difference in forced vibration and how it depends on frequency and damping.

Objective

Understand the concept of phase difference between the external force and the system response in forced vibration, and how it varies with frequency and damping.

Activity

▪ Define the *phase difference* (ϕ) as the angle by which the response of the system lags or leads the external forcing function.

▪ Explain that in forced vibration, the system's displacement response can be out of phase with the external force. This phase difference depends on the

- forcing frequency (ω)
- natural frequency (ω_n)
- damping ratio (ζ)

▪ The phase angle ϕ is given by

$$tan\phi = \frac{2\zeta\omega_n\omega}{\omega_n^2 - \omega^2}$$

▪ Discuss how

- at frequencies much lower than the natural frequency $(\omega \ll \omega_n)$, the response is in phase with the force $(\phi \approx 0°)$
- at resonance $(\omega = \omega_n)$, the phase difference is $90°$.
- at frequencies much higher than the natural frequency $(\omega \gg \omega_n)$, the response lags the force by $180°$ $(\phi \approx 180°)$

Numerical Example

A mass-spring-damper system has a mass $m = 2$ kg, stiffness $k = 800$ N/m, and damping coefficient $c = 20$ Ns/m. It is subjected to a harmonic force with a frequency $f = 5$ Hz. Calculate the phase difference between the external force and the system response.

Solution

1. Calculate the natural angular frequency ω_n:

$$\omega_n = \sqrt{\frac{k}{m}} = \sqrt{\frac{800}{2}} = 20 \, rad/s$$

2. Calculate the forcing angular frequency ω.

$$\omega = 2\pi f = 2\pi \times 5 = 10\pi \, rad/s \approx 31.42 \, rad/s$$

3. Calculate the damping ratio ζ:

$$\zeta = \frac{c}{2\sqrt{km}} = \frac{20}{2\sqrt{800 \times 2}} = 0.25$$

4. Calculate $\tan \phi$:

$$tan\phi = \frac{2\zeta\omega_n\omega}{\omega_n^2 - \omega^2} = \frac{2 \times 0.25 \times 20 \times 31.42}{20^2 - (31.42)^2} = -\frac{314.2}{587.22} = -0.535$$

5. Calculate the phase angle ϕ:

$$\phi = \tan^{-1}(-0.535) = -28.15°$$

Since the phase angle is negative, the system response lags the external force by approximately $28°$.

ChatGPT Integration

- Ask ChatGPT to explain how phase difference varies with the frequency and damping and verify your calculations.
- Use ChatGPT to explore how changing the damping ratio or forcing frequency affects the phase difference, possibly by plotting ϕ versus ω.

PROMPT 227

Discuss the frequency response of an SDOF system and how the amplitude of vibration changes with varying excitation frequency.

Objective

Understand how the amplitude of vibration in a forced SDOF system changes with the excitation frequency and how to interpret the system's frequency response.

Activity

- Define the *frequency response* as the system's steady-state amplitude response to an external force as a function of the excitation frequency.
- Explain that the amplitude varies significantly with excitation frequency, especially near the system's natural frequency.
- Key points:
  - At low frequencies $(\omega \ll \omega_n)$, the system moves almost in phase with the external force, with a small amplitude.
  - At resonance $(\omega \approx \omega_n)$, the amplitude increases dramatically, especially if the system is lightly damped.

- At high frequencies $(\omega \gg \omega_n)$, the amplitude decreases as the system's response lags the force.

Numerical Example

Consider a system with $m = 5$ kg, $k = 400$ N/m, and $c = 20$ Ns/m. Calculate the amplitude response at different forcing frequencies: $f = 1$ Hz, 5 Hz, and 10 Hz.

Solution

1. First, calculate the natural frequency:

$$\omega_n = \sqrt{\frac{k}{m}} = \sqrt{\frac{400}{5}} = \sqrt{80} = 8.94 \ \text{rad/s} \ .$$

2. The forcing frequencies are:

$$\omega_1 = 2\pi \times 1 = 6.28 \ \text{rad/s},$$

$$\omega_2 = 2\pi \times 5 = 31.42 \ \text{rad/s},$$

$$\omega_3 = 2\pi \times 10 = 62.83 \ \text{rad/s}$$

3. For each frequency, calculate the amplitude response using the formula:

$$X = \frac{F_0/m}{\sqrt{\left(\omega_n^2 - \omega^2\right)^2 + \left(2\zeta\omega_n\omega\right)^2}}$$

(Use specific values of F_0 and ζ as needed to compute the amplitude at each frequency).

ChatGPT Integration

- Ask ChatGPT to explain how to interpret frequency response plots and how damping affects the amplitude at different frequencies.
- Use ChatGPT to explore the effect of changing stiffness or damping on the system's frequency response.

PROMPT 228

Solve a problem involving the amplitude of forced vibration at a frequency close to resonance.

Objective

Understand how to calculate the amplitude of a forced vibration system when the excitation frequency is near the system's natural frequency, accounting for resonance effects.

Activity

- Provide a mass-spring-damper system with specified parameters.
- Instruct the reader to calculate the amplitude of vibration at an excitation frequency slightly below, at, and above the natural frequency to observe the effects of resonance.

Numerical Example

A system has a mass $m = 3$ kg, stiffness $k = 200$ N/m, and damping coefficient $c = 10$ N.s/m. It is subjected to a harmonic force with an amplitude of $F_0 = 30$ N.

Tasks:

1. Calculate the natural frequency ω_n.
2. Determine the amplitude of vibration at excitation frequencies

$$\omega = \omega_n - 0.5 \text{ rad}/s$$

$$\omega = \omega_n$$

$$\omega = \omega_n + 0.5 \text{ rad}/s$$

Solution

1. The natural frequency is

$$\omega_n = \sqrt{\frac{k}{m}} = \sqrt{\frac{200}{3}} = 8.16 \text{ rad}/s$$

2. The amplitude at different frequencies is

At $\omega = \omega_n - 0.5 \text{ rad}/s$, we get

$$X = \frac{30/3}{\sqrt{\left(8.16^2 - 7.66^2\right)^2 + \left(2 \times 0.204 \times 8.16 \times 7.66\right)^2}} = 0.374 \text{ m}$$

The amplitude at $\omega = \omega_n$.

At resonance, when the excitation frequency ω equals the natural frequency ω_n, the amplitude X is given by

$$X = \frac{F_0}{m\sqrt{\left(\omega_n^2 - \omega^2\right)^2 + \left(2\zeta\omega_n\omega\right)^2}}$$

For $\omega = \omega_n$, the equation simplifies due to the resonance:

$$X = \frac{F_0}{2\zeta m\omega_n}$$

where

- F_0 = 30 N
- m = 3 kg
- $\zeta = \dfrac{c}{2\sqrt{km}} = \dfrac{10}{2\sqrt{200 \cdot 3}} = 0.204$
- ω_n = 8.16 rad/s

By substituting values, we obtain

$$X = \frac{30}{2 \times 0.204 \times 3 \times 8.16} = 3 \text{ m}$$

So, the amplitude at $\omega = \omega_n$ is $X = 3$ m.

For $\omega = \omega_n + 0.5$, $\omega = 8.16 + 0.5 = 8.66$ rad/s

The amplitude X is calculated using the following:

$$X = \frac{F_0}{m\sqrt{\left(\omega_n^2 - \omega^2\right)^2 + \left(2\zeta\omega_n\omega\right)^2}}$$

By substituting the values, we obtain

$$X = \frac{30}{3\sqrt{\left(8.16^2 - 8.66^2\right)^2 + \left(2 \times 0.204 \times 8.16 \times 8.66\right)^2}}$$

By calculating the terms, we obtain

1. $\omega_n^2 = 66.58$
2. $\omega^2 = 8.66^2 = 75.03$

3. $\left(\omega_n^2 - \omega^2\right)^2 = \left(66.58 - 75.03\right)^2 = \left(-8.45\right)^2 = 71.4$

$$\left(2\zeta\omega_n\omega\right)^2 = \left(2 \cdot 0.204 \cdot 8.16 \cdot 8.66\right)^2 = 831.26$$

Thus,

$$X = \frac{30}{3\sqrt{71.4 + 831.26}} = 0.333 \text{ m}$$

So, the amplitude at $\omega = \omega_n + 0.5$ is approximately $X = 0.333$ m.

The amplitude of vibration increases dramatically near resonance (≈ 3.00 m), while it remains significantly lower at nearby frequencies (≈ 0.374 m and ≈ 0.333 m). This highlights the sensitivity of forced response to the excitation frequency in lightly damped systems.

PROMPT 229

Explain the concept of transmissibility in forced vibration and its application in vibration isolation systems.

Objective

Understand the concept of transmissibility in forced vibration and how it applies to vibration isolation.

Activity

- Define transmissibility in forced vibration as the ratio of the amplitude of the output (or transmitted) vibration to the amplitude of the input (or driving) vibration. It is given by the formula

$$T = \frac{X}{X_0} = \frac{1}{\sqrt{\left(1 - \frac{\omega^2}{\omega_n^2}\right)^2 + \left(2\xi\frac{\omega}{\omega_n}\right)^2}}$$

where
- ω is the forcing frequency.
- ω_n is the natural frequency.
- $\zeta = \dfrac{c}{2\sqrt{km}}$ is the damping ratio.

- Explain that transmissibility helps determine how much vibration is transmitted through a system, especially for frequencies above and below the natural frequency.
- Discuss its application in *vibration isolation*, where systems are designed to reduce transmitted vibrations (e.g., in machinery mounts or building foundations).

Numerical Example

A vibration isolation system has a mass m = 50 kg, stiffness k = 10,000 N/m, and damping coefficient c = 100 Ns/m. The system is subjected to a forcing frequency f = 12 Hz.

Tasks:

1. Calculate the natural frequency ω_n.
2. Determine the transmissibility at the forcing frequency.

Solution

1. natural frequency: $\omega_n = \sqrt{\dfrac{k}{m}} = \sqrt{\dfrac{10000}{50}} = 14.14$ rad/s

2. forcing angular frequency: $\omega = 2\pi f = 2\pi \times 12 = 75.4$ rad/s

3. damping ratio: $\xi = \dfrac{c}{2\sqrt{km}} = \dfrac{100}{2\sqrt{1000 \times 50}} = 0.071$

4. transmissibility: $T = \dfrac{1}{\sqrt{\left(1 - \dfrac{75.4^2}{14.14^2}\right)^2 + \left(2 \times 0.071 \times \dfrac{75.4}{14.14}\right)^2}} = 0.036$

The transmissibility at the forcing frequency is approximately 0.036, indicating that only 3.6% of the vibration is transmitted to the structure.

ChatGPT Integration

- Ask ChatGPT to verify your understanding of transmissibility and its application in vibration isolation systems.

■ Use ChatGPT to explore how changing the damping ratio or forcing frequency affects the transmissibility of the system.

PROMPT 230

Discuss real-world examples of forced vibration in mechanical systems and how engineers mitigate harmful resonance effects.

Objective

Understand how forced vibration affects mechanical systems in real-world applications and explore methods engineers use to mitigate harmful resonance.

Activity

■ Discuss several real-world examples where forced vibration is significant, such as

- Rotating machinery: Engines, turbines, and pumps often experience forced vibrations due to imbalances, leading to potential resonance issues.

- Bridges and buildings: Large structures like bridges and skyscrapers can experience forced vibrations due to wind forces, seismic activity, or traffic. If the external force frequency matches the structure's natural frequency, resonance can lead to failure.

- Vehicles: Cars, trains, and airplanes are subjected to various external forces, leading to forced vibrations in the suspension systems, wheels, and fuselage.

■ Explain how engineers design systems to mitigate resonance, including

- damping: adding dampers to absorb energy and reduce amplitude near resonance

- tuned mass dampers: devices used in structures like skyscrapers to counteract resonance by oscillating out of phase with the structure's vibrations

- isolation mounts: systems designed to prevent the transmission of vibrations to sensitive equipment or structures

Numerical Example

A building is equipped with a tuned mass damper (TMD) to counteract wind-induced forced vibrations. The building has a natural frequency of 1.5 Hz, and the wind exerts a periodic force with a frequency of 1.45 Hz. Explain how the TMD helps reduce the effects of forced vibration and estimate the impact on amplitude reduction.

Solution

- The TMD is designed to oscillate out of phase with the building's vibrations, effectively canceling out the forces that would otherwise cause the building to resonate.

- The small difference between the forcing frequency (1.45 Hz) and the natural frequency (1.5 Hz) means the system is near resonance, so the TMD plays a critical role in keeping the vibrations under control. While the specific amplitude reduction depends on the system's design, it is common for TMDs to reduce vibration amplitudes by up to 50% or more.

ChatGPT Integration

- Ask ChatGPT to explore real-world cases where engineers successfully mitigated resonance issues, such as in bridges or tall buildings.

- Use ChatGPT to investigate other engineering solutions that minimize forced vibrations and ensure structural safety.

CHAPTER 24

DAMPING IN VIBRATING SYSTEMS

Prompt 231 Define damping and explain its significance in vibrating systems.

Prompt 232 Discuss the different types of damping. viscous, Coulomb (dry friction), and structural damping.

Prompt 233 Derive the equation of motion for an SDOF system with viscous damping.

Prompt 234 Explain the concept of the damping ratio and how it affects the system's response to vibration.

Prompt 235 Solve a problem involving the calculation of the damping ratio for a given system.

Prompt 236 Discuss the concept of critical damping and its importance in avoiding oscillations.

Prompt 237 Explain the difference between underdamped, critically damped, and overdamped systems with examples.

Prompt 238 Solve a numerical problem involving the transient response of an underdamped system.

Prompt 239 Discuss the role of damping in real-world applications such as shock absorbers, building structures, and machinery mounts.

Prompt 240 Explore advanced damping techniques used in engineering systems, such as tuned mass dampers and viscoelastic materials.

PROMPT 231

Define damping and explain its significance in vibrating systems.

Objective

Understand the concept of damping and its role in controlling vibrations in mechanical systems.

Activity

- Define *damping* as the mechanism by which vibrational energy is dissipated, usually due to friction, resistance, or material deformation. Damping prevents a system from vibrating indefinitely by converting mechanical energy into other forms, such as heat.
- Explain the significance of damping in reducing the amplitude of vibrations over time. In real-world systems, damping is essential for avoiding excessive vibrations that can lead to system damage or failure.
- Provide examples of systems where damping is critical:
  - Car suspension systems: Damping helps smooth out the ride by reducing oscillations after the car goes over a bump.
  - Building structures: Damping helps prevent tall buildings from excessive vibrations due to wind or seismic activity.

Numerical Example

A spring-mass system has a damping coefficient $c = 15$ Ns/m and a mass $m = 5$ kg. The system is displaced and released from rest. Calculate the initial energy dissipation rate due to damping.

Solution

The initial energy dissipation rate (power dissipated) is given by

$$P = c \times v^2$$

where v is the velocity of the mass when it is released. If the initial velocity is 2m/s, the power dissipated is

$$P = 15 \times (2)^2 = 15 \times 4 = 60 \text{ Watts}$$

Thus, the system dissipates 60 W of energy initially.

ChatGPT Integration

- Ask ChatGPT to explain why damping is important in various mechanical systems.
- Use ChatGPT to explore additional real-world examples where damping plays a crucial role in controlling vibrations.

PROMPT 232

Discuss the different types of damping: viscous, Coulomb (dry friction), and structural damping.

Objective

Understand the different types of damping mechanisms and their applications in mechanical systems.

Activity

- Define and explain the following types of damping:
  - Viscous Damping: Damping that is proportional to velocity, where the damping force is given by $F_d = -c\dot{x}$. It occurs in systems where the motion is resisted by a fluid, such as shock absorbers in vehicles or hydraulic dampers.
  - Coulomb (Dry Friction) Damping: Damping those results from the constant friction force between surfaces in contact. The damping force is independent of velocity and is given by $F_d = \pm\mu N$, where μ is the coefficient of friction and N is the normal force.
  - Structural Damping: Damping that occurs due to internal friction within a material as it deforms. It is often present in solid materials and is significant in structures subjected to vibrations.
- Discuss real-world applications for each type of damping:
  - Viscous damping: used in automotive suspensions and earthquake-resistant building designs
  - Coulomb damping: found in mechanical joints and sliding components
  - Structural damping: relevant in materials such as metals and polymers, where internal friction limits excessive vibrations

Numerical Example

A mechanical system is subjected to Coulomb damping with a friction force and a normal force of $N = 100\,\text{N}$. The system slides with a coefficient of friction $\mu = 0.1$. Calculate the damping force acting on the system.

Solution

The damping force is given by

$$F_d = \pm\mu N = 0.1 \times 100 = 10\ \text{N}$$

Thus, the damping force is 10 N.

ChatGPT Integration

- Ask ChatGPT to explain the differences between viscous, Coulomb, and structural damping, and explore additional examples where each type is used.
- Use ChatGPT to explore how engineers select damping types based on the application and material properties.

PROMPT 233

Derive the equation of motion for an SDOF system with viscous damping.

Objective

Understand how to derive the equation of motion for a single-degree-of-freedom (SDOF) system with viscous damping.

Activity

- Start with Newton's second law, applied to a mass-spring-damper system. The forces acting on the system are
  - Restoring force from the spring: $F_s = -kx$
  - Damping force: $F_d = -c\dot{x}$
  - External force (if any): $F_{ext}(t)$
- The equation of motion for the system is $m\ddot{x}(t) + c\dot{x}(t) + kx(t) = F_{ext}(t)$
- If no external force is applied (free vibration), the equation simplifies to $m\ddot{x}(t) + c\dot{x}(t) + kx(t) = 0$

- Explain the significance of each term in the equation:
  - $m\ddot{x}(t)$: inertial force (mass times acceleration)
  - $c\dot{x}(t)$: damping force proportional to velocity
  - $kx(t)$: restoring force proportional to displacement

Numerical Example

A system has a mass $m = 2$ kg, damping coefficient $c = 5$ Ns/m, and stiffness $k = 200$ N/m. Write the equation of motion for this system in free vibration.

Solution

The equation of motion is

$$2\ddot{x}(t) + 5\dot{x}(t) + 200x(t) = 0$$

This represents the damped free vibration of the system.

ChatGPT Integration

- Ask ChatGPT to verify your derivation of the equation of motion for a damped system and explore different cases, such as forced vibration.
- Use ChatGPT to analyze how changing the damping coefficient or stiffness affects the system's response over time.

PROMPT 234

Explain the concept of the damping ratio and how it affects the system's response to vibration.

Objective

Understand the damping ratio and its influence on the behavior of vibrating systems.

Activity

- Define the *damping ratio* ζ as a dimensionless parameter that describes how a system's damping compares to critical damping. It is given by the formula: $\zeta = \dfrac{c}{2\sqrt{km}}$

where
- c is the damping coefficient
- k is the stiffness
- m is the mass

- Explain how the damping ratio affects the system's behavior:
  - Underdamped systems $\zeta < 1$: The system oscillates with gradually decreasing amplitude.
  - Critically damped systems $\zeta = 1$: The system returns to equilibrium as quickly as possible without oscillating.
  - Overdamped systems $\zeta > 1$: The system returns to equilibrium without oscillating, but more slowly than in the critically damped case.

Numerical Example

A mass-spring-damper system has a mass $m = 3$ kg, damping coefficient $c = 8$ Ns/m, and stiffness $k = 100$ N/m. Calculate the damping ratio ζ.

Solution

$$\zeta = \frac{c}{2\sqrt{km}} = \frac{8}{2\sqrt{100 \times 3}} = \frac{8}{2 \times 17.32} = \frac{8}{34.64} \approx 0.231$$

The damping ratio $\zeta \approx 0.231$, which indicates that the system is underdamped.

ChatGPT Integration

- Ask ChatGPT to explore the effects of varying damping ratios on system response, such as underdamped, critically damped, and overdamped behavior.
- Use ChatGPT to verify the damping ratio calculation and explore additional examples with different parameters.

PROMPT 235

Solve a problem involving the calculation of the damping ratio for a given system.

Objective

Learn how to calculate the damping ratio for a mechanical system with known mass, damping coefficient, and stiffness.

Activity

- Provide the formula for calculating the damping ratio, $\zeta = \dfrac{c}{2\sqrt{km}}$ where
  - c is the damping coefficient
  - k is the stiffness
  - m is the mass
- Emphasize that the damping ratio helps determine whether the system is underdamped, critically damped, or overdamped, and explain its relevance in controlling system response to vibrations.

Numerical Example

A mechanical system consists of a mass $m = 4$ kg, stiffness $k = 300$ N/m, and damping coefficient $c = 12$ Ns/m. Calculate the damping ratio ζ for this system and classify the system as underdamped, critically damped, or overdamped.

Solution

Calculate the damping ratio:

$$\zeta = \frac{c}{2\sqrt{km}} = \frac{12}{2\sqrt{300 \times 4}} = \frac{12}{2 \times \sqrt{1200}} = \frac{12}{2 \times 34.64} = \frac{12}{69.28} \approx 0.173$$

Since $\zeta = 0.173$, the system is underdamped because $\zeta < 1$.

ChatGPT Integration

- Ask ChatGPT to solve additional examples involving the calculation of damping ratios for different mechanical systems.
- Use ChatGPT to verify the calculation and explore how varying parameters like damping coefficient or mass affect the damping ratio.

PROMPT 236

Discuss the concept of critical damping and its importance in avoiding oscillations.

Objective

Understand the concept of critical damping and its role in preventing oscillations in mechanical systems.

Activity

- Define *critical damping* as the specific amount of damping that causes a system to return to equilibrium as quickly as possible without oscillating. For a critically damped system, the damping ratio $\zeta = 1$.
- Explain the significance of critical damping:
  - In systems with underdamping ($\zeta < 1$), the system oscillates with gradually decreasing amplitude.
  - In systems with critical damping ($\zeta = 1$), the system returns to equilibrium without oscillation.
  - In systems with overdamping ($\zeta > 1$), the system returns to equilibrium without oscillation, but more slowly than in the critically damped case.
- Discuss applications where critical damping is crucial, such as in car suspension systems, doors with hydraulic closers, and seismic protection devices.

Numerical Example

A system has a mass $m = 10$ and stiffness $k = 500$ N/m. Calculate the damping coefficient c required for critical damping.

Solution

The damping coefficient for critical damping is given by

$$c_{\text{crit}} = 2\sqrt{km}$$

By substituting values, we obtain

$$c_{\text{crit}} = 2\sqrt{500 \times 10} = 2\sqrt{5000} = 2 \times 70.71 = 141.42\,\text{Ns/m}$$

The damping coefficient required for critical damping is 141.42 Ns/m.

ChatGPT Integration

- Ask ChatGPT to explain why critical damping is important in various engineering applications.
- Use ChatGPT to explore additional examples where critical damping is applied, such as in shock absorbers or doors with hydraulic closers.

PROMPT 237

Explain the difference between underdamped, critically damped, and overdamped systems with examples.

Objective

Understand and compare the behaviors of underdamped, critically damped, and overdamped systems.

Activity

- Define the three types of damping:
  - Underdamped systems $\zeta < 1$: The system oscillates with gradually decreasing amplitude as it returns to equilibrium.
  - Critically damped systems $\zeta = 1$: The system returns to equilibrium as quickly as possible without oscillating.
  - Overdamped systems $\zeta > 1$: The system returns to equilibrium without oscillation but more slowly than a critically damped system.
- Provide examples for each type:
  - Underdamped: A car's suspension system is often underdamped to provide a smooth ride, where some oscillation is acceptable.
  - Critically damped: Hydraulic door closers are designed to return to a closed position quickly without oscillating.
  - Overdamped: Heavy industrial machinery or seismic dampers in buildings may be overdamped to ensure no oscillations occur, even if it takes longer to return to equilibrium.

Numerical Example

Consider a system with mass $m = 6$ kg, stiffness $k = 240$ N/m, and three different damping coefficients:

- $c_1 = 10$ Ns/m (underdamped)
- $c_2 = 76$ Ns/m (critically damped
- $c_3 = 80$ Ns/m (overdamped)

Classify each system as underdamped, critically damped, or overdamped.

Solution

Calculate the critical damping coefficient:

$$c_{\text{crit}} = 2\sqrt{km} = 2\sqrt{240 \times 6} = 2 \times \sqrt{1440} = 75.9\,\text{Ns/m}$$

Compare the damping coefficients:

- $c_1 = 10$ Ns/m $< c_{\text{crit}} \rightarrow$ Underdamped
- $c_2 = 76$ Ns/m $\approx c_{\text{crit}} \rightarrow$ Critically damped
- $c_3 = 80$ Ns/m $> c_{\text{crit}} \rightarrow$ Overdamped

ChatGPT Integration

- Ask ChatGPT to explain the differences between underdamped, critically damped, and overdamped systems with real-world examples.
- Use ChatGPT to explore the effects of changing the damping coefficient on system behavior.

PROMPT 238

Solve a numerical problem involving the transient response of an underdamped system.

Objective

Learn how to solve for the transient response of an underdamped system using the system's parameters.

Activity

- Define the transient response for an underdamped system as the part of the motion that decays over time, while the system oscillates before eventually returning to equilibrium.
- The displacement for an underdamped system is given by the equation $x(t) = X_0 e^{-\zeta \omega_n t} \cos(\omega_d t + \phi)$

where
- X_0 is the initial amplitude
- ζ is the damping ratio
- ω_n is the natural angular frequency
- $\omega_d = \omega_n \sqrt{1-\zeta^2}$ is the damped angular frequency
- ϕ is the phase angle

Numerical Example

A mass-spring-damper system has a mass $m = 4$ kg, stiffness $k = 250$ N/m, damping coefficient $c = 10$ Ns/m, and initial displacement $X_0 = 0.1$ m.

Calculate the displacement at time $t = 2$ seconds.

Solution

1. Calculate the natural frequency.

$$\omega_n = \sqrt{\frac{k}{m}} = \sqrt{\frac{250}{4}} = \sqrt{62.5} \approx 7.91 \text{ rad/s}$$

2. Calculate the damping ratio.

$$\zeta = \frac{c}{2\sqrt{km}} = \frac{10}{2 \times \sqrt{250 \times 4}} = \frac{10}{2 \times 31.62} = 0.158$$

3. Calculate the damped frequency.

$$\omega_d = \omega_n \sqrt{1-\zeta^2} = 7.91 \times \sqrt{1-0.158^2} \approx 7.81 \text{ rad/s}$$

4. Calculate the displacement at $t = 2$ seconds.

$$x(2) = 0.1e^{-0.158 \times 7.81 \times 2} \cos(7.81 \times 2 + 0)$$

- $e^{-0.158 \times 7.91 \times 2} \approx e^{-2.5} \approx 0.082$
- $\cos(15.62) \approx -0.996$

$$x(2) = 0.1 \times 0.082 \times (-0.996) \approx -0.0082 \text{ m}$$

ChatGPT Integration

- Ask ChatGPT to help solve similar transient response problems for underdamped systems.
- Use ChatGPT to explore how different damping ratios affect the transient response and decay rates.

PROMPT 239

Discuss the role of damping in real-world applications such as shock absorbers, building structures, and machinery mounts.

Objective

Understand how damping is applied in various real-world systems to control vibrations and protect structures or equipment.

Activity

- Explain how damping is used in important real-world applications:
  - Shock absorbers: In vehicles, damping helps control the motion of the suspension system, ensuring a smooth ride by reducing the oscillations after encountering a bump.
  - Building structures: Damping systems, such as tuned mass dampers or base isolators, help absorb energy from wind forces or seismic activity, protecting the building from excessive vibrations and potential damage.
  - Machinery mounts: In industrial settings, machines are mounted on damping systems to reduce vibrations that could damage both the machine and the surrounding equipment or structures.
- Discuss how the appropriate amount and type of damping are selected for each application based on factors such as the mass of the system, the type of forces it encounters, and the desired response.

Numerical Example

A shock absorber in a car suspension system has a damping coefficient of $c = 120$ Ns/m and the mass of the car over the suspension is $m = 250$ kg. The stiffness of the suspension is $k = 15,000$ N/m. Calculate the damping ratio ζ for the suspension system and explain if the system is underdamped, critically damped, or overdamped.

Solution

Calculate the damping ratio:

$$\zeta = \frac{c}{2\sqrt{km}} = \frac{120}{2\sqrt{15,000 \times 250}} = \frac{120}{2 \times 193.65} = \frac{120}{387.3} \approx 0.31$$

Since ζ = 0.31, the system is underdamped, which is appropriate for a suspension system to provide both a smooth ride and controlled response.

ChatGPT Integration

- Ask ChatGPT to explore how damping is applied in other real-world applications, such as in aerospace systems or sports equipment.
- Use ChatGPT to investigate how changes in mass, damping coefficient, or stiffness affect the damping ratio in different applications.

PROMPT 240

Explore advanced damping techniques used in engineering systems, such as tuned mass dampers and viscoelastic materials.

Objective

Understand advanced damping techniques and their applications in controlling vibrations in engineering systems.

Activity

- Discuss two key advanced damping techniques:
  - Tuned Mass Dampers (TMDs): A secondary mass is attached to the main structure to absorb vibrational energy. TMDs are commonly used in skyscrapers and bridges to counteract wind-induced or seismic vibrations by oscillating out of phase with the structure.
  - Viscoelastic Materials: These materials exhibit both viscous and elastic behavior, dissipating energy through internal molecular movement. Viscoelastic damping is often used in aerospace and automotive applications to absorb vibrations while maintaining structural integrity.

- Explain the principles behind each technique and provide real-world examples of where they are applied:
  - TMDs are used in buildings like the Taipei 101 tower and the Millennium Bridge in London.
  - Viscoelastic materials are used in aircraft panels, automotive parts, and sporting equipment to reduce vibrations.

Numerical Example

A building with a mass $m = 20 \times 10^3 \, kg$ is equipped with a tuned mass damper with a mass $m_{tmd} = 500$ kg. The building experiences vibrations at a frequency of 0.8 Hz. Calculate the natural frequency of the tuned mass damper to minimize vibrations in the building.

Solution

Calculate the natural frequency of the TMD. The natural frequency of the TMD should match the forcing frequency of the building: $\omega_{tmd} = 2\pi f = 2\pi \times 0.8 \approx 5.03 \, rad/s$. The TMD is tuned to 0.8 Hz to effectively counteract the building's vibrations.

ChatGPT Integration

- Ask ChatGPT to explain how advanced damping techniques like TMDs and viscoelastic materials are applied in various industries, such as aerospace, civil engineering, and automotive.
- Use ChatGPT to explore more case studies involving advanced damping techniques and how they have successfully mitigated vibrations in complex systems.

CHAPTER 25

VIBRATION OF MULTI-DEGREE-OF-FREEDOM (MDOF) SYSTEMS

Prompt 241 Define a multi-degree-of-freedom (MDOF) system and provide examples in mechanical systems.

Prompt 242 Explain how the equations of motion are derived for an MDOF system using Newton's second law.

Prompt 243 Discuss the concept of mode shapes in MDOF systems and their significance in vibration analysis.

Prompt 244 Explain the procedure for determining natural frequencies and mode shapes of an MDOF system.

Prompt 245 Solve a problem involving the calculation of natural frequencies for a 2-DOF system.

Prompt 246 Discuss the role of matrix methods in solving MDOF vibration problems.

Prompt 247 Explain the significance of coupling in MDOF systems and how it affects system behavior.

Prompt 248 Solve a problem involving the transient response of a 2-DOF system.

Prompt 249 Discuss the use of numerical methods, such as finite element analysis (FEA), for analyzing MDOF systems.

Prompt 250 Explore real-world applications of MDOF vibration analysis in engineering, such as in bridges, vehicles, and buildings.

PROMPT 241

Define a multi-degree-of-freedom (MDOF) system and provide examples in mechanical systems.

Objective

Understand the concept of multi-degree-of-freedom (MDOF) systems and recognize their importance in complex mechanical systems.

Activity

- Define an *MDOF system* as a mechanical system that can move or vibrate in more than one independent direction or coordinate. Each degree of freedom represents a distinct mode of motion or displacement.
- Provide examples of MDOF systems:
  - A two-mass-spring system: Each mass can move independently, leading to two degrees of freedom.
  - Building structures: High-rise buildings can experience multiple degrees of freedom due to sway in different directions and torsion.
  - Vehicles: Car suspensions are modeled as MDOF systems, with vertical, lateral, and angular displacements for each wheel.

Numerical Example

Consider a simple system with two masses connected by springs. Each mass can move in one direction, making this a 2-DOF system. If mass $m_1 = 5$ kg, mass $m_2 = 3$ kg, and the spring constants are $k_1 = 200$ N/m and $k2 = 150$ N/m, describe how the system would exhibit two degrees of freedom.

Solution

The two masses can move independently along the axis, with each mass experiencing a restoring force from the connected springs. These independent movements define the two degrees of freedom for the system.

ChatGPT Integration

- Ask ChatGPT to explore additional examples of MDOF systems in engineering.

- Use ChatGPT to investigate how MDOF systems are analyzed in more complex mechanical structures, such as aircraft or mechanical linkages.

PROMPT 242

Explain how the equations of motion are derived for an MDOF system using Newton's second law.

Objective

Understand the process of deriving the equations of motion for an MDOF system based on Newton's second law of motion.

Activity

- Start by applying Newton's second law to each mass in an MDOF system. For each mass, the sum of forces acting on the mass equals the product of the mass and its acceleration: $m_i\ddot{x}_i = F_{\text{spring},i} + F_{\text{damping},i} + F_{\text{external},i}$, where m_i is the mass, $\ddot{x}_i$ is the acceleration, and the right-hand side includes spring forces, damping forces, and any external forces.
- For a two-degree-of-freedom system, the system of equations would look like $m_2\ddot{x}_2 = -k_2(x_2 - x_1)$.
- Explain how these equations describe the motion of each mass in relation to the forces acting on them.

Numerical Example

Consider a 2-DOF system with masses $m_1 = 4$ kg and $m_2 = 2$ kg, and spring constants $k_1 = 100$ N/m and $k_2 = 50$ N/m. Derive the equations of motion for this system.

Solution

1. For m_1: $4\ddot{x}_1 = -100x_1 + 50(x_2 - x_1)$
2. For m_2: $2\ddot{x}_2 = -50(x_2 - x_1)$

These two equations represent the coupled motion of the two masses.

ChatGPT Integration

- Ask ChatGPT to verify the derived equations of motion and explore other examples of MDOF systems.

- Use ChatGPT to investigate how additional forces, such as damping or external driving forces, affect the equations of motion for MDOF systems.

PROMPT 243

Discuss the concept of mode shapes in MDOF systems and their significance in vibration analysis.

Objective

Understand the concept of mode shapes in multi-degree-of-freedom (MDOF) systems and their importance in vibration analysis.

Activity

- Define *mode shapes* as the specific patterns of displacement that a system undergoes when it vibrates at one of its natural frequencies. In an MDOF system, each natural frequency corresponds to a unique mode shape.

- Explain that in each mode shape, the relative motion of each degree of freedom is determined by the system's properties (mass, stiffness, and damping).

- Discuss the significance of mode shapes:
  - Natural frequencies and mode shapes are critical in predicting the dynamic response of a system.
  - Mode shapes help engineers understand how different parts of the system will move relative to each other when excited by external forces.
  - In some vibration modes, certain degrees of freedom may have zero displacement. These stationary points or regions are called nodal points (in 1D) or nodal lines/surfaces (in higher dimensions).

Numerical Example

Consider a 2-DOF system with masses $m_1 = 3$ kg, $m_2 = 2$ kg, and spring constants $k_1 = 100$ N/m and $k_2 = 50$ N/m. Describe the mode shapes for this system at its two natural frequencies.

Solution

For this system, the mode shapes indicate how m_1 and m_2 move relative to each other. At the first natural frequency, both masses may move in the same direction (in-phase motion). At the second natural frequency, the masses could move in opposite directions (out-of-phase motion).

ChatGPT Integration

- Ask ChatGPT to explore the calculation of mode shapes in more complex MDOF systems.
- Use ChatGPT to investigate how mode shapes are used in real-world applications, such as the design of buildings, bridges, and vehicles.

PROMPT 244

Explain the procedure for determining natural frequencies and mode shapes of an MDOF system.

Objective

Learn the steps involved in determining the natural frequencies and mode shapes for an MDOF system.

Activity

- Outline the procedure for calculating natural frequencies and mode shapes in MDOF systems:

  1. Set up the equations of motion. Begin by deriving the equations of motion for the MDOF system, using Newton's second law.
  2. Write in matrix form. Express the equations in matrix form:
     $$[M]\{\ddot{x}\} + [K]\{x\} = 0$$
     where
     - $[M]$ is the mass matrix
     - $[K]$ is the stiffness matrix
     - $\{x\}$ is the displacement vector
  3. Assume harmonic motion. Assume a harmonic solution of the form $\{x(t)\} = \{X\} e^{i\omega t}$, where ω is the natural frequency.

4. Solve the eigenvalue problem. Substitute the harmonic motion assumption into the equation of motion to get the eigenvalue problem: $([K] - \omega^2[M])\{X\} = 0$

5. Solving this equation gives the natural frequencies ω (eigenvalues) and mode shapes {X} (eigenvectors).

Numerical Example

Consider a 2-DOF system with the mass matrix $[M] = \begin{bmatrix} 3 & 0 \\ 0 & 2 \end{bmatrix}$ and stiffness matrix $[K] = \begin{bmatrix} 100 & -50 \\ -50 & 50 \end{bmatrix}$.

Calculate the natural frequencies of this system.

Solution

Solve the eigenvalue problem, by letting the determinant to be zero: $\left| [K] - \omega^2[M] \right| = 0$.

After solving, the natural frequencies are found to be $\omega_1 \approx 2.89$ rad/s and $\omega_2 \approx 7.07$ rad/s.

ChatGPT Integration

- Ask ChatGPT to verify the eigenvalue problem solution and explore more complex MDOF systems.

- Use ChatGPT to investigate how natural frequencies and mode shapes are used in engineering design to prevent resonance and system failure.

PROMPT 245

Solve a problem involving the calculation of natural frequencies for a 2-DOF system.

Objective

Learn how to calculate the natural frequencies for a simple two-degree-of-freedom (2-DOF) system.

Activity

Outline the steps for calculating the natural frequencies of a 2-DOF system:

1. Write the equations of motion for the system based on Newton's second law.
2. Formulate the mass and stiffness matrices for the system.
3. Solve the eigenvalue problem $([K]-\omega^2[M])\{X\}=0$ to find the natural frequencies.

Numerical Example

A 2-DOF system has the following mass and stiffness matrices:

$$[M]=\begin{bmatrix} 2 & 0 \\ 0 & 1 \end{bmatrix} \quad [K]=\begin{bmatrix} 100 & -50 \\ -50 & 50 \end{bmatrix}$$

Calculate the natural frequencies of the system.

Solution

Set up the eigenvalue equation $\left(\begin{bmatrix} 100 & -50 \\ -50 & 50 \end{bmatrix}-\omega^2\begin{bmatrix} 2 & 0 \\ 0 & 1 \end{bmatrix}\right)\{X\}=0$.

This leads to the characteristic equation, by letting determinant to be zero: $\begin{vmatrix} 100 & -50 \\ -50 & 50 \end{vmatrix}-\omega^2\begin{bmatrix} 2 & 0 \\ 0 & 1 \end{bmatrix} = \begin{vmatrix} 100-2\omega^2 & -50 \\ -50 & 50-\omega^2 \end{vmatrix}=0.$

Solving the characteristic equation gives the natural frequencies: $\omega_1 = 3.83\,rad/s$, $\omega_2 = 9.24\,rad/s$.

ChatGPT Integration

- Ask ChatGPT to verify the solution and explore the calculation of natural frequencies for more complex systems.
- Use ChatGPT to analyze how changing mass or stiffness values affects the natural frequencies of a 2-DOF system.

PROMPT 246

Discuss the role of matrix methods in solving MDOF vibration problems.

Objective

Understand the importance of matrix methods in analyzing and solving vibration problems in MDOF systems.

Activity

- Explain that matrix methods are essential for solving MDOF vibration problems because they simplify the process of managing multiple degrees of freedom, which would be too complex to solve manually using scalar equations.
- Discuss how matrix methods allow for the compact representation of mass, stiffness, and damping properties, and how they facilitate solving the equations of motion efficiently using linear algebra techniques.
- Provide examples where matrix methods are used:
  - The mass matrix [M] describes how mass is distributed across the system's degrees of freedom.
  - The stiffness matrix [K] represents how stiffness is distributed across the system.
  - The damping matrix [C] represents the damping forces in the system.
- Matrix methods are especially useful for systems with multiple coupled degrees of freedom, where simultaneous equations need to be solved for natural frequencies, mode shapes, and system response.

Numerical Example

For a 2-DOF system with the following matrices, we find

$$[M] = \begin{bmatrix} 3 & 0 \\ 0 & 2 \end{bmatrix} \quad [K] = \begin{bmatrix} 200 & -100 \\ -100 & 100 \end{bmatrix}$$

Explain how the mass and stiffness matrices are used to set up the eigenvalue problem for finding natural frequencies.

Solution

The natural frequencies are found by solving the eigenvalue problem:

$$\det\left([K] - \omega^2 [M]\right) = 0$$

which involves subtracting the scaled mass matrix from the stiffness matrix and solving for the eigenvalues, which correspond to the natural frequencies.

ChatGPT Integration

- Ask ChatGPT to verify the setup and solution of the eigenvalue problem for the MDOF system.
- Use ChatGPT to explore how matrix methods are applied in real-world systems, such as structural engineering or aerospace design.

PROMPT 247

Explain the significance of coupling in MDOF systems and how it affects system behavior.

Objective

Understand the concept of coupling in MDOF systems and its influence on the dynamic behavior of the system.

Activity

- Define *coupling* in an MDOF system as the interaction between degrees of freedom where the motion of one part of the system affects the motion of another. This is represented mathematically by off-diagonal terms in the mass, stiffness, or damping matrices.
- Explain that in a coupled system, the displacement or velocity of one degree of freedom influences the motion of others, making the system behave as an integrated whole rather than as independent parts.
- Discuss how coupling leads to complex motion patterns and affects natural frequencies and mode shapes. In contrast, uncoupled systems can be solved independently for each degree of freedom.

- Provide examples of coupling in real-world systems:
  - Building structures: Coupling occurs when different parts of a building sway or rotate in response to forces such as wind or earthquakes.
  - Mechanical linkages: In vehicles or machinery, the movement of one component often influences the movement of connected components, especially in multi-link suspensions.

Numerical Example

Consider a 2-DOF system with mass matrix $[M] = \begin{bmatrix} 3 & 0 \\ 0 & 2 \end{bmatrix}$ and stiffness matrix $[K] = \begin{bmatrix} 100 & -50 \\ -50 & 100 \end{bmatrix}$.

Explain how coupling is represented in the stiffness matrix and how it affects the system's natural frequencies.

Solution

The off-diagonal terms (-50) in the stiffness matrix indicate coupling between the two degrees of freedom. This means that the displacement of one mass affects the forces on the other mass. Solving the eigenvalue problem will reveal that the natural frequencies are influenced by this coupling.

ChatGPT Integration

- Ask ChatGPT to further explain the impact of coupling on MDOF systems and explore uncoupled versus coupled systems.
- Use ChatGPT to investigate real-world applications where coupling plays a critical role, such as in structural design or vehicle dynamics.

PROMPT 248

Solve a problem involving the transient response of a 2-DOF system.

Objective

Learn how to solve for the transient response of a 2-degree-of-freedom (2-DOF) system subjected to an initial displacement or force.

Activity

- Explain the concept of *transient response* in an MDOF system as the part of the system's motion that decays over time, eventually leaving only the steady-state response.
- The equations of motion for a 2-DOF system can be written as follows: $[M]\{\ddot{x}\} + [C]\{\dot{x}\} + [K]\{x\} = 0$, where $[M]$, $[C]$, and $[K]$ are the mass, damping, and stiffness matrices, and $\{x\}$ is the displacement vector.
- Explain that the transient response can be solved by assuming an initial displacement or velocity for each degree of freedom, then using numerical methods or an eigenvalue approach to find the system's response over time.

Numerical Example

Consider a 2-DOF system with the following mass and stiffness matrices:

$$[M] = \begin{bmatrix} 2 & 0 \\ 0 & 1 \end{bmatrix} \quad [K] = \begin{bmatrix} 100 & -50 \\ -50 & 50 \end{bmatrix}$$

The system is initially displaced by $x_1(0) = 0.1$ m, and $x_2(0) = 0.05$ m, with zero initial velocities, the transient response can be computed using modal analysis (eigenvalue decomposition) or numerical integration methods for a chosen time t.

Solution

1. Set up the system of equations using the initial conditions.
2. Apply numerical methods (e.g., Runge-Kutta or eigenvalue methods) to find the displacement of both masses over time.
3. The transient response decays as the system's energy is dissipated, and the response at time ttt depends on the system's natural frequencies and mode shapes.

ChatGPT Integration

- Ask ChatGPT to assist with the numerical solution of transient response problems for MDOF systems.
- Use ChatGPT to explore how the transient response of an MDOF system changes with different initial conditions or damping values.

PROMPT 249

Discuss the use of numerical methods, such as finite element analysis (FEA), for analyzing MDOF systems.

Objective

Understand how numerical methods, particularly finite element analysis (FEA), are used to analyze complex MDOF systems in engineering applications.

Activity

- Explain how finite element analysis (FEA) works by breaking down complex systems into smaller, simpler elements (finite elements) and then solving for the system's dynamic behavior using matrix methods.
- Discuss how FEA is particularly useful for solving MDOF vibration problems, where complex geometries, boundary conditions, and material properties make analytical solutions difficult.
- Mention how FEA involves discretizing the structure into elements, forming the global mass, stiffness, and damping matrices, and solving the equations of motion for natural frequencies, mode shapes, and transient or steady-state responses.

Numerical Example

Consider a structural beam modeled as a continuous system. The beam is discretized into 10 finite elements, each represented by its own stiffness and mass properties. The natural frequencies and mode shapes of the beam are calculated using FEA. Explain the steps involved in setting up the FEA model for this beam and calculating its natural frequencies.

Solution

1. Discretize the beam into finite elements and define the mass and stiffness matrices for each element.

2. Assemble the global mass $[M]$ and stiffness $[K]$ matrices for the entire beam.

3. Solve the eigenvalue problem: $([K] - \omega^2 [M])\{\phi\} = \{0\}$ to determine the natural frequencies ω and mode shapes $\{\phi\}$ of the beam.

ChatGPT Integration

- Ask ChatGPT to provide an overview of FEA and how it applies to solving MDOF vibration problems.
- Use ChatGPT to explore case studies where FEA was used to analyze complex structures, such as bridges or vehicle components.

PROMPT 250

Explore real-world applications of MDOF vibration analysis in engineering, such as in bridges, vehicles, and buildings.

Objective

Understand how MDOF vibration analysis is applied in various engineering fields to solve complex problems related to dynamic behavior.

Activity

- Discuss the application of MDOF vibration analysis in important engineering areas:
  - Bridges: Bridges, especially long-span suspension bridges, are analyzed as MDOF systems to account for vibrations caused by traffic, wind, or seismic activity. Engineers use MDOF analysis to ensure stability and avoid resonance.
  - Vehicles: In automotive engineering, vehicles are modeled as MDOF systems to analyze the suspension system's response to road irregularities. This helps in designing suspensions that provide comfort and safety.
  - Buildings: Skyscrapers and other large structures are modeled as MDOF systems to assess their behavior under wind forces and earthquakes. MDOF analysis helps in designing damping systems, such as tuned mass dampers (TMDs), to control vibrations.
- Explain that MDOF systems are used in engineering to predict how structures will behave dynamically and to ensure that natural frequencies do not coincide with the frequencies of external forces (avoiding resonance).

Numerical Example

A vehicle is modeled as a 4-DOF system, with each wheel having independent suspension and damping. The natural frequencies of the system are found to be 2.5 Hz, 3.2 Hz, 4.1 Hz, and 5.0 Hz. Explain how MDOF analysis helps in improving vehicle ride quality by tuning the suspension system.

Solution

MDOF analysis provides insight into how different parts of the vehicle—such as wheels, suspension, and chassis—interact dynamically. By identifying the natural frequencies and associated mode shapes, engineers can tune the suspension stiffness and damping properties to:

– Avoid resonance with road-induced excitation frequencies,
– Improve passenger comfort by reducing vibration transmission to the cabin, and
– Enhance stability and handling by controlling the dynamic response of each wheel.

This analysis enables targeted suspension tuning for better ride quality across a range of driving conditions.

ChatGPT Integration

- Ask ChatGPT to explore additional real-world applications where MDOF vibration analysis is crucial, such as in aerospace, mechanical, and civil engineering projects.

- Use ChatGPT to investigate case studies where MDOF analysis has led to improved designs and safer structures.

CHAPTER 26

VIBRATION ISOLATION AND ABSORBERS

Prompt 251 Define vibration isolation and explain its significance in mechanical systems.

Prompt 252 Discuss the concept of transmissibility and how it is used to measure the effectiveness of vibration isolation.

Prompt 253 Explain how vibration isolators are designed to reduce transmitted forces in mechanical systems.

Prompt 254 Solve a problem involving the calculation of transmissibility for a given vibration isolation system.

Prompt 255 Explain the role of damping in vibration isolation and how it affects system performance.

Prompt 256 Discuss the difference between passive and active vibration isolation systems with examples.

Prompt 257 Explain the concept of vibration absorbers and how they are used to reduce vibrations in mechanical systems.

Prompt 258 Solve a problem involving the design of a tuned mass damper for vibration absorption in a structure.

Prompt 259 Discuss real-world applications of vibration isolation and absorbers in engineering, such as in buildings, vehicles, and machinery.

Prompt 260 Explore advanced vibration isolation and absorption techniques used in aerospace, automotive, and civil engineering industries.

PROMPT 251

Define vibration isolation and explain its significance in mechanical systems.

Objective

Understand the concept of vibration isolation and why it is important in various mechanical systems.

Activity

- Define *vibration isolation* as the process of reducing the transmission of vibrations from a vibrating source to a surrounding structure or equipment. The primary goal is to protect sensitive systems from unwanted vibrations that can cause damage or reduce performance.

- Explain that vibration isolation is essential in various applications to ensure the longevity of machinery and structures and to enhance user comfort in systems like vehicles and buildings.

- Provide examples:
  - Machinery: Vibration isolation helps prevent vibrations from heavy machinery affecting the structure or nearby equipment.
  - Buildings: Vibration isolators are used in tall buildings to reduce the transmission of ground vibrations caused by earthquakes or traffic.
  - Vehicles: Car suspensions are designed to isolate passengers from road-induced vibrations.

Numerical Example

Consider a machine that produces vibrations at a frequency of 10 Hz, and you want to isolate these vibrations from reaching the supporting structure. A vibration isolator with a natural frequency of 3 Hz is selected. Explain how this isolator helps reduce the transmission of vibrations.

Solution

The effectiveness of a vibration isolator depends on the ratio of the excitation frequency to the system's natural frequency. In this case, the excitation frequency (10 Hz) is significantly higher than the isolator's natural frequency (3 Hz), resulting in a frequency ratio greater than $\sqrt{2}$. Under such conditions, the isolator operates in the isolation region, where transmitted force is reduced. Since the system avoids resonance (which occurs when the excitation frequency is near the natural frequency), and operates in a range where transmissibility drops below 1, the isolator effectively attenuates the vibration energy reaching the support structure.

ChatGPT Integration

- Ask ChatGPT to provide additional examples of where vibration isolation is critical in engineering systems.
- Use ChatGPT to explore how the choice of isolator frequency affects system performance in different applications.

PROMPT 252

Discuss the concept of transmissibility and how it is used to measure the effectiveness of vibration isolation.

Objective

Understand the concept of transmissibility and how it relates to the effectiveness of vibration isolation systems.

Activity

- Define *transmissibility* as the ratio of the transmitted response (such as force, displacement, or acceleration) to the input excitation in a vibration isolation system. It quantifies how much of the input vibration is transmitted through the isolator to the supporting structure.

- The transmissibility T is given by $T = \dfrac{1 + \left(2\zeta\,\dfrac{\omega}{\omega_n}\right)^2}{\sqrt{\left(1 - \left(\dfrac{\omega}{\omega_n}\right)^2\right)^2 + \left(2\zeta\,\dfrac{\omega}{\omega_n}\right)^2}}$

  where
  - ω is the forcing frequency
  - ω_n is the natural frequency of the isolation system
  - ζ is the damping ratio

- Discuss how transmissibility helps engineers assess the effectiveness of a vibration isolation system by determining the degree to which vibrations are reduced or amplified based on the relationship between the forcing frequency and the natural frequency of the isolator.

Numerical Example

A machine produces vibrations at a frequency of f = 15 Hz, and the vibration isolator has a natural frequency of f_n = 5, with a damping ratio of ζ = 0.05. Calculate the transmissibility and explain the isolation effectiveness.

Solution

1. Calculate the angular frequencies.

$$\omega = 2\pi \times 15 = 94.25 \, \text{rad/s}, \omega_n = 2\pi \times 5 = 31.42 \, \text{rad/s}$$

2. Calculate the transmissibility.

$$T = \frac{\sqrt{1 + (2 \times 0.05 \times 3)^2}}{\sqrt{(1 - 3^2)^2 + (2 \times 0.05 \times 3)^2}} = \frac{\sqrt{1 + (0.3)^2}}{\sqrt{(1 - 9^2)^2 + (0.3)^2}}$$

$$T = \frac{\sqrt{1 + 0.09}}{\sqrt{64 + 0.09}} = \frac{\sqrt{1.09}}{\sqrt{64.09}} \approx \frac{1.044}{8.005} \approx 0.1305$$

The transmissibility is approximately 0.13, meaning only 13% of the vibration is transmitted through the isolator. This indicates that the isolator provides very effective vibration isolation in this scenario.

ChatGPT Integration

- Ask ChatGPT to explain how transmissibility changes as the damping ratio or forcing frequency varies.
- Use ChatGPT to explore additional real-world examples where transmissibility is critical for vibration isolation design.

PROMPT 253

Explain how vibration isolators are designed to reduce transmitted forces in mechanical systems.

Objective

Learn the principles behind the design of vibration isolators and how they minimize transmitted forces.

Activity

▪ Explain that vibration isolators are designed to decouple a vibrating source from a structure, reducing the amount of vibrational energy transmitted to the surrounding environment. They achieve this by introducing a flexible element between the source and the structure, which acts as a spring-damper system.

▪ Discuss key design elements:

• Stiffness: The stiffness of the isolator affects its natural frequency. A lower stiffness results in a lower natural frequency, which improves isolation performance at higher vibration frequencies.

• Damping: Damping helps to absorb vibrational energy, preventing excessive oscillations and improving the effectiveness of isolation across a wider range of frequencies.

• Resonance: Isolators are designed to ensure that the system's natural frequency is much lower than the operating frequency, minimizing the chance of resonance, where vibrations would be amplified instead of reduced.

▪ Provide examples of vibration isolators used in various applications:

• Rubber mounts are often used in machinery to isolate vibrations and absorb shocks.

• Spring isolators are commonly used in HVAC systems to isolate large equipment from transmitting vibrations to building structures.

Numerical Example

A machine producing vibrations at 20 Hz is mounted on a spring isolator with a stiffness of $k = 500$ N/m and a mass of $m = 10$ kg. Calculate the natural frequency of the isolator and explain its effectiveness in reducing transmitted forces.

Solution

Calculate the natural frequency:

$$\omega_n = \sqrt{\frac{k}{m}} = \sqrt{\frac{500}{10}} = \sqrt{50} \approx 7.07 \, \text{rad/s}$$

We converting this to Hz: $f_n = \frac{\omega_n}{2\pi} \approx 1.12 \, \text{Hz}$.

The natural frequency of the isolator (1.12 Hz) is much lower than the vibration frequency (20 Hz), Since isolation is effective when the operating frequency is at least $\sqrt{2}$ times higher than the natural frequency, this isolator provides excellent vibration isolation, significantly reducing transmitted forces.

ChatGPT Integration

- Ask ChatGPT to explain how stiffness and damping values influence the design of vibration isolators.
- Use ChatGPT to explore how different industries, such as automotive or aerospace, design vibration isolators to protect equipment from vibrational forces.

PROMPT 254

Solve a problem involving the calculation of transmissibility for a given vibration isolation system.

Objective

Learn how to calculate the transmissibility for a given vibration isolation system and understand how it reflects the effectiveness of vibration isolation.

Activity

- Provide the equation for transmissibility

$$T = \frac{\sqrt{1 + \left(2\zeta \dfrac{\omega}{\omega_n} \right)^2}}{\sqrt{\left(1 - \left(\dfrac{\omega}{\omega_n} \right)^2 \right)^2 + \left(2\zeta \dfrac{\omega}{\omega_n} \right)^2}}$$

where
- ω is the forcing frequency
- ω_n is the natural frequency of the isolator
- ζ is the damping ratio
- Explain that transmissibility values less than 1 indicate effective isolation, while values greater than 1 indicate amplification of vibrations.

Numerical Example

A machine produces vibrations at a frequency of $f = 18$ Hz, and the isolator has a natural frequency of $f_n = 6$ Hz with a damping ratio of $\zeta = 0.1$. Calculate the transmissibility of the system.

Solution

1. Calculate the angular frequencies.

$$\omega = 2\pi \times 18 = 113.1 \, \text{rad/s}, \, \omega_n = 2\pi \times 6 = 37.7 \, \text{rad/s}$$

2. Calculate the transmissibility.

$$T = \frac{\sqrt{1 + \left(2 \times 0.1 \times 3\right)^2}}{\sqrt{\left(1 - 3^2\right)^2 + \left(2 \times 0.1 \times 3\right)^2}} = \frac{\sqrt{1 + \left(0.6\right)^2}}{\sqrt{\left(1 - 9\right)^2 + \left(0.6\right)^2}}$$

$$T = \frac{\sqrt{1 + 0.36}}{64 + 0.36} = \frac{1.36}{64.36} \approx \frac{1.166}{8.021} \approx 0.1454$$

3. The transmissibility is approximately $T \approx 0.145$, which means that only 14.5% of the vibration is transmitted. This indicates effective vibration isolation.

ChatGPT Integration

- Ask ChatGPT to verify the transmissibility calculation and explore how different parameters, such as the damping ratio and natural frequency, affect the results.
- Use ChatGPT to explore real-world applications where transmissibility is crucial for vibration isolation, such as in machinery or structural engineering.

PROMPT 255

Explain the role of damping in vibration isolation and how it affects system performance.

Objective

Understand the influence of damping on the performance of vibration isolation systems and how it affects the transmission of vibrations.

Activity

- Discuss the role of damping in vibration isolation systems:
  - Damping helps absorb and dissipate vibrational energy, reducing the amplitude of oscillations in both the system and the structure it protects.
  - In vibration isolation, damping is critical to controlling the response near the system's resonance frequency, preventing large oscillations.
- Explain that damping improves isolation performance by reducing the system's transmissibility across a wider range of frequencies.
- Provide examples of how too little or too much damping can affect isolation performance:
  - Underdamping may lead to excessive vibrations near the natural frequency.
  - Critical damping provides optimal isolation performance without oscillation.
  - Overdamping can reduce the isolation system's responsiveness and lead to slower recovery times.

Numerical Example

A vibration isolation system has a damping ratio $\zeta = 0.3$, and the system is subjected to a forcing frequency $f = 12$ Hz. The natural frequency of the isolator is $f_n = 5$ Hz. Calculate the transmissibility and explain how damping affects the system's performance.

Solution

1. Calculate the angular frequencies.

$$\omega = 2\pi \times 12 = 75.4 \text{ rad/s}, \omega_n = 2\pi \times 5 = 31.4 \text{ rad/s}$$

2. Calculate the transmissibility.

$$T = \frac{\sqrt{1 + \left(2 \times 0.3 \times \frac{12}{5}\right)^2}}{\sqrt{\left(1 - \left(\frac{12}{5}\right)^2\right)^2 + \left(2 \times 0.3 \times \frac{12}{5}\right)^2}} = \frac{\sqrt{1 + \left(\frac{36}{25}\right)^2}}{\sqrt{\left(1 - \left(\frac{144}{25}\right)^2\right) + \left(\frac{36}{25}\right)^2}}$$

About 35% of the vibration is transmitted, indicating effective vibration isolation with the given damping.

ChatGPT Integration

- Ask ChatGPT to explore how increasing or decreasing the damping ratio affects the performance of vibration isolators.
- Use ChatGPT to investigate real-world systems where damping plays a critical role in preventing excessive vibrations, such as in automotive suspensions or building foundations.

PROMPT 256

Discuss the difference between passive and active vibration isolation systems with examples.

Objective

Understand the difference between passive and active vibration isolation systems and how each is used in engineering applications.

Activity

- Define and explain *passive vibration isolation*:
  - Passive isolation relies on physical components, such as springs and dampers, to absorb and reduce vibrations. These systems do not require external power or feedback mechanisms.
  - Passive isolation is typically simpler, more cost-effective, and easier to implement in most mechanical systems.
- Define and explain *active vibration isolation*:
  - Active isolation uses sensors and actuators to detect and counteract vibrations. Feedback control systems monitor the vibrations and adjust the actuators to cancel out unwanted motion.
  - Active isolation systems are more complex and expensive but provide superior performance in controlling vibrations, especially in high-precision environments.
- Provide examples:
  - passive isolation: rubber mounts in machinery, car suspensions, and building foundations
  - active isolation: precision equipment in laboratories, optical systems, and spacecraft

Numerical Example

A laboratory optical table requires isolation from external vibrations. A passive isolator has a transmissibility of 0.1, and an active isolator achieves a transmissibility of 0.01. Compare the performance of the passive and active isolation systems.

Solution

The passive isolator transmits 10% of the external vibrations, while the active isolator transmits only 1%. This demonstrates the significantly improved isolation performance of the active system, which is crucial in applications where even minor vibrations can affect precision, such as in optical or semiconductor manufacturing.

ChatGPT Integration

- Ask ChatGPT to explain when passive and active isolation systems are best suited and explore examples of their application in different industries.
- Use ChatGPT to explore how active isolation systems use feedback control to cancel out vibrations in high-performance environments.

PROMPT 257

Explain the concept of vibration absorbers and how they are used to reduce vibrations in mechanical systems.

Objective

Understand how vibration absorbers work and their applications in reducing unwanted vibrations in mechanical systems.

Activity

- Define *vibration absorbers* as devices that are specifically tuned to absorb energy at certain frequencies, reducing the amplitude of vibrations in the primary system.
- Explain the principle behind a vibration absorber:
  - A vibration absorber is typically a mass-spring system attached to the primary system. It is designed to have a natural frequency that matches the forcing frequency, allowing it to absorb the vibrational energy and prevent it from affecting the main system.

- Discuss applications of vibration absorbers:
    - Tuned mass dampers (TMDs) are used in skyscrapers and bridges to reduce the effects of wind or seismic vibrations.
    - Dynamic vibration absorbers are found in rotating machinery, automotive engines, and turbine blades to reduce resonant vibrations.

Numerical Example

A dynamic vibration absorber with a mass $m = 5$ kg and stiffness $k = 800$ N/m is attached to a system that vibrates at a frequency of 4 Hz. Calculate the natural frequency of the absorber and determine if it is effectively tuned to reduce vibrations.

Solution

Calculate the natural frequency of the absorber:

$$\omega_n = \sqrt{\frac{k}{m}} = \sqrt{\frac{800}{5}} = \sqrt{160} \approx 12.65 \text{ rad/s}$$

We convert this to Hz: $f_n = \dfrac{\omega_n}{2\pi} \approx 2.01$ Hz .

The absorber's natural frequency is significantly lower than the forcing frequency (4 Hz), meaning it is not optimally tuned for this application. The absorber should be adjusted to match the forcing frequency more closely for effective vibration reduction.

ChatGPT Integration

- Ask ChatGPT to explain how vibration absorbers are used in various industries to prevent resonant vibrations.
- Use ChatGPT to explore how engineers tune vibration absorbers for specific applications, such as in tall buildings or rotating machinery.

PROMPT 258

Solve a problem involving the design of a tuned mass damper for vibration absorption in a structure.

Objective

Learn how to design a tuned mass damper (TMD) to reduce vibrations in a structure.

Activity

- Explain the concept of a tuned mass damper (TMD):
  - A TMD is a mass-spring-damper system attached to a structure that vibrates. The TMD is designed to oscillate out of phase with the structure's vibrations, effectively canceling them out by absorbing vibrational energy.
  - The TMD is tuned to the natural frequency of the structure or the frequency of the external forces acting on it.
- Discuss the design considerations for TMDs:
  - The mass of the damper should be a significant portion of the mass of the structure.
  - The spring and damping values should be selected to tune the TMD to the desired frequency.
  - TMDs are commonly used in large structures like skyscrapers and bridges to reduce wind or earthquake-induced vibrations.

Numerical Example

A building with a mass $M = 10^5 kg$ experiences vibrations at a frequency of 0.5 Hz. A TMD with a mass $m = 5,000$ kg is designed to reduce these vibrations. Calculate the required stiffness of the TMD spring to tune it to the building's vibration frequency.

Solution

1. Calculate the angular frequency of the building's vibrations:

$$\omega = 2\pi \times 0.5 = 3.14 \text{ rad/s}$$

2. Use the TMD tuning equation:

$$\omega = \sqrt{\frac{k}{m}} \quad \Rightarrow \quad k = \omega^2 m$$

3. We substitute the values to obtain

$$k = (3.14)^2 \times 5{,}000 = 9.86 \times 5{,}000 = 49{,}300 \text{ N/m}$$

The required stiffness for the TMD spring is k = 49,300 N/m.

ChatGPT Integration

▪ Ask ChatGPT to explore real-world examples of tuned mass dampers in structures and how they have successfully reduced vibrations.

▪ Use ChatGPT to investigate how varying the mass, stiffness, and damping properties of a TMD affects its performance in vibration control.

PROMPT 259

Discuss real-world applications of vibration isolation and absorbers in engineering, such as in buildings, vehicles, and machinery.

Objective

Understand how vibration isolation and absorbers are applied in various engineering fields to control vibrations and protect systems.

Activity

▪ Discuss the use of vibration isolation and absorbers in key engineering applications:

• Buildings: Vibration isolators and tuned mass dampers (TMDs) are used in skyscrapers to reduce vibrations from wind and seismic forces. Notable examples include the Taipei 101 tower and the Burj Khalifa.

• Vehicles: Automotive suspensions are designed to isolate passengers from road vibrations. Dynamic absorbers are also used in engines and drive shafts to reduce harmful vibrations.

• Machinery: Heavy industrial machinery is often mounted on vibration isolators to prevent vibrations from affecting the surrounding environment or equipment. Dynamic absorbers are used in turbines and pumps to minimize resonant vibrations.

▪ Explain how vibration isolation and absorbers improve system performance by reducing wear and tear, enhancing comfort, and increasing the lifespan of structures and equipment.

Numerical Example

In a vehicle, a dynamic vibration absorber is installed to reduce vibrations in the engine operating at 3,000 RPM. The absorber has a mass of $m = 3$ kg, and the stiffness of the spring is $k = 600$ N/m. Calculate the natural frequency of the absorber and determine if it effectively reduces vibrations at the engine speed.

Solution

1. Convert RPM to Hz:

$$f = \frac{3,000}{60} = 50 \text{ Hz}$$

2. Calculate the natural frequency of the absorber:

$$\omega_n = \sqrt{\frac{k}{m}} = \sqrt{\frac{600}{3}} = \sqrt{200} \approx 14.14 \text{ rad/s}$$

3. Convert this to Hz:

$$f_n = \frac{\omega_n}{2\pi} \approx 2.25 \text{ Hz}$$

The absorber's natural frequency (2.25 Hz) is much lower than the engine's operating frequency (50 Hz), indicating it is not effectively tuned for this specific application.

ChatGPT Integration

▪ Ask ChatGPT to explore other real-world applications where vibration isolation and absorbers are used, such as in aerospace or electronics.

▪ Use ChatGPT to investigate how engineers fine-tune vibration isolation systems to ensure optimal performance in different industries.

PROMPT 260

Explore advanced vibration isolation and absorption techniques used in aerospace, automotive, and civil engineering industries.

Objective

Understand the advanced techniques used for vibration isolation and absorption in cutting-edge engineering applications.

Activity

- Discuss advanced vibration isolation techniques used in modern industries:
  - Magnetic levitation isolators are used in high-precision systems such as satellites and cleanroom environments, magnetic levitation provides isolation without mechanical contact, reducing friction and wear.
  - Active vibration control systems: These systems use sensors, actuators, and feedback control to actively counteract vibrations. They are commonly used in aerospace applications where high precision is needed, such as in spacecraft and jet engines.
  - Smart materials: Materials like piezoelectric and viscoelastic materials are used to absorb and dampen vibrations. These materials change their properties in response to external stimuli, offering adaptive vibration control.
- Discuss advanced vibration absorbers:
  - Tuned liquid dampers (TLDs) are used in buildings and bridges to reduce large-scale vibrations from wind or earthquakes. The movement of liquid inside a tank is tuned to counteract the structure's motion.
  - Semi-active suspension systems are found in modern vehicles, these systems use sensors to adjust the damping characteristics in real-time, providing better control over vehicle vibrations on varying road conditions.

Numerical Example

A satellite uses a magnetic levitation isolator to protect sensitive equipment from vibrations caused by onboard motors. The isolator's stiffness is $k = 100$ N/m, and the mass of the isolated system is $m = 2$ kg. Calculate the natural frequency of the isolator and determine its effectiveness in isolating vibrations at a frequency of 15 Hz.

Solution

1. Calculate the natural frequency:

$$\omega_n = \sqrt{\frac{k}{m}} = \sqrt{\frac{100}{2}} = \sqrt{50} \approx 7.07 \text{ rad/s.}$$

2. We convert this to Hz:

$$f_n = \frac{\omega_n}{2\pi} \approx 1.12 \text{ Hz}$$

The natural frequency (1.12 Hz) is much lower than the operating frequency (15 Hz), indicating effective vibration isolation.

ChatGPT Integration

- Ask ChatGPT to explore how advanced techniques such as magnetic levitation and smart materials are applied in high-precision fields like aerospace and robotics.

- Use ChatGPT to investigate how future innovations in vibration isolation and absorbers could further improve system performance in demanding environments.

CHAPTER 27

CONTINUOUS SYSTEMS: STRINGS, BEAMS, AND SHAFTS

Prompt 261 Explain the concept of continuous systems and how they differ from discrete systems in vibration analysis.

Prompt 262 Discuss the vibration of strings and derive the equation for the natural frequencies of a vibrating string.

Prompt 263 Explain the concept of beam vibration and derive the equation for the natural frequencies of a simply supported beam.

Prompt 264 Solve a problem involving the calculation of the fundamental frequency of a vibrating string or beam.

Prompt 265 Discuss the role of boundary conditions in determining the mode shapes and natural frequencies of continuous systems.

Prompt 266 Explain the concept of torsional vibrations in rotating shafts and derive the equation for the natural frequencies of a shaft.

Prompt 267 Solve a problem involving the calculation of the natural frequencies of a rotating shaft.

Prompt 268 Discuss the use of energy methods in analyzing the vibrations of continuous systems.

Prompt 269 Compare the methods used to analyze continuous systems with those used for discrete systems in vibration analysis.

Prompt 270 Explore real-world applications of continuous systems in mechanical and structural engineering, such as bridges, turbines, and cables.

PROMPT 261

Explain the concept of continuous systems and how they differ from discrete systems in vibration analysis.

OBJECTIVE

Understand the key differences between continuous and discrete systems in vibration analysis.

Activity

- Define *continuous systems* as systems where mass and elasticity are distributed continuously along the length of the structure. In contrast, *discrete systems* consist of distinct masses connected by springs or other elements that vibrate at specific points.
- Provide examples of continuous systems:
  - strings: vibrating strings in musical instruments or mechanical structures like cables
  - beams: structural beams in buildings or bridges that experience vibration along their entire length
  - shafts: rotating shafts in machinery that experience torsional and bending vibrations
- Discuss how the vibration of continuous systems is described by partial differential equations (PDEs), whereas discrete systems are modeled by ordinary differential equations (ODEs).

Numerical Example

A string of length L = 2 m is fixed at both ends and experiences transverse vibrations. The wave speed in the string is v = 100 m/s. Calculate the fundamental frequency of vibration for the string.

Solution

$$f_1 = \frac{v}{2L} = \frac{100}{2 \times 2} = 25\,\text{Hz}$$

The fundamental frequency of the string is 25 Hz.

ChatGPT Integration

- Ask ChatGPT to explain the differences between continuous and discrete systems in terms of their vibration behavior and provide additional examples of continuous systems.
- Use ChatGPT to explore how boundary conditions (e.g., fixed or free ends) affect the vibration frequencies of continuous systems.

PROMPT 262

Discuss the vibration of strings and derive the equation for the natural frequencies of a vibrating string.

OBJECTIVE

Understand the behavior of vibrating strings and how their natural frequencies are calculated.

Activity

- Explain the concept of string vibration, where the string is modeled as a continuous system that oscillates transversely when displaced from its equilibrium position.
- The equation of motion for a vibrating string is given by the wave equation: $\dfrac{\partial^2 y(x,t)}{\partial t^2} = v^2 \dfrac{\partial^2 y(x,t)}{\partial x^2}$

   where
   - $y(x, t)$ is the transverse displacement of the string at position x and time t
   - v: wave speed (depends on string tension T and linear mass density μ: $v = \sqrt{\frac{T}{\mu}}$
- Derive the equation for the natural frequencies of a string fixed at both ends: $f_n = \dfrac{nv}{2L}$

   where
   - n is the mode number (such as 1 for the fundamental frequency and 2 for the second harmonic)
   - L is the length of the string
   - v is the wave speed

Numerical Example

A string of length $L = 1$ is fixed at both ends and has a wave speed of $v = 120$ m/s. Calculate the first two natural frequencies of the string.

Solution

▪ For $n = 1$ (fundamental frequency): $f_1 = \dfrac{1 \times 120}{2 \times 1} = 60\,\text{Hz}$

▪ For $n = 2$ (second harmonic): $f_2 = \dfrac{2 \times 120}{2 \times 1} = 120\,\text{Hz}$

The first two natural frequencies are $60\,\text{Hz}$ and $120\,\text{Hz}$.

ChatGPT Integration

▪ Ask ChatGPT to explore the derivation of the natural frequencies for other boundary conditions (e.g., one end fixed and one end free).

▪ Use ChatGPT to investigate real-world applications of vibrating strings in engineering and physics.

PROMPT 263

Explain the concept of beam vibration and derive the equation for the natural frequencies of a simply supported beam.

OBJECTIVE

Understand the behavior of beams under vibration and how to calculate their natural frequencies.

Activity

▪ Explain that a beam is a continuous system that can experience transverse vibrations when subjected to dynamic forces. The equation governing the transverse vibrations of a beam is the Euler-Bernoulli beam equation: $\dfrac{\partial^2}{\partial x^2}\left(EI\dfrac{\partial^2 y(x,t)}{\partial x^2}\right) + \rho A \dfrac{\partial^2 y(x,t)}{\partial t^2} = 0$

where
- E is the modulus of elasticity
- I is the moment of inertia of the beam's cross-section
- ρ is the density

- A is the cross-sectional area
- $y(x, t)$ is the transverse displacement
- For a simply supported beam (with both ends pinned), the natural frequencies are given by $f_n = \dfrac{n^2\pi}{2L^2}\sqrt{\dfrac{EI}{\rho A}}$

  where
  - n is the mode number
  - L is the length of the beam

Numerical Example

A simply supported beam has a length L = 2 m, modulus of elasticity E = 210 GPa, moment of inertia I = 50×10⁻⁶ m⁴, density ρ = 7800 kg/m³, and cross-sectional area A = 0.01 m². Calculate the fundamental frequency of the beam.

Solution

Calculate the natural frequency for n = 1:

$$f_1 = \frac{\pi}{2(2)^2}\sqrt{\frac{210\times10^9\times50\times10^{-6}}{7800\times0.01}} \approx 1441 \text{ Hz.}$$

The fundamental frequency of the beam is approximately 1441 Hz.

ChatGPT Integration

- Ask ChatGPT to explore how different boundary conditions (e.g., clamped and free) affect the natural frequencies of a beam.
- Use ChatGPT to investigate how beam vibration analysis is applied in structural engineering, such as in bridges or mechanical components.

PROMPT 264

Solve a problem involving the calculation of the fundamental frequency of a vibrating string or beam.

Objective

Practice solving problems related to the calculation of the fundamental frequency of vibrating strings or beams.

Activity

- Provide a problem scenario for calculating the fundamental frequency of either a vibrating string or a beam, based on the system's parameters.

Numerical Example

A string of length $L = 3$ m is fixed at both ends and has a wave speed of $v = 200$ m/s. Calculate the fundamental frequency of the string.

Solution

$$f_1 = \frac{v}{2L} = \frac{200}{2 \times 3} = 33.33 \text{ Hz}$$

The fundamental frequency of the string is 33.33 Hz.

Numerical Example

A simply supported beam has a length $L = 5$ m, modulus of elasticity $E = 200$ GPa, moment of inertia $I = 100 \times 10^{-6}$ m^4, density $\rho = 7850$ kg/m^3, and cross-sectional area $A = 0.015$ m^2.

Calculate the fundamental frequency of the beam.

Solution

$$f_1 = \frac{\pi^2}{2L^2} \sqrt{\frac{EI}{\rho A}} = 25.89 \text{ Hz}$$

The fundamental frequency of the beam is approximately 25.89 Hz.

ChatGPT Integration

- Ask ChatGPT to verify the solutions and explore similar problems involving different system parameters.
- Use ChatGPT to investigate the factors that most significantly influence the fundamental frequency in real-world applications, such as tension, material properties, or geometry.

PROMPT 265

Discuss the role of boundary conditions in determining the mode shapes and natural frequencies of continuous systems.

Objective

Understand how boundary conditions affect the mode shapes and natural frequencies of continuous systems such as strings, beams, and shafts.

Activity

- Explain the concept of boundary conditions, which refer to the constraints applied at the ends or interfaces of a continuous system. These constraints strongly influence the system's vibrational behavior—specifically, the mode shapes (how the system deforms in each mode) and the natural frequencies (the frequencies at which these vibrations occur without external forcing).

- Discuss common boundary conditions:
  - Fixed-Fixed: Both ends are fixed, allowing no displacement or rotation (e.g., beams in structural frameworks).
  - Fixed-Free: One end is fixed, and the other is free to move (e.g., cantilever beams or trees swaying in the wind).
  - Free-Free: Both ends are free, allowing full displacement (e.g., cables or ropes).
  - Clamped-Free: One end is clamped (no displacement or rotation), and the other end is free.

- Explain that boundary conditions define the mathematical form of the solution, determine the admissible mode shapes, and strongly affect the system's natural frequencies. For instance, a clamped beam has a higher fundamental frequency than a simply supported beam of the same dimensions and material, due to greater stiffness from the clamped boundary.

- Provide contrasting examples to illustrate how different boundary conditions lead to different frequencies and deformation patterns.

Numerical Example

A uniform beam has a length $L = 4$ m, modulus of elasticity $E = 210$ GPa, and moment of inertia $I = 60 \times 10^{-6}$ m^4. Mass density $\rho = 7800$ kg/m^3 Cross-sectional area $A = 0.02$ m^2 Calculate the fundamental frequency for two boundary conditions: (a) Fixed-Fixed, and (b) Free-Free.

Solution

▪ For the Fixed-Fixed case, the fundamental frequency will be

$$f_1 = \frac{\beta_1^2}{2\pi L^2}\sqrt{\frac{EI}{\rho A}}$$

where

▪ $\beta_1 L \approx 4.73004 \Rightarrow \beta_1 = \dfrac{4.73004}{L}$

$$f_1 = \frac{4.73004^2}{2\pi \cdot 16} \cdot \frac{\sqrt{210 \times 10^9 \cdot 60 \times 10^{-6}}}{70.4}$$

Step by Step:

▪ $\dfrac{4.73004^2}{2\pi \cdot 16} \approx \dfrac{22.35}{100.53} \approx 0.2223$

▪ $\dfrac{210 \times 10^9 \cdot 60 \times 10^{-6}}{70 \cdot 4} \approx \dfrac{12.6 \times 10^6}{70.4} \approx 179,545.45$

▪ $\sqrt{179545.45} \approx 423.7$

Final:

$$f_1 \approx 0.2223 \cdot 423.7 \approx \boxed{94.2\,\text{Hz}}$$

ChatGPT Integration

▪ Ask ChatGPT to explain how different boundary conditions affect the vibration response of strings, beams, and shafts.

▪ Use ChatGPT to explore real-world engineering examples where boundary conditions play a significant role, such as in bridge designs or architectural structures.

PROMPT 266

Explain the concept of torsional vibrations in rotating shafts and derive the equation for the natural frequencies of a shaft.

Objective

Understand the concept of torsional vibrations in rotating shafts and how to calculate their natural frequencies.

Activity

- Explain that torsional vibrations occur when a rotating shaft oscillates about its axis due to applied torques. These vibrations can lead to excessive stresses and eventual failure if not properly analyzed.
- The equation governing torsional vibrations in a shaft is given by

$$\frac{\partial^2 \theta(x, t)}{\partial t^2} = \frac{GJ}{I} \cdot \frac{\partial^2 \theta(x,t)}{\partial x^2},$$

where
- I is the mass moment of inertia per unit length of the shaft (kg·m)
- G is the shear modulus
- J is the polar moment of inertia
- $\theta(x, t)$ is the angular displacement

- Derive the equation for the natural frequencies of torsional vibrations for a shaft fixed at both ends: $f_n = \frac{n}{2L}\sqrt{\frac{GJ}{I}}$, where n is the mode number and L is the length of the shaft.

Numerical Example

A rotating shaft has a length $L = 1.5$ m, shear modulus $G = 80$ GPa, and polar moment of inertia $J = 2 \times 10^{-6} \text{m}^4$. Calculate the fundamental frequency of torsional vibration for the shaft, assuming $I = 0.05$ kg.m.

Solution

$$f_1 = \frac{1}{2L}\sqrt{\frac{GJ}{I}} = \frac{1}{2 \times 1.5}\sqrt{\frac{80 \times 10^9 \times 2 \times 10^{-6}}{0.05}} = 596.3 \text{ Hz}$$

The fundamental torsional frequency depends on the moment of inertia of the shaft.

ChatGPT Integration

- Ask ChatGPT to explain how torsional vibrations can affect machinery and structures, particularly in rotating equipment.
- Use ChatGPT to explore real-world applications of torsional vibration analysis in engineering, such as in crankshafts, turbines, and drive shafts.

PROMPT 267

Solve a problem involving the calculation of the natural frequencies of a rotating shaft.

Objective

Practice calculating the natural frequencies of a rotating shaft experiencing torsional vibrations.

Activity

- Provide a problem scenario for calculating the natural frequency of a rotating shaft based on the system's parameters.

Numerical Example

A rotating steel shaft has the following properties:

- Length $L = 2$ m
- Shear modulus $G = 80$ GPa
- Polar moment of inertia $J = 1.5 \times 10^{-6}$ m^4
- Moment of inertia $I = 0.02$ kg·m

Calculate the fundamental torsional frequency of the shaft.

Solution

Calculate the natural frequency:

$$f_1 = \frac{1}{2L}\sqrt{\frac{GJ}{I}} = \frac{1}{2 \times 2}\sqrt{\frac{80 \times 10^9 \times 1.5 \times 10^{-6}}{0.02}}$$

$$f_1 = \frac{1}{4}\sqrt{\frac{120 \times 10^3}{0.02}} = \frac{1}{4} \times \sqrt{6 \times 10^6} \approx 612.4\,\text{Hz}$$

The fundamental torsional frequency is approximately 612.4 Hz.

ChatGPT Integration

- Ask ChatGPT to verify the solution and explore similar problems involving rotating shafts with different material properties or dimensions.

▪ Use ChatGPT to explore how engineers design rotating shafts to minimize torsional vibrations in applications like engines, turbines, or windmills.

PROMPT 268

Discuss the use of energy methods in analyzing the vibrations of continuous systems.

Objective

Understand how energy methods, such as the Rayleigh-Ritz method, are used to analyze the vibrations of continuous systems like strings, beams, and shafts.

Activity

▪ Explain the concept of energy methods in vibration analysis. These methods involve calculating the total potential and kinetic energy of the system to approximate the natural frequencies and mode shapes.

▪ Discuss the Rayleigh-Ritz method, which is commonly used to approximate the natural frequencies of continuous systems. This method involves

1. Assuming an approximate mode shape $y(x)$ that satisfies geometric boundary conditions.

2. Calculating the system's total kinetic energy T and potential energy U.

3. Using the Rayleigh quotient to estimate the natural frequency:

$$\omega^2 \approx \frac{U}{T} \Rightarrow f \approx \frac{1}{2\pi}\sqrt{\frac{U}{T}}$$

For beams transverse vibration, this becomes:

$$\omega^2 \approx \frac{\int_0^L EI\left(\dfrac{d^2 y}{dx^2}\right)dx}{\int_0^L \rho A\left(\dfrac{dy}{dt}\right)^2 dx}$$

where EI is flexural rigidity, ρA is mass per unit length.

- Mention that energy methods are particularly useful when solving complex vibration problems that are difficult to handle with exact analytical solutions.

Numerical Example

Consider a simply supported beam of length $L = 3$ m, modulus of elasticity $E = 200$ GPa, and moment of inertia $I = 80 \times 10^{-6}$ m^4.
Use the Rayleigh-Ritz method to approximate the fundamental frequency of the beam.

Solution

1. **Assume an admissible mode shape:**

   For a simply supported beam, a reasonable choice is:

   $$y(x,t) = \phi(x)\cos(\omega t), \quad \text{with} \quad \phi(x) = \sin\left(\frac{\pi x}{L}\right)$$

2. **Compute potential energy** (strain energy):

   $$U = \frac{1}{2}\int_0^L EI\left(\frac{d^2\phi}{dx^2}\right)^2 dx$$

   $$\frac{d^2\phi}{dx^2} = -\left(\frac{\pi}{L}\right)^2 \sin\left(\frac{\pi x}{L}\right) \Rightarrow \left(\frac{d^2\phi}{dx^2}\right)^2 = \left(\frac{\pi^2}{L^2}\right)\sin^2\left(\frac{\pi x}{L}\right)$$

   $$U = \frac{1}{2}EL\left(\frac{\pi^4}{L^4}\right)\int_0^L \sin^2\left(\frac{\pi x}{dx^2}\right)^2 dx = \frac{1}{2}EI\left(\frac{\pi^4}{L^4}\right)\cdot\frac{L}{2} = \frac{EL\pi^2}{4L^3}$$

3. **Computer kinetic energy:**

   $$T = \frac{1}{2}\int_0^L \rho A\left(\frac{dy}{dt}\right)^2 dx = \frac{1}{2}\omega^2\int_0^L \rho A\phi^2(x)\,dx = \frac{1}{2}\omega^2\rho A\int_0^L \sin^2\left(\frac{\pi x}{L}\right)$$

   $$dx = \frac{1}{2}\omega^2\rho A\cdot\frac{L}{2} = \frac{\omega^2\rho AL}{4}$$

4. **Apply Rayleigh quotient:**

   $$\omega^2 = \frac{U}{T} = \frac{\dfrac{EI\pi^4}{4L^3}}{\dfrac{\omega^2\rho AL}{4}} \Rightarrow \omega^2 = \frac{EI\pi^4}{\rho AL^4} \Rightarrow f = \frac{\omega}{2\pi} - \frac{1}{2\pi}\sqrt{\frac{EI\pi^4}{\rho AL^4}}$$

Insert values:

$$f = \frac{1}{2\pi}\sqrt{\frac{(200\times10^9)(80\times10^{-6})\pi^4}{12\cdot(3)^4}} = \frac{1}{2\pi}\sqrt{\frac{16\times10^3\cdot\pi^4}{972}} \approx \frac{1}{2\pi}\sqrt{\frac{16.97.4091}{972}}\cdot10^3$$

$$\approx \frac{1}{2\pi}\sqrt{1.602\times10^3} = \frac{1}{2\pi}\cdot40.02 \approx \boxed{6.37\ \text{Hz}}$$

ChatGPT Integration

- Ask ChatGPT to explain how energy methods like the Rayleigh-Ritz method are applied to continuous systems in engineering.
- Use ChatGPT to explore real-world applications of energy methods, particularly in structural analysis, where exact solutions are challenging to obtain.

PROMPT 269

Compare the methods used to analyze continuous systems with those used for discrete systems in vibration analysis.

Objective

Understand the differences between the methods used to analyze continuous systems and those used for discrete systems in vibration analysis.

Activity

- Compare the analysis techniques for continuous systems and discrete systems:
  - Continuous systems: These involve distributed parameters like mass and stiffness along the length of the structure, requiring the use of partial differential equations (PDEs). Continuous systems are analyzed using methods like the wave equation and beam theory.
  - Discrete systems: These are modeled using lumped masses and springs, where the motion is governed by ordinary differential equations (ODEs). Discrete systems are typically analyzed using matrix methods or Newton's second law.
- Explain that continuous systems require more complex mathematical treatments due to the distributed nature of their properties,

whereas discrete systems are simpler and involve solving sets of algebraic equations.

- Discuss how both types of systems can be approximated using finite element analysis (FEA), which discretizes continuous systems into smaller, manageable parts for numerical solution.

Numerical Example

Consider a 2-meter-long string with a mass per unit length of $\mu = 0.05$ kg/m and a tension of $T = 500$ N, and a discrete system with a mass $m = 0.1$ kg connected to a spring with stiffness $k = 500$ N/m. Compare the approaches for calculating the natural frequencies of these two systems.

Solution

- For the continuous string, the wave equation is used: $f_n = \dfrac{n}{2L}\sqrt{\dfrac{T}{M/L}}$.

For the discrete system, the natural frequency is calculated using:

$$f_n = \frac{1}{2\pi}\sqrt{\frac{k.L}{m}} \Rightarrow f_1 = \frac{1}{2.2}\sqrt{\frac{500}{0.05}} = \frac{1}{4}\sqrt{10,000} = \frac{1}{4}\cdot 100 = 25\,\text{Hz}.$$

ChatGPT Integration

- Ask ChatGPT to explain how both continuous and discrete systems are analyzed using finite element methods (FEM).
- Use ChatGPT to explore examples of complex systems that require a combination of continuous and discrete methods for accurate vibration analysis.

PROMPT 270

Explore real-world applications of continuous systems in mechanical and structural engineering, such as bridges, turbines, and cables.

Objective

Understand how continuous systems are applied in various engineering fields, particularly in the design and analysis of structures and machines.

Activity

- Discuss real-world applications of continuous systems:
  - Bridges: Continuous beams in bridges must be analyzed for vibrations caused by traffic, wind, and earthquakes. Engineers use beam theory to ensure that the structure can withstand dynamic loads without experiencing excessive deflection or resonance.
  - Turbines: Turbine blades experience continuous vibrations during operation, especially at high rotational speeds. Torsional vibration analysis is crucial to prevent failure due to fatigue.
  - Cables: Cables in suspension bridges or power lines behave as continuous systems and are analyzed for their natural frequencies to avoid resonance due to wind or dynamic loading.
- Explain how engineers apply the principles of continuous systems to predict and control vibrations, improving the safety and longevity of these structures and machines.

Numerical Example

Consider a bridge with a simply supported beam of length L = 10 m, modulus of elasticity E = 200 GPa, and moment of inertia I = 200×10^{-6}m^4. Calculate the fundamental frequency of the bridge beam and explain its significance in the design.

Solution

$$f_1 = \frac{\pi}{2L^2}\sqrt{\frac{EI}{\rho A}} = \frac{\pi}{2.10^2}\sqrt{\frac{200\times10^9 \cdot 200\times10^{-6}}{7850\cdot0.12}} = 324 \text{ Hz}$$

The fundamental frequency informs engineers whether the structure can avoid resonating with environmental or operational forces.

ChatGPT Integration

- Ask ChatGPT to provide additional real-world examples of continuous systems in engineering, including their analysis and design considerations.
- Use ChatGPT to explore how continuous systems are designed to minimize the risk of resonant vibrations in critical infrastructure like bridges, skyscrapers, or aircraft components.

CHAPTER 28

MODAL ANALYSIS AND MODE SHAPES

Prompt 271 Define modal analysis and explain its significance in vibration analysis of mechanical systems.

Prompt 272 Explain the concept of mode shapes and how they relate to the natural frequencies of a system.

Prompt 273 Describe the procedure for conducting a modal analysis of a multi-degree-of-freedom (MDOF) system.

Prompt 274 Solve a problem involving the calculation of natural frequencies and mode shapes for a 2-degree-of-freedom (2-DOF) system.

Prompt 275 Discuss the importance of orthogonality of mode shapes in modal analysis.

Prompt 276 Explain the use of experimental modal analysis (EMA) in identifying mode shapes and natural frequencies in real-world systems.

Prompt 277 Discuss the role of damping in modal analysis and how it affects the mode shapes and natural frequencies of a system.

Prompt 278 Solve a problem involving the calculation of mode shapes for a continuous system, such as a beam or string.

Prompt 279 Discuss the use of modal analysis in practical applications, such as in aerospace, automotive, and civil engineering.

Prompt 280 Explore advanced topics in modal analysis, including the use of computational tools like finite element analysis (FEA) for determining mode shapes in complex structures.

PROMPT 271

Define modal analysis and explain its significance in vibration analysis of mechanical systems.

Objective

Understand the concept of modal analysis and its importance in studying the dynamic behavior of mechanical systems.

Activity

- Define *modal analysis* as a method used to determine the natural frequencies, mode shapes, and damping characteristics of a mechanical system. It helps predict how a system will respond to dynamic loads and external forces.
- Explain the significance of modal analysis:
  - It provides insights into the dynamic behavior of complex systems by identifying their resonant frequencies.
  - Engineers use modal analysis to prevent resonance, which can lead to excessive vibrations and potential failure.
  - Modal analysis is crucial for designing structures and machines that are resistant to dynamic forces.
- Discuss the steps involved in performing modal analysis:
  - Model the system using its mass, stiffness, and damping properties.
  - Solve the eigenvalue problem to find the natural frequencies and corresponding mode shapes.
  - Use the results to design systems that can avoid harmful resonance.

Numerical Example

A two-mass system with masses $m_1 = 2$ kg and $m_2 = 3$ kg, and spring constants $k_1 = 100$ N/m and $k_2 = 150$ N/m. Perform a basic modal analysis to calculate the natural frequencies of the system.

Solution

Set up the equations of motion, form the mass and stiffness matrices, and solve the eigenvalue problem to find the natural frequencies.

Step 1: Mass and Stiffness Matrices

$$M = \begin{bmatrix} 2 & 0 \\ 0 & 3 \end{bmatrix}, \quad K = \begin{bmatrix} 250 & -150 \\ -150 & 150 \end{bmatrix}$$

Step 2: Eigenvalue Equation

$$\det(K - \omega^2 M) = 0 \Rightarrow \det = \begin{bmatrix} 250 - 2\omega^2 & -150 \\ -150 & 150 - 3\omega^2 \end{bmatrix} = 0$$

$$(250 - 2\omega^2)(150 - 3\omega^2) - 22500 = 0 \Rightarrow 6\omega^4 - 1050\omega^2 + 15000 = 0$$

Step 3: Solve Quadratic

Let $x = \omega^2$:

$$2x^2 - 350x + 5000 = 0 \Rightarrow x_{1,2} = \frac{350 \pm \sqrt{82500}}{4} \approx 159.31, \ 15.69$$

Step 4: Natural Frequencies

$$\omega_1 \approx \sqrt{15.69} = 3.96 \text{ rad/s}, \quad \omega_2 \approx \sqrt{159.31} = 12.62 \text{ rad/s},$$

$$f_1 \approx \frac{3.96}{2\pi} = 0.63 \text{ Hz}, \quad f_2 \approx \frac{12.62}{2\pi} = 2.01 \text{ Hz}$$

ChatGPT Integration

- Ask ChatGPT to explain the steps of modal analysis in more detail and explore how it is applied in various engineering fields.
- Use ChatGPT to investigate how modal analysis helps prevent resonance in systems like buildings, bridges, or vehicles.

PROMPT 272

Explain the concept of mode shapes and how they relate to the natural frequencies of a system.

Objective

Understand the concept of mode shapes and their relationship to the natural frequencies of mechanical systems.

Activity

- Define *mode shapes* as the specific patterns of motion that a system undergoes when vibrating at one of its natural frequencies. Each mode shape corresponds to a distinct natural frequency.
- Explain that when a system vibrates at its natural frequency, the entire structure moves in a predictable, repeatable pattern called a *mode shape*.
- Discuss the significance of mode shapes:
  - Mode shapes provide insights into which parts of the system move the most and which remain stationary (nodal points).
  - The natural frequencies determine the specific frequencies at which these mode shapes occur.
- Illustrate how higher mode shapes correspond to higher natural frequencies, and how mode shapes differ for different types of structures (e.g., beams, plates, or rotating systems).

Numerical Example

Consider a two-mass system with masses m_1 = 4 kg and m_2 = 6 kg, and spring constants k_1 = 200 N/m and k_2 = 150 N/m. Solve for the mode shapes of the system at its two natural frequencies.

Solution

- Set up the equations of motion.
- Solve the eigenvalue problem to find the natural frequencies. The eigenvalues correspond to the square of the natural frequencies of the system, representing the fundamental vibration characteristics of the structure.
- Use the eigenvectors to determine the mode shapes for each frequency. The eigenvectors correspond to the mode shapes of the system, describing the relative displacement patterns of the structure during vibration at each natural frequency.

Step 1: Define Matrices

Mass matrix:

$$M = \begin{bmatrix} 4 & 0 \\ 0 & 6 \end{bmatrix}$$

Stiffness matrix:

$$M = \begin{bmatrix} k_1 + k_2 & -k_2 \\ -k_2 & k_2 \end{bmatrix} = \begin{bmatrix} 350 & -150 \\ -150 & 150 \end{bmatrix}$$

Step 2: Solve the generalized eigenvalue problem

$$(K - \omega^2 M)x = 0 \Rightarrow \omega^2 = \text{eigenvalues of } M^{-1}K$$

The computed natural frequencies are:

▪ $\omega_1 = \sqrt{12.5} = 3.536$ rad/s

▪ $\omega_2 = \sqrt{100} = 10.000$ rad/s

Step 3: Mode Shapes (Normalized w.r.t. second mass)

$$\text{Mode 1}: \begin{bmatrix} -3 \\ 1 \end{bmatrix}, \text{Mode 2}: \begin{bmatrix} 0.5 \\ 1 \end{bmatrix}$$

This represent the relative displacements of the two masses at each natural frequency.

ChatGPT Integration

▪ Ask ChatGPT to explore the relationship between mode shapes and natural frequencies in more complex systems.

▪ Use ChatGPT to investigate how mode shapes are used in the design of buildings, bridges, and machines to control vibrations.

PROMPT 273

Describe the procedure for conducting a modal analysis of a multi-degree-of-freedom (MDOF) system.

Objective

Learn the steps involved in performing a modal analysis on a multi-degree-of-freedom (MDOF) system.

Activity

▪ Explain the procedure for conducting modal analysis of an MDOF system:

1. Set up the equations of motion: Begin by writing the equations of motion for the system using Newton's second law or Lagrangian methods.

2. Formulate the mass and stiffness matrices: The system of equations can be represented in matrix form, where [M] is the mass matrix and [K] is the stiffness matrix: $[M]\{\ddot{x}\} + [K]\{x\} = 0$.

3. Solve the eigenvalue problem: Assume harmonic motion of the form $\{x(t)\} = \{X\}e^{i\omega t}$ and substitute this into the equations of motion. Solve the resulting eigenvalue problem: $([K] - \omega^2 [M])\{X\} = 0$.

4. Find the natural frequencies and mode shapes: The eigenvalues of this equation correspond to the natural frequencies ω, and the eigenvectors $\{X\}$ represent the mode shapes.

5. Interpret the results: Use the natural frequencies and mode shapes to analyze the dynamic behavior of the system.

Numerical Example

Consider a system of three masses $m_1 = 2$ kg, $m_2 = 3$ kg, and $m_3 = 4$ kg, connected in series by springs with spring constants $k_1 = 100$ N/m, $k_2 = 150$ N/m, and $k_3 = 200$ N/m. Determine the natural frequencies and mode shapes of the system.

Solution

1. Set up the equations of motion using Newton's second law for each mass:
   - For m_1: $m_1 \ddot{x}_1 = -k_1 x_1 + k_2 (x_2 - x_1)$
   - For m_2: $m_2 \ddot{x}_2 = -k_2 (x_2 - x_1) + k_3 (x_3 - x_2)$
   - For m_3: $m_3 \ddot{x}_3 = -k_3 (x_3 - x_2)$

2. Formulate the matrix equation in the form $M\ddot{x} + Kx = 0$, where M is the mass matrix, K is the stiffness matrix, and x is the displacement vector:

$$[M] = \begin{bmatrix} 2 & 0 & 0 \\ 0 & 3 & 0 \\ 0 & 0 & 4 \end{bmatrix}, \quad [K] = \begin{bmatrix} k_1 + k_2 & -k_2 & 0 \\ -k_2 & k_2 + k_3 & -k_3 \\ 0 & -k_3 & k_3 \end{bmatrix} = \begin{bmatrix} 250 & -150 & 0 \\ -150 & 350 & -200 \\ 0 & -200 & 200 \end{bmatrix}$$

3. Solve the eigenvalue problem $det\left[\left(K-\omega^2 M\right)\right]=0$ to find the natural frequencies $\omega_1, \omega_2,$ and ω_3 of the system, $\omega_1 = 4.27,$ $\omega_2=8.63,$ $\omega_3=14.35$ rad/s.

4. Determine the mode shapes by finding the eigenvectors corresponding to each natural frequency. Each eigenvector will represent a mode shape, showing the relative motion of $x_1, x_2,$ and x_3 at each natural frequency.

   This gives three eigenvectors, representing mode shapes, e.g.:

   Mode 1 (low frequency): all masses move in phase

   Mode 2: middle mass moves out of phase

   Mode 3: outer masses out of phase, node at center

ChatGPT Integration

- Ask ChatGPT to explore how modal analysis is conducted in more complex systems with multiple degrees of freedom.
- Use ChatGPT to investigate how modal analysis is applied in industries like aerospace or automotive engineering to design stable, vibration-resistant systems.

PROMPT 274

Solve a problem involving the calculation of natural frequencies and mode shapes for a 2-degree-of-freedom (2-DOF) system.

Objective

Practice solving problems related to the calculation of natural frequencies and mode shapes for a multi-degree-of-freedom (MDOF) system.

Activity

- Provide a scenario for solving the natural frequencies and mode shapes of a 2-degree-of-freedom (2-DOF) system.

Numerical Example

A two-degree-of-freedom system has the following mass and stiffness matrices:

$$[M] = \begin{bmatrix} 2 & 0 \\ 0 & 1 \end{bmatrix} \quad [K] = \begin{bmatrix} 200 & -100 \\ -100 & 100 \end{bmatrix}$$

Calculate the natural frequencies and mode shapes of the system.

Solution

1. Set up the eigenvalue problem: $[K] - \omega^2 [M] = 0$.
2. Solve for the eigenvalues (natural frequencies) and eigenvectors (mode shapes).

Step 1: Solve

$$\det \left([K] - \omega^2 [M] = 0 \right)$$

$$\det \begin{bmatrix} 200 - 2\omega^2 & -100 \\ -100 & 100 - \omega^2 \end{bmatrix} = 0$$

$$(200 - 2\omega^2)(100 - \omega^2) - 100^2 = 0 \Rightarrow \omega^4 - 200\omega^2 + 5000 = 0$$

$$\omega^2 = 100 \pm 50\sqrt{2} \Rightarrow \omega_1 \approx 5.41, \quad \omega_2 = 13.06 \text{ rad/s}$$

Step 2: Mode Shapes

For $\omega_1^2 \approx 29.29$:

$$141.42 x_1 = 100 x_2 \Rightarrow \frac{x_1}{x_2} = 0.707 \Rightarrow X^{(1)} = \begin{bmatrix} 1 \\ \sqrt{2} \end{bmatrix}$$

For $\omega_2^2 \approx 170.71$:

$$-141.42 x_1 = 100 x_2 \Rightarrow \frac{x_1}{x_2} = -0.707 \Rightarrow X^{(2)} = \begin{bmatrix} -1 \\ \sqrt{2} \end{bmatrix}$$

ChatGPT Integration

- Ask ChatGPT to verify the solution and explore more complex problems involving multi-degree-of-freedom systems.
- Use ChatGPT to investigate real-world applications where modal analysis of MDOF systems is critical, such as in structural engineering or mechanical design.

PROMPT 275

Discuss the importance of orthogonality of mode shapes in modal analysis.

Objective

Understand the concept of orthogonality in modal analysis and why it is important for solving vibration problems in mechanical systems.

Activity

- Explain the concept of *orthogonality of mode shapes*, which means that the mode shapes of a system are mutually orthogonal with respect to the mass and stiffness matrices. This property simplifies the solution of vibration problems in multi-degree-of-freedom (MDOF) systems.
- The mode shapes $\{X_i\}^T[M]\{X_j\} = 0$ for $i \neq j$ of an MDOF system satisfy the following orthogonality conditions:

$$\{X_i\}^T[K]\{X_j\} = 0 \quad \text{for} \quad i \neq j$$

- Discuss why orthogonality is important:
  - It allows for the decoupling of the equations of motion, simplifying the analysis of the system's response to external forces.
  - Orthogonality enables the use of modal superposition to express the system's overall response as a combination of individual mode shapes.
- Provide examples of how orthogonality is used in practice to solve complex vibration problems.

Numerical Example

Consider a 2-DOF system with the mass matrix of

$$[M] = \begin{bmatrix} 3 & 0 \\ 0 & 2 \end{bmatrix}$$

and stiffness matrix

$$[K] = \begin{bmatrix} 100 & -50 \\ -500 & 100 \end{bmatrix}$$

Verify the orthogonality of the mode shapes for the system.

Solution

1. Solve for the mode shapes $\{X_1\}$, $\{X_2\}$.
2. Check the orthogonality condition by calculating $\{X_1\}^T[M]\{X_2\}$ and $\{X_1\}^T[K]\{X_2\}$.

If both expressions equal zero, the mode shapes are orthogonal.

1. **Solve the eigenvalue problem:**

$$KX = \omega^2 MX$$

2. **Natural frequencies (rad/s):**

$$\omega_1 \approx 4.43 \quad \text{and} \quad \omega_2 \approx 7.98$$

3. **Corresponding mode shapes** (unnormalized eigenvectors):

$$X_1 = \begin{bmatrix} -0.77 \\ -0.64 \end{bmatrix}, \quad X_2 = \begin{bmatrix} 0.48 \\ -0.88 \end{bmatrix}$$

4. **Orthogonality check:**

$$X_1^T MX_2 \approx 2.76 \times 10^{-16} \approx 0 \text{ (mass orthogonality)}$$
$$X_1^T KX_2 \approx 1.09 \times 10^{-14} \approx 0 \text{ (stiffness orthogonality)}$$

ChatGPT Integration

- Ask ChatGPT to explain how orthogonality is applied in more complex systems with multiple degrees of freedom.
- Use ChatGPT to explore how engineers use orthogonality in modal analysis for structural and mechanical design.

PROMPT 276

Explain the use of experimental modal analysis (EMA) in identifying mode shapes and natural frequencies in real-world systems.

Objective

Understand the role of experimental modal analysis (EMA) in determining the dynamic characteristics of real-world systems.

Activity

- Define experimental modal analysis (EMA) as a technique used to identify the natural frequencies, mode shapes, and damping characteristics of a structure through physical testing.
- Explain the procedure for performing EMA:
  1. Excitation of the system: The structure is excited using a force input, such as an impact hammer or shaker, to induce vibrations.
  2. Measurement of response: Sensors like accelerometers or displacement transducers are placed at different locations on the structure to measure the vibration response.
  3. Analysis of data: The vibration data is processed using frequency response functions (FRFs) to identify the system's natural frequencies and mode shapes.
- Discuss the advantages of EMA:
  - Captures the actual dynamic behavior of the physical structure, including effects from damping, material non-uniformity, and boundary conditions.
  - Useful for validating theoretical or simulation-based modal analysis (e.g., Finite Element Analysis) by comparing measured vs. predicted mode shapes and frequencies.
  - Can detect structural damage, changes in stiffness, or loosened joints—making it a powerful diagnostic and validation tool.

Numerical Example

A cantilever beam is subjected to experimental modal analysis using an impact hammer. The measured natural frequencies are $f_1 = 10\ H_z$ and $f_2 = 25\ H_z$

Based on the measured data, how would you interpret the mode shapes of the beam?

Solution

1. First Mode ($f_1 = 10$ Hz):

   This is the fundamental bending mode of a cantilever beam. The beam bends in a smooth, single-curve shape. Maximum displacement occurs at the free end, with zero displacement at the fixed end. No internal nodes (zero-crossings) along the beam.

2. Second Mode (f_2 = 25 Hz):

   This represents the second bending mode. The beam shows a more complex shape with at least one node (point of zero displacement) along its span. The free end still moves, but not with maximum amplitude. Indicates higher energy and shorter wavelength deformation pattern.

ChatGPT Integration

- Ask ChatGPT to explore the steps involved in conducting experimental modal analysis on different types of structures.
- Use ChatGPT to investigate real-world applications where EMA is critical, such as in aerospace, automotive, and civil engineering for validating design performance.

PROMPT 277

Discuss the role of damping in modal analysis and how it affects the mode shapes and natural frequencies of a system.

Objective

Understand how damping influences the results of modal analysis and its effect on the dynamic behavior of mechanical systems.

Activity

- Explain that damping is a measure of how energy is dissipated in a vibrating system. In real-world systems, damping reduces the amplitude of vibrations over time, and it can affect the natural frequencies and mode shapes.
- Discuss the two main types of damping used in modal analysis:
  - Proportional damping: Assumes that the damping matrix is a linear combination of the mass and stiffness matrices. This is common in analytical models because it simplifies the solution process.
  - Non-proportional damping: Damping forces are not directly related to the mass and stiffness properties of the system, making the analysis more complex.

- Explain how damping affects natural frequencies:
  - With light damping, the natural frequencies remain nearly unchanged.
  - With heavy damping, the system's natural frequencies decrease, and the vibration response becomes more subdued.
- Discuss how mode shapes are affected:
  - In lightly damped systems, the mode shapes are close to those in the undamped case.
  - In systems with significant damping, the mode shapes may change due to the influence of damping forces.

Numerical Example

A two-degree-of-freedom system has the following mass and stiffness matrices:

$$[M] = \begin{bmatrix} 2 & 0 \\ 0 & 1 \end{bmatrix}, \quad [K] = \begin{bmatrix} 100 & -50 \\ -50 & 50 \end{bmatrix}$$

The damping matrix is proportional:

$$[C] = 0.1[M] + 0.05[K]$$

Calculate the damped natural frequencies and discuss the influence of damping on the mode shapes.

Solution

1. Set up the equations of motion with damping.
2. Solve for the damped natural frequencies using the Rayleigh damping model, which assumes that damping is a linear combination of mass and stiffness proportional damping coefficients to approximate energy dissipation in the system.
3. Compare the mode shapes for the damped and undamped systems to observe the influence of damping.

Steps:
1. Calculate C:

$$[C] = \begin{bmatrix} 5.2 & 2.5 \\ 2.5 & 2.6 \end{bmatrix}$$

2. Find undamped natural frequencies ω_i by solving

$$\det \left([K] - \omega^2 [M]\right) = 0$$

Result:

$$\omega_1 = 9.24 \text{ rad/s}, \qquad \omega_2 = 3.83 \text{ rad/s}$$

3. Determine mode shapes

$$\Phi_1 = \begin{bmatrix} 1 \\ 1.414 \end{bmatrix}, \quad \Phi_2 = \begin{bmatrix} 1 \\ -1.414 \end{bmatrix}$$

4. Calculate model damping ratios ζ_i:

$$\zeta_i = \frac{1}{2\omega_i} \frac{\Phi_i^T [C] \Phi_i}{\Phi_i^T [M] \Phi_i}$$

Results:

$$\zeta_i = 0.236, \qquad \zeta_2 = 0.109$$

5. Calculate damped natural frequencies:

$$\omega_{d,i} = \omega_i \sqrt{1 - \zeta_i^2}$$

Results:

$$\omega_{d,1} = 8.97 \text{ rad/s}, \qquad \omega_{d,2} = 3.80 \text{ rad/s}$$

Interpretation:

Damping reduces the natural frequencies slightly.

Mode shapes remain nearly unchanged due to proportional damping.

Mode 1 is more heavily damped than mode 2.

ChatGPT Integration

- Ask ChatGPT to explain how engineers incorporate damping into modal analysis to predict the real-world behavior of structures.
- Use ChatGPT to explore the importance of damping in specific industries, such as automotive or civil engineering, where vibrations must be controlled to ensure safety and performance.

PROMPT 278

Solve a problem involving the calculation of mode shapes for a continuous system, such as a beam or string.

Objective

Practice solving problems related to the calculation of mode shapes for continuous systems.

Activity

- Provide a scenario involving a continuous system, such as a beam or string, and solve for the mode shapes based on the system's boundary conditions and physical properties.

Numerical Example

Consider a simply supported beam of length L = 4 m, modulus of elasticity E = 210 GPa, moment of inertia $I = 50 \times 10^{-6}$ m^4, and mass per unit length μ = 0.1 kg/m. Calculate the first two mode shapes for the beam.

Solution

1. Determine the natural frequencies.

   The natural frequencies for a simply supported beam are given by

   $$f_n = \frac{n^2 \pi}{2L^2}\sqrt{\frac{EI}{\mu}}$$

   For n = 1 and n = 2, calculate the first two natural frequencies.
2. Calculate the mode shapes.

   The mode shapes for a simply supported beam are given by

   $$y_n(x) = \sin\left(\frac{n\pi x}{L}\right)$$

   The first mode shape corresponds to a half sine wave along the length of the beam, while the second mode shape corresponds to a full sine wave.

Step 1: Use formula

$$f_n = \frac{n^2\pi}{2L^2}\sqrt{\frac{EL}{\mu}}$$

Step 2: Compute constant term

$$\sqrt{\frac{EL}{\mu}} = \sqrt{\frac{210\times10^9 \times 4.5\times10^{-6}}{0.1}} = \sqrt{9.45\times10^6} \approx 3074.8 \text{ rad/s}$$

Step 3: Compute frequencies

For $n = 1$:

$$f_1 = \frac{1^2\pi}{2\times4^2}\times3074.8 = \frac{\pi}{32}\times3074.8 \approx 0.09817\times3074.8 \approx \boxed{301.9\,\text{Hz}}$$

For $n = 2$:

$$f_2 = \frac{4\pi}{32}\times3074.8 = \frac{\pi}{8}\times3074.8 \approx 0.3927\times3074.8 \approx \boxed{1207.5\,\text{Hz}}$$

Step 4: Mode shapes

For a simply beam, the mode shapes are:

$$\phi_n(x) = \sin\left(\frac{n\pi x}{L}\right)$$

- $\phi_1(x) = \sin\left(\frac{nx}{L}\right)$

- $\phi_2(x) = \sin\left(\frac{2\pi x}{4}\right) = \sin\left(\frac{\pi x}{2}\right)$

ChatGPT Integration

- Ask ChatGPT to verify the solution and explore more advanced problems involving the mode shapes of continuous systems.
- Use ChatGPT to investigate how engineers use the mode shapes of continuous systems to design structures that avoid resonance, such as in bridges or skyscrapers.

PROMPT 279

Discuss the use of modal analysis in practical applications, such as in aerospace, automotive, and civil engineering.

Objective

Understand how modal analysis is applied in real-world engineering fields to improve the performance and safety of structures and machines.

Activity

- Explore the practical applications of modal analysis in various industries:
  - Aerospace: Modal analysis is critical in the design of aircraft and spacecraft, ensuring that the structures can withstand vibrations caused by aerodynamic forces, engine operations, and other dynamic loads. Engineers use modal analysis to avoid resonance, which could cause structural failure in flight.
  - Automotive: In the automotive industry, modal analysis is used to design vehicles that are comfortable and safe by controlling vibrations from engines, road conditions, and wind. It is applied in the design of suspensions, exhaust systems, and frames.
  - Civil engineering: Modal analysis is essential for ensuring the structural integrity of buildings, bridges, and towers. It helps engineers predict how structures will respond to dynamic loads like wind, earthquakes, or traffic, and ensures that the natural frequencies of the structures do not align with environmental forces.
- Provide examples of how modal analysis has led to significant improvements in safety and performance across these industries.

Numerical Example

A civil engineering team is analyzing a bridge to ensure it can withstand dynamic loads from traffic. The bridge is modeled as a continuous beam with a length of $L = 100$ m, modulus of elasticity $E = 200$ GPa, moment of inertia $I = 300 \times 10^{-6} \text{ m}^4$, and mass per unit length $\mu = 500$ kg/m. Perform a modal analysis to calculate the first natural frequency of the bridge and discuss its implications for the design.

Solution

1. Use the natural frequency formula for a simply supported beam:

$$f_n = \frac{n^2 \pi}{2L^2} \sqrt{\frac{EI}{\mu}} \text{ (in Hz)}$$

2. Calculate the natural frequency and interpret how it affects the design to prevent resonance with traffic-induced vibrations.

Step 1: Correct formula for natural frequency (in Hz)

For a simply supported team (continuous system), the n-th natural frequency is:

$$f_n = \frac{n^2 \pi}{2L^2} \sqrt{\frac{EI}{\mu}} \text{ (in Hz)}$$

Step 2: Compute $\sqrt{\frac{EI}{\mu}}$

$$EI = 200 \times 10^9 \times 3.0 \times 10^{-4} = 60 \times 10^6 \text{ Nm}^2$$

$$\frac{EI}{\mu} = \frac{60 \times 10^6}{500} = 120000 \Rightarrow \sqrt{\frac{EI}{\mu}} = \sqrt{120000} \approx 346.41$$

Step 3: Compute f_1 (first mode)

$$f_1 = \frac{1^2 \pi}{2 \times 100^2} \times 346.41 = \frac{\pi}{20000} \times 346.41 \simeq 0.000157 \times 346.41 \approx \boxed{0.0545\,\text{Hz}}$$

Step 4: Interpretation for bridge design

- The first natural frequency is very low ($\sim$ 0.0545 Hz), which is expected for long, flexible structures like bridges.
- Traffic-induced loads (from vehicles or pedestrains) often contain frequency components up to several Hz.
- Since 0.0545 Hz is well below typical traffic excitation frequencies, resonance is unlikely from normal traffic.
- However, very slow, heavy vehicles or synchronized pedestrain motion (e.g., troops marching) could excite low modes. Designers may need to:
  - Use damping systems
  - Avoid resonant pacing (e.g., signs for troops)
  - Perform dynamic load analysis for unusual events

ChatGPT Integration

* Ask ChatGPT to explore real-world case studies where modal analysis played a crucial role in improving the safety and performance of structures or vehicles.

* Use ChatGPT to investigate how different industries use modal analysis to address specific vibration-related challenges.

PROMPT 280

Explore advanced topics in modal analysis, including the use of computational tools like finite element analysis (FEA) for determining mode shapes in complex structures.

Objective

Understand the advanced topics in modal analysis and how computational tools like finite element analysis (FEA) are used to analyze complex systems.

Activity

* Discuss the limitations of analytical solutions in modal analysis for complex systems. For intricate structures like aircraft components, bridges, or machinery with irregular geometry, analytical methods become infeasible.

* Introduce finite element analysis (FEA) as a powerful computational tool used to model and solve modal analysis problems for complex structures:

  * FEA breaks down a structure into smaller elements, allowing the natural frequencies and mode shapes to be calculated for each element and then assembled for the entire system.

  * FEA accounts for variations in material properties, boundary conditions, and geometric complexities that are difficult to handle analytically.

* Explain how FEA software tools (e.g., ANSYS and COMSOL) are used in industries such as aerospace, automotive, and civil engineering to conduct modal analysis.

Numerical Example

An engineer is tasked with performing modal analysis on a turbine blade with complex geometry. The blade is modeled using finite elements in an FEA software tool. Describe the process of using FEA to determine the natural frequencies and mode shapes of the turbine blade.

Solution

1. Discretize the blade into finite elements and apply the boundary conditions.

2. Define the material properties for each element (e.g., Young's modulus and density).

3. Solve the eigenvalue problem for the system using the FEA tool to obtain the natural frequencies and corresponding mode shapes.

4. Interpret the results to identify critical areas of the blade that are prone to excessive vibration.

ChatGPT Integration

- Ask ChatGPT to explain how FEA is applied in modal analysis and explore case studies where FEA was used to optimize designs in various industries.

- Use ChatGPT to investigate how advances in computational tools have expanded the capabilities of modal analysis in handling larger and more complex structures.

CHAPTER 29

VIBRATION MEASUREMENT AND INSTRUMENTATION

Prompt 281 Define vibration measurement and explain the significance of measuring vibrations in engineering systems.

Prompt 282 Discuss the different types of sensors used to measure vibrations, such as accelerometers, velocity sensors, and displacement sensors.

Prompt 283 Explain how an accelerometer works and how it is used to measure vibration levels in machinery.

Prompt 284 Solve a problem involving the calculation of vibration amplitude using sensor data.

Prompt 285 Discuss the role of signal conditioning in vibration measurement and how it improves the accuracy of sensor readings.

Prompt 286 Explain the concept of frequency analysis and how tools like the Fast Fourier Transform (FFT) are used in vibration monitoring.

Prompt 287 Discuss the use of data acquisition systems in vibration measurement and how they collect and process vibration data in real time.

Prompt 288 Explain the process of calibrating vibration sensors to ensure accurate measurements.

Prompt 289 Explore real-world applications of vibration measurement in industries such as aerospace, automotive, and manufacturing.

Prompt 290 Discuss the role of vibration measurement in predictive maintenance and how it helps prevent equipment failures.

PROMPT 281

Define vibration measurement and explain the significance of measuring vibrations in engineering systems.

Objective

Understand the importance of measuring vibrations in engineering systems and how it contributes to maintaining performance and safety.

Activity

- Define vibration measurement as the process of detecting, recording, and analyzing the vibratory motion of mechanical systems.
- Explain the significance of vibration measurement:
  - It helps detect excessive vibrations that can lead to equipment wear, fatigue, and failure.
  - Monitoring vibrations provides insights into the health of machinery and structures, allowing for preventive maintenance.
  - Vibration data is critical for ensuring safety, especially in industries where mechanical failure could be catastrophic (e.g., aerospace and automotive).
- Discuss how engineers use vibration measurement to assess system performance, ensure structural integrity, and optimize designs for vibration reduction.

Numerical Example

A machine experiences vibrations at a frequency of f = 50 Hz with an amplitude of A = 0.05 m. Calculate the velocity and acceleration of the vibration.

Solution

Velocity: $v_{max} = 2\pi f A = 2\pi \times 50 \times 0.05 = 15.71$ m/s

Acceleration: $a_{max} = (2\pi f)^2 A = (2\pi \times 50)^2 \times 0.05 = 4934.8$ m/s^2

ChatGPT Integration

- Ask ChatGPT to provide additional examples of how vibration measurement is used in different industries and explore the consequences of ignoring excessive vibrations in machinery.
- Use ChatGPT to investigate the role of vibration measurement in improving the lifespan and safety of engineering systems.

PROMPT 282

Discuss the different types of sensors used to measure vibrations, such as accelerometers, velocity sensors, and displacement sensors.

Objective

Understand the various types of sensors used in vibration measurement and their applications in different engineering systems.

Activity

- Explain the three main types of sensors used in vibration measurement:
  - Accelerometers measure the acceleration of vibrating systems. They are the most common type of sensor used in vibration monitoring, providing data on how quickly a system's velocity is changing.
  - Velocity sensors measure the velocity of vibrating systems. These sensors are used when engineers are interested in knowing how fast the system is oscillating at a given moment.
  - Displacement sensors measure the displacement or position of a vibrating system. Displacement sensors are useful when engineers need to know the amplitude of vibrations, such as in machinery experiencing significant deflection.
- Discuss the specific applications of each sensor:
  - Accelerometers are used in machinery, automobiles, and aerospace systems to monitor dynamic forces and vibrations.
  - Velocity sensors are used in rotating equipment like motors, turbines, and pumps.
  - Displacement sensors are commonly used in precision engineering, such as in CNC machines and high-precision instruments.

Numerical Example

An accelerometer placed on a vibrating system records an acceleration of $a = 5$ m/s^2 at a frequency of $f = 60$ Hz. Calculate the corresponding displacement.

Solution

▪ The relationship between displacement and acceleration is given by

$$A = \frac{a}{\left(2\pi f\right)^2}$$

▪ By substituting the values, we obtain

$$A = \frac{5}{\left(2\pi \times 60\right)^2} \approx 3.52 \times 10^{-5} \text{ m} = 0.0352 \text{ mm}$$

ChatGPT Integration

▪ Ask ChatGPT to explain how these sensors are used together in real-world applications to provide a complete picture of a system's vibration profile.

▪ Use ChatGPT to investigate the advantages and limitations of each type of sensor in different industrial applications.

PROMPT 283

Explain how an accelerometer works and how it is used to measure vibration levels in machinery.

Objective

Understand the working principle of an accelerometer and its application in vibration measurement for mechanical systems.

Activity

▪ Explain the basic working principle of an accelerometer:
  ● An accelerometer typically consists of a seismic mass suspended by flexible supports (springs) within a housing. When the device experiences acceleration (e.g., from vibration), the mass tends to resist the motion (due to inertia), causing a relative displacement between the mass and the housing.
  ● This relative motion is detected using various transduction mechanisms:
    – Piezoelectric: converts mechanical strain into voltage.

- Capacitive: changes in capacitance due to distance variation.
- Piezoresistive or MEMS-based: detect deformation electronically.
- The measured displacement or force is proportional to acceleration, allowing computation via Newton's second law $F = ma$.

- Discuss how accelerometers are used to measure vibration levels in machinery:
  - They provide real-time data on acceleration, which can be converted into velocity and displacement to give a full picture of the vibration profile.
  - Accelerometers are widely used in industries to monitor machine health, detect imbalances, misalignments, or mechanical wear that could lead to failure.
  - Explain the common placement of accelerometers on rotating machinery, engines, and structural components to measure vibration in different axes.

Numerical Example

An accelerometer is installed on a machine to monitor vibrations at a frequency of $f = 80$ Hz. The maximum recorded acceleration is $a = 10$ m/s^2. Calculate the corresponding velocity and displacement of the vibrations.

Solution

- Velocity: $v_{max} = \dfrac{a}{2\pi f} = \dfrac{10}{2\pi \times 80} \approx 0.02$ m/s

- Displacement: $A = \dfrac{a}{\left(2\pi f\right)^2} = \dfrac{10}{\left(2\pi \times 80\right)^2} \approx 0.0396$ mm

ChatGPT Integration

- Ask ChatGPT to explain how accelerometers are used in specific industries, such as automotive or aerospace, to monitor vibration levels.
- Use ChatGPT to explore different types of accelerometers, such as piezoelectric and MEMS accelerometers, and their applications.

PROMPT 284

Solve a problem involving the calculation of vibration amplitude using sensor data.

Objective

Practice calculating vibration amplitude from sensor data, such as acceleration or velocity measurements.

Activity

- Provide a scenario where the user needs to calculate the amplitude of vibration using data from a sensor (e.g., an accelerometer or velocity sensor).

Numerical Example

An accelerometer records a peak acceleration of a_max=12 m/s² for a machine vibrating at a frequency of f = 100 Hz. Calculate the corresponding vibration amplitude.

Solution

1. The relationship between acceleration and displacement amplitude is

$$A = \frac{a_{\max}}{(2\pi f)^2}$$

2. By substituting the given values, we obtain

$$A = \frac{12}{(2\pi \times 100)^2} = \frac{12}{(628.32)^2} \approx 3.04 \times 10^{-5} \text{ m} = 0.0304 \text{ mm}$$

The vibration amplitude is approximately 0.0304 mm.

ChatGPT Integration

- Ask ChatGPT to explain how to use different sensor data (e.g., velocity or displacement) to calculate the amplitude of vibration.
- Use ChatGPT to explore how engineers use amplitude data to assess the severity of vibrations and predict potential equipment failures.

PROMPT 285

Discuss the role of signal conditioning in vibration measurement and how it improves the accuracy of sensor readings.

Objective

Understand the importance of signal conditioning in vibration measurement and how it enhances the quality and accuracy of sensor data.

Activity

- Explain that *signal conditioning* refers to the process of modifying or filtering sensor signals to improve their quality and ensure accurate measurements. It is especially important in vibration measurement, where raw sensor data can be noisy or distorted.
- Discuss key components of signal conditioning:
  - Amplification boosts weak signals from sensors, making them easier to analyze.
  - Filtering removes noise from the sensor signals, such as high-frequency or low-frequency noise that is not relevant to the measurement.
  - Analog-to-Digital Conversion (ADC) converts analog signals from sensors into digital data that can be processed and analyzed.
- Explain how signal conditioning is crucial for obtaining reliable vibration data, ensuring that engineers can accurately assess the condition of machines or structures.

Numerical Example

A displacement sensor outputs a voltage signal that is very small and includes high-frequency noise. Describe how signal conditioning (amplification and filtering) could improve the quality of the sensor's output for use in vibration analysis.

Solution

- Amplification: The weak sensor signal is amplified to a level that can be accurately read and analyzed by the data acquisition system.
- Filtering: A low-pass filter is applied to remove high-frequency noise that may interfere with the true vibration signal, allowing for more accurate vibration data.

ChatGPT Integration

- Ask ChatGPT to explain how signal conditioning systems are used in different industries to improve vibration measurement accuracy.
- Use ChatGPT to explore the types of filters used in vibration analysis, such as low-pass and band-pass filters, and how they affect the final sensor readings.

PROMPT 286

Explain the concept of frequency analysis and how tools like the Fast Fourier Transform (FFT) are used in vibration monitoring.

Objective

Understand the concept of frequency analysis and how the Fast Fourier Transform (FFT) is applied in vibration measurement.

Activity

- Define *frequency analysis* as the process of analyzing the frequency components of a vibration signal to identify dominant frequencies and amplitudes.
- Explain how Fast Fourier Transform (FFT) is used in vibration monitoring:
  - The FFT converts a time-domain signal (such as a vibration signal recorded by a sensor) into the frequency domain, allowing engineers to see the amplitude of vibrations at different frequencies.
  - FFT is used to detect resonant frequencies, identify faults in machinery, and monitor system health.
- Discuss the significance of frequency spectra obtained from FFT:
  - Peaks in the frequency spectrum indicate dominant frequencies at which the system vibrates.
  - By analyzing these peaks, engineers can identify the sources of vibration, such as imbalance, misalignment, or wear in rotating equipment.

Numerical Example

A machine generates a vibration signal with a time-domain representation. Using FFT, the frequency components are identified, and the dominant frequency is $f = 120$ Hz with an amplitude of 0.02 m/s². How can the information obtained from FFT be used to assess the condition of the machine?

Solution

The FFT reveals that the dominant frequency in the vibration spectrum is 120 Hz, with a peak acceleration amplitude of 0.02 m/s². This frequency corresponds to the component of vibration with the highest energy.

Interpretation:

A dominant peak at 120 Hz may suggest a mechanical issue such as:

Rotating imbalance (if 120 Hz matches a shaft's rotational frequency or its multiples)

Misalignment or bearing defect (if consistent with characteristic fault frequencies)

Application:

By comparing 120 Hz to known operating speeds, gear mesh frequencies, or bearing defect frequencies, engineers can pinpoint the source of the vibration.

The amplitude (0.02 m/s²) gives a measure of vibration severity at this frequency. Though relatively low, continuous monitoring is essential to detect any increase over time, which could indicate worsening conditions.

ChatGPT Integration

- Ask ChatGPT to explain how frequency analysis helps detect specific vibration issues in machinery and structures.
- Use ChatGPT to explore other tools besides FFT used in frequency analysis, such as wavelet transforms, and their applications in real-time vibration monitoring.

PROMPT 287

Discuss the use of data acquisition systems in vibration measurement and how they collect and process vibration data in real time.

Objective

Understand the role of data acquisition systems (DAQ) in vibration measurement and how they enable real-time monitoring and analysis of vibration data.

Activity

- Explain that a data acquisition system (DAQ) collects, processes, and analyzes signals from vibration sensors (e.g., accelerometers, velocity sensors, and displacement sensors) in real time.
- Describe the components of a typical DAQ system:
  - Sensors convert physical vibrations into electrical signals.
  - Signal conditioning amplifies and filters the sensor signals to improve data quality.
  - Analog-to-Digital Converter (ADC) converts the analog signals into digital data for analysis.
  - Data processing units store and analyze the digital data, often using tools like FFT for frequency analysis.
  - A software interface provides real-time visualization of vibration data, enabling engineers to monitor the system's behavior and identify potential issues immediately.
- Discuss how DAQ systems are used in real-time vibration monitoring to detect faults, prevent equipment failure, and optimize performance in industries like aerospace, automotive, and manufacturing.

Numerical Example

A DAQ system records the vibration data of a rotating machine in real time. The system measures an acceleration of $a = 5$ m/s^2 at a frequency of $f = 80$ Hz. Calculate the displacement amplitude using the recorded data.

Solution

$$A = \frac{\alpha}{\left(2\pi f\right)^2}$$

$$A = \frac{5}{\left(2\pi \times 80\right)^2} = 1.98 \times 10^{-5} \, \text{m}$$

The displacement amplitude is 0.0198 mm.

ChatGPT Integration

▪ Ask ChatGPT to explain how DAQ systems are set up for specific vibration monitoring applications, such as in vehicles or industrial machinery.

▪ Use ChatGPT to explore different DAQ systems and their features for vibration measurement, including wireless DAQ systems and cloud-based data storage.

PROMPT 288

Explain the process of calibrating vibration sensors to ensure accurate measurements.

Objective

Learn the process of calibrating vibration sensors and why calibration is essential for ensuring accurate and reliable vibration measurements.

Activity

▪ Define *calibration* as the process of adjusting and verifying the accuracy of a sensor by comparing its readings with a known reference standard.

▪ Explain the importance of calibrating vibration sensors:

• Calibration ensures that the sensor provides accurate readings, which is crucial for detecting small changes in vibration levels.

• It prevents measurement errors that could lead to incorrect assessments of machine health, potentially causing unexpected equipment failures.

- Regular calibration helps maintain the reliability and consistency of vibration data, especially in critical applications like aerospace or industrial machinery.
- Discuss the steps involved in calibrating a vibration sensor:
  1. Set up a reference source: Use a vibration calibrator or a reference device that generates a known vibration frequency and amplitude.
  2. Compare sensor readings: Place the sensor on the reference source and compare the sensor's output with the known vibration values.
  3. Adjust sensor output: If discrepancies are found, adjust the sensor's settings (e.g., gain or sensitivity) to match the reference values.
  4. Document results: Record the calibration data and verify that the sensor meets the required accuracy standards.

Numerical Example

A displacement sensor is calibrated using a reference device that vibrates at $f = 50$ Hz with an amplitude of $A = 0.02$ m. The sensor initially reads an amplitude of 0.018 m. By what percentage should the sensor's output be adjusted to match the reference value?

Solution

1. Calculate the adjustment factor. Adjustment factor = $0.02/0.018 \approx 1.11$.
2. The sensor's output should be increased by approximately 11%.

ChatGPT Integration

- Ask ChatGPT to explain how calibration procedures differ for various types of vibration sensors (e.g., accelerometers and displacement sensors).
- Use ChatGPT to explore the importance of regular calibration in industries with high vibration sensitivity, such as semiconductor manufacturing or aviation.

PROMPT 289

Explore real-world applications of vibration measurement in industries such as aerospace, automotive, and manufacturing.

Objective

Understand how vibration measurement is applied in various industries to ensure safety, performance, and reliability of systems.

Activity

- Discuss the real-world applications of vibration measurement in different industries:
  - Aerospace: Vibration measurement is used to monitor the health of aircraft engines, wings, and fuselage structures. Early detection of abnormal vibrations can prevent fatigue failure in critical components, ensuring the safety of flights.
  - Automotive: Vibration monitoring is crucial in detecting issues like imbalance in wheels, misalignment in engine components, or loose parts in the vehicle body. By measuring and analyzing vibrations, automotive engineers can improve ride comfort, safety, and vehicle longevity.
  - Manufacturing: In industrial machinery, excessive vibrations can indicate mechanical wear, imbalance, or misalignment in rotating equipment like motors, pumps, and compressors. Vibration measurement allows for predictive maintenance, preventing unplanned downtime and extending equipment lifespan.
- Provide examples of how vibration measurement improves performance and reduces risk in these industries.
- *Performance improvement:* Fine-tuning machine operation by reducing vibration-related losses.
- *Safety enhancement:* Preventing catastrophic failures by detecting early signs of damage.
- *Cost reduction:* Lowering repair and maintenance costs via early diagnosis.
- *Product quality:* Ensuring parts and systems operate within acceptable vibration limits.

Numerical Example

In an automotive application, an accelerometer records abnormal vibrations in an engine at a frequency of $f = 150$ Hz and an amplitude of $A = 0.02$ m. How can this vibration data be used to diagnose engine issues?

Solution

The recorded frequency may correspond to a specific engine component, such as a rotating shaft or imbalance in the engine. Engineers can use this information to locate and address the source of the vibration, preventing further damage or failure.

ChatGPT Integration

- Ask ChatGPT to explore additional real-world applications of vibration measurement in industries such as wind energy, railways, or power plants.
- Use ChatGPT to investigate how industries use vibration data to improve predictive maintenance strategies and optimize machinery performance.

PROMPT 290

Discuss the role of vibration measurement in predictive maintenance and how it helps prevent equipment failures.

Objective

Understand how vibration measurement is used in predictive maintenance programs to prevent unexpected equipment failures and improve system reliability.

Activity

- Define *predictive maintenance* as a maintenance strategy that uses real-time data to predict when equipment is likely to fail, allowing for timely repairs or replacements before breakdowns occur.
- Explain the role of vibration measurement in predictive maintenance:
  - Vibration data provides insights into the condition of rotating machinery, such as motors, pumps, and turbines.

- Abnormal vibrations can indicate issues like misalignment, imbalance, or bearing wear. By continuously monitoring vibrations, engineers can detect early signs of equipment deterioration and schedule maintenance before a critical failure occurs.

- Predictive maintenance reduces downtime, minimizes repair costs, and extends the lifespan of equipment by addressing problems before they become severe.

- Discuss the use of vibration monitoring systems in industries such as manufacturing, aerospace, and energy generation.

Numerical Example

A pump experiences increasing vibration levels over time, with a measured acceleration of $a = 8\,\text{m/s}^2$ at a frequency of $f = 120$ Hz. How can this data be used in a predictive maintenance program?

Solution

The increasing vibration levels suggest potential issues like bearing wear or imbalance. By monitoring the trend in vibration data, engineers can predict when the pump will require maintenance, preventing an unexpected failure and minimizing costly downtime.

ChatGPT Integration

- Ask ChatGPT to explain how predictive maintenance systems are implemented in various industries using vibration data.

- Use ChatGPT to explore case studies where predictive maintenance based on vibration analysis has successfully reduced equipment failure rates and improved operational efficiency.

CHAPTER 30

REAL-WORLD APPLICATIONS IN MECHANICAL VIBRATIONS

Prompt 291 Discuss the role of vibration analysis in the design of bridges and how it helps prevent structural failures.

Prompt 292 Explain how vibration analysis is applied in the aerospace industry to ensure the safety and performance of aircraft.

Prompt 293 Explore the use of vibration measurement in wind turbines and how it helps improve their reliability and efficiency.

Prompt 294 Discuss how vibration control systems are implemented in buildings to reduce the effects of earthquakes and wind forces.

Prompt 295 Explain how vibration isolation systems are used in precision instruments, such as microscopes and optical devices.

Prompt 296 Discuss the challenges of controlling vibrations in marine engineering, particularly in ships and submarines.

Prompt 297 Explain how vibrations are measured and controlled in automotive engines to enhance performance and comfort.

Prompt 298 Explore the use of modal analysis in the design and testing of railway vehicles to ensure stability and safety.

Prompt 299 Discuss the role of vibration damping materials in reducing noise and vibration in electronic devices.

Prompt 300 Explore advanced techniques used in vibration analysis and control in robotics and automation systems.

PROMPT 291

Discuss the role of vibration analysis in the design of bridges and how it helps prevent structural failures.

Objective

Understand how vibration analysis is used in bridge design to ensure structural safety and stability.

Activity

- Explain that vibration analysis plays a critical role in bridge design by predicting how a bridge will respond to dynamic forces such as traffic, wind, and earthquakes.
- Discuss how engineers use vibration analysis to
  - identify natural frequencies of the bridge structure to avoid resonance, which could lead to excessive vibrations or failure
  - analyze the mode shapes of the bridge to ensure that key structural elements are properly designed to handle dynamic loads
  - evaluate the impact of external forces, such as wind or seismic activity, on the bridge's dynamic behavior
- Provide examples of famous bridge failures caused by poor vibration management, such as the Tacoma Narrows Bridge collapse, and how modern vibration analysis techniques have prevented similar incidents.

Lessons from Past Failures:

Tacoma Narrows Bridge (1940)

Collapsed due to aeroelastic flutter, a form of self-excited vibration.

The designers had not accounted for the bridge's torsional mode coupling with wind excitation, which caused oscillations to grow uncontrollably.

This incident led to the modern integration of wind tunnel testing and computational fluid-structure interaction (FSI) modeling in long-span bridge design.

London Millennium Bridge (2000)

Experienced lateral synchronous pedestrian excitation shortly after opening.

Solved using damping devices and better understanding of human-structure interaction via vibration analysis.

Modern Tools and Preventive: Practices

Use of Finite Element Analysis (FEA) and Modal Testing to simulate and measure dynamic responses.

Incorporation of tuned mass dampers, base isolators, and aerodynamic fairings to mitigate vibration effects.

Regular structural health monitoring (SHM) using sensors to track vibrations in real time.

Numerical Example

A bridge is modeled as a simply supported beam with a length of $L = 100$ m, modulus of elasticity $E = 200$ GPa, and moment of inertia $I = 250 \times 10^{-6}$ m^4; $\mu =$ mass per unit length. Calculate the first natural frequency of the bridge and discuss its significance in preventing resonance.

Solution

Use the formula for the natural frequency of a simply supported beam: $f_1 = \dfrac{\pi^2}{L^2} \sqrt{\dfrac{EI}{\mu}}$, where μ is the mass per unit length. Calculate the first natural frequency and discuss how engineers use this information to design the bridge to avoid resonance with traffic or environmental forces.

Step 1: Use the formula for the 1st natural frequency:

$$f_1 = \frac{\pi^2}{2L^2} \sqrt{\frac{EI}{\mu}}$$

Step 2: Plug in values:

$$f_1 = \frac{\pi^2}{2 \cdot (100)^2} \sqrt{\frac{200 \times 10^9 \cdot 4.25 \times 10^{-2}}{2500}}$$

Step 3: Simlify inside the square root:

$$\frac{EI}{\mu} = \frac{200 \times 10^9 \cdot 4.25 \times 10^{-2}}{2500} = \frac{8.5 \times 10^9}{2500} = 3.4 \times 10^6$$

$$\sqrt{3.4 \times 10^6} \approx 1843.91$$

Step 4: Compute frequency:

$$f_1 = \frac{3.1416}{2 \cdot 10^4} \cdot 1843.91 = \frac{3.1416}{20000} \cdot 1843.91$$

$$f_1 = 0.00015708 \cdot 1843.91 \approx 0.2896 \text{ Hz}$$

ChatGPT Integration

- Ask ChatGPT to provide additional case studies of how vibration analysis has been applied to modern bridges.
- Use ChatGPT to explore the tools and methods used in the field of bridge engineering for vibration analysis, such as finite element analysis (FEA).

PROMPT 292

Explain how vibration analysis is applied in the aerospace industry to ensure the safety and performance of aircraft.

Objective

Understand the role of vibration analysis in aerospace engineering and how it helps improve the safety and performance of aircraft.

Activity

- Explain that vibration analysis in the aerospace industry is used to monitor and control the vibrations experienced by aircraft during flight. Excessive vibrations can cause fatigue in critical components, leading to failure.
- Discuss how vibration analysis is applied in key areas:
  - Engines: Vibration analysis ensures that rotating components, such as turbines and compressors, operate without experiencing dangerous levels of vibration that could lead to engine failure.

- Airframes: The fuselage and wings are analyzed to ensure they can withstand dynamic forces such as air turbulence and aerodynamic loads.

- Flight comfort: Vibration analysis helps improve passenger comfort by reducing cabin vibrations from engines and airflow.

- Mention how modal analysis is used to determine the natural frequencies and mode shapes of aircraft structures to avoid resonance during flight.

Numerical Example

An aircraft wing experiences vibrations at a frequency of f = 20 Hz with a displacement amplitude of A = 0.01 m. Calculate the maximum velocity and acceleration of the vibration.

Solution

- Velocity: $v_{max} = 2\pi f A = 2\pi \times 20 \times 0.01 = 1.26$ m/s
- Acceleration: $a_{max} = (2\pi f)^2 A = (2\pi \times 20)^2 \times 0.01 = 158$ m/s^2

ChatGPT Integration

- Ask ChatGPT to explain how vibration analysis has been applied to improve the performance of specific aircraft models.

- Use ChatGPT to explore the tools and techniques used by aerospace engineers to monitor and control vibrations in flight, such as onboard vibration monitoring systems.

PROMPT 293

Explore the use of vibration measurement in wind turbines and how it helps improve their reliability and efficiency.

Objective

Understand the role of vibration measurement in wind turbines and how it enhances the reliability and operational efficiency of these systems.

Activity

- Explain that vibration measurement is critical in wind turbines to monitor the health of rotating components such as blades, shafts, and gearboxes. Excessive vibrations can lead to mechanical failures, reducing the reliability and efficiency of the turbine.

- Discuss the key areas of vibration measurement in wind turbines:
  - Blades: Monitoring vibrations in the blades helps detect issues such as blade imbalance, structural defects, or wear over time.
  - Gearbox: Vibration analysis of the gearbox can reveal early signs of gear misalignment, tooth damage, or bearing wear.
  - Tower and foundation: Measuring vibrations in the tower and foundation ensures that the turbine structure remains stable and free from excessive dynamic forces caused by wind loads.
- Mention how vibration monitoring systems enable predictive maintenance, helping operators detect and address potential issues before they cause costly downtime or damage.

Numerical Example

A wind turbine blade experiences vibrations with an amplitude of $A = 0.015$ m at a frequency of $f = 5$ Hz. Calculate the maximum velocity and acceleration of the blade's vibration.

Solution

- Velocity: $v_{max} = 2\pi f A = 2\pi \times 5 \times 0.015 = 0.47$ m/s
- Acceleration: $a_{max} = (2\pi f)^2 A = (2\pi \times 5)^2 \times 0.015 = 14.8$ m/s^2

ChatGPT Integration

- Ask ChatGPT to explore how vibration measurement is applied in wind farms to optimize turbine performance and reduce operational costs.
- Use ChatGPT to investigate the advanced vibration monitoring systems used in modern wind turbines, including real-time data collection and remote diagnostics.

PROMPT 294

Discuss how vibration control systems are implemented in buildings to reduce the effects of earthquakes and wind forces.

Objective

Understand how vibration control systems are used in buildings to mitigate the effects of dynamic forces such as earthquakes and wind.

Activity

- Explain that vibration control systems in buildings are designed to reduce the impact of dynamic forces, protecting the structure from damage during earthquakes and high winds.
- Discuss the different types of vibration control systems:
  - Tuned mass dampers (TMDs): A mass attached to the building that moves in opposition to the building's vibrations, reducing the amplitude of vibrations. TMDs are commonly used in skyscrapers to mitigate wind-induced vibrations.
  - Base isolation systems: The building's foundation is isolated from the ground using flexible bearings, allowing the structure to move independently of ground motion during an earthquake.
  - Viscoelastic dampers: These materials absorb and dissipate vibrational energy, reducing the amplitude of vibrations throughout the structure.
- Provide examples of iconic buildings with vibration control systems, such as the Taipei 101 skyscraper, which uses a large tuned mass damper to withstand typhoon winds and seismic activity.

Numerical Example

A building with a mass of $m = 500{,}000$ kg is equipped with a tuned mass damper that oscillates at a frequency of $f = 0.5$ Hz. The damper's mass is $m_{damper} = 5{,}000$ kg, and it reduces the building's vibration amplitude by 20%. Calculate the maximum displacement of the damper if the original amplitude of the building's vibration was $A = 0.1$ m.

Solution

- The reduced amplitude of the building's vibration is
  $A_{reduced} = 0.1\,\text{m} \times 0.8 = 0.08\,\text{m}$.
- The damper's displacement will match the building's reduced amplitude, so its maximum displacement is $A_{damper} = 0.08\,\text{m}$.

ChatGPT Integration

- Ask ChatGPT to explore more examples of buildings that use advanced vibration control systems for earthquake protection.
- Use ChatGPT to investigate how engineers design these systems to ensure structural safety and comfort in high-rise buildings.

PROMPT 295

Explain how vibration isolation systems are used in precision instruments, such as microscopes and optical devices.

Objective

Understand the role of vibration isolation systems in ensuring the accuracy and performance of precision instruments like microscopes and optical devices.

Activity

- Explain that vibration isolation systems are critical in precision instruments where even the smallest vibrations can disrupt performance or accuracy. Instruments such as electron microscopes and optical devices require stable environments to function properly.
- Discuss the different types of vibration isolation systems:
  - Passive isolation systems use materials like rubber pads, springs, or air cushions to absorb and dampen vibrations.
  - Active isolation systems use sensors and actuators to detect vibrations and generate forces that counteract them, providing real-time vibration control.
  - Magnetic levitation systems provide isolation without mechanical contact by suspending the instrument using magnetic forces, eliminating vibrations caused by friction.
- Mention specific applications in laboratories and industrial settings where precision instruments are isolated from environmental vibrations (e.g., buildings, traffic, and equipment).

Numerical Example

A microscope is placed on a vibration isolation table with a spring constant $k = 1,000$ N/m and a mass of $m = 50$ kg. Calculate the natural frequency of the isolated system to determine whether it effectively reduces vibrations at a frequency of 10Hz.

Solution

- The natural frequency is given by $f_n = \dfrac{1}{2\pi}\sqrt{\dfrac{k}{m}} = \dfrac{1}{2\pi}\sqrt{\dfrac{1000}{50}} \approx 0.71\,\text{Hz}.$
- Since the isolation system's natural frequency is much lower than the external vibration frequency (10 Hz), it will effectively isolate the microscope from vibrations.

ChatGPT Integration

- Ask ChatGPT to explore how vibration isolation systems are designed for use in ultra-precision instruments, such as in semiconductor manufacturing.
- Use ChatGPT to investigate the advancements in active isolation technologies that enable precise control of vibrations in sensitive equipment.

PROMPT 296

Discuss the challenges of controlling vibrations in marine engineering, particularly in ships and submarines.

Objective

Understand the challenges of vibration control in marine engineering and how engineers mitigate these issues in ships and submarines.

Activity

- Explain that vibration control in marine engineering is critical to the performance, safety, and comfort of vessels like ships and submarines. Excessive vibrations can affect the structural integrity of the vessel, the performance of onboard machinery, and crew comfort.
- Discuss the sources of vibrations in marine environments:

- Propeller-induced vibrations: Cavitation and imbalances in the propeller can generate significant vibrations.
- Engine vibrations: The engines, especially in submarines, must be carefully isolated to prevent vibrations that could be detected by sonar.
- Hull vibrations: External forces such as waves can induce vibrations in the ship's hull, leading to fatigue and structural failure.

- Explain how engineers use vibration isolation systems, damping materials, and advanced monitoring techniques to control vibrations in marine vessels.

Numerical Example

A submarine's propulsion system generates vibrations at a frequency of $f = 20$ Hz, with a displacement amplitude of $A = 0.005$ m. Calculate the maximum velocity and acceleration of the vibrations.

Solution

Velocity: $v_{max} = 2\pi f A = 2\pi \times 20 \times 0.005 = 0.63$ m/s
Acceleration: $a_{max} = (2\pi f)^2 A = (2\pi \times 20)^2 \times 0.005 = 79.0$ m/s^2

ChatGPT Integration

- Ask ChatGPT to explore how vibration control is implemented in modern submarines to reduce noise and increase stealth.
- Use ChatGPT to investigate the role of advanced damping materials in reducing vibrations in ship hulls and critical components in the marine industry.

PROMPT 297

Explain how vibrations are measured and controlled in automotive engines to enhance performance and comfort.

Objective

Understand how vibrations are measured and controlled in automotive engines to improve vehicle performance, durability, and passenger comfort.

Activity

- Explain that vibration measurement and control in automotive engines are essential to reducing wear, improving fuel efficiency, and enhancing passenger comfort. Vibrations in engines are caused by factors like imbalance, misfiring, and resonance.

- Discuss how engineers measure vibrations in engines using the following:

  - Accelerometers are placed on key engine components to monitor vibrations and detect issues such as misalignment or imbalance.

  - Frequency analysis involves tools like FFT that are used to analyze the vibration spectrum and identify abnormal frequencies that indicate mechanical problems.

- Explain vibration control methods:

  - Engine mounts: Rubber or hydraulic engine mounts isolate the engine from the vehicle frame, reducing the transmission of vibrations to the cabin.

  - Dynamic dampers are installed on the crankshaft or other rotating components to reduce torsional vibrations.

  - Balance shafts are used in some engines to counteract vibration forces, improving engine smoothness.

Numerical Example

An engine vibrates at a frequency of f = 75 Hz with an amplitude of A = 0.01 m.

Calculate the maximum velocity and acceleration of the engine's vibration.

Solution

- Velocity: $v_{max} = 2\pi f A = 2\pi \times 75 \times 0.01 = 4.71\,\text{m/s}$
- Acceleration: $a_{max} = (2\pi f)^2 A = (2\pi \times 75)^2 \times 0.01 = 2220.66\ \text{m/s}^2$

ChatGPT Integration

- Ask ChatGPT to explore the advanced techniques used in modern cars to reduce engine vibrations and improve driving comfort.

- Use ChatGPT to investigate how automotive manufacturers use vibration analysis to enhance vehicle reliability and reduce maintenance costs.

PROMPT 298

Explore the use of modal analysis in the design and testing of railway vehicles to ensure stability and safety.

Objective

Understand how modal analysis is applied in the design and testing of railway vehicles to ensure they remain stable and safe during operation.

Activity

- Explain that modal analysis is critical in railway vehicle design to assess the dynamic behavior of the vehicle as it moves over tracks. The analysis identifies natural frequencies and mode shapes to ensure that the vehicle remains stable and avoids dangerous vibrations.
- Discuss how engineers use modal analysis to
  - design railway vehicles that avoid resonance caused by the interaction between the vehicle and track
  - optimize the suspension system to provide a smooth ride while maintaining vehicle stability at high speeds
  - test new designs to ensure that the vehicle can withstand dynamic forces from track irregularities, curves, and braking
- Mention how modal testing is performed on both the vehicle structure and its components, such as wheels, axles, and suspension systems.

Numerical Example

A railway car experiences vertical vibrations at a frequency of $f = 5$ Hz with a displacement amplitude of $A = 0.02$ m. Calculate the maximum velocity and acceleration of the vibration.

Solution

- Velocity: $v_{max} = 2\pi f A = 2\pi \times 5 \times 0.02 = 0.63$ m/s
- Acceleration: $a_{max} = (2\pi f)^2 A = (2\pi \times 5)^2 \times 0.02 = 19.74$ m/s^2

ChatGPT Integration

- Ask ChatGPT to explore how modal analysis is used to improve the performance and safety of modern high-speed trains.
- Use ChatGPT to investigate case studies where modal analysis helped solve specific vibration problems in railway vehicle design.

PROMPT 299

Discuss the role of vibration damping materials in reducing noise and vibration in electronic devices.

Objective

Understand how vibration damping materials are used to reduce noise and vibration in electronic devices, improving performance and user experience.

Activity

- Explain that vibration damping materials are essential in electronic devices to minimize vibrations that can cause noise, affect performance, or damage sensitive components.
- Discuss the different types of damping materials used in electronic devices:
  - Viscoelastic materials: These materials absorb vibrational energy and convert it into heat, reducing the amplitude of vibrations. They are commonly used in smartphones, computers, and other consumer electronics.
  - Foam and rubber: These materials are often placed between electronic components to prevent direct transmission of vibrations, reducing noise and improving durability.
  - Constrained layer damping (CLD): A technique where a damping layer is sandwiched between two stiff layers, commonly used in audio devices and hard drives to reduce unwanted vibrations.
- Provide examples of how damping materials improve the performance of sensitive components such as hard drives, microphones, and speakers by reducing vibration-induced noise.

Numerical Example

A smartphone experiences vibrations with a frequency of f = 200 Hz and an amplitude of A = 0.001 m. Calculate the maximum velocity and acceleration of the vibration and explain how damping materials could reduce the impact on sensitive components.

Solution

- Velocity: $v_{max} = 2\pi f A = 2\pi \times 200 \times 0.001 = 1.26$ m/s
- Acceleration: $a_{max} = (2\pi f)^2 A = (2\pi \times 200)^2 \times 0.001 = 1580$ m/s^2

Damping materials can absorb these vibrations and reduce the impact on the sensitive electronic components within the smartphone.

ChatGPT Integration

- Ask ChatGPT to explore how specific damping materials are used in electronic devices to reduce vibrations and extend component life.
- Use ChatGPT to investigate advancements in vibration damping technology in modern consumer electronics.

PROMPT 300

Explore advanced techniques used in vibration analysis and control in robotics and automation systems.

Objective

Understand the advanced techniques used to analyze and control vibrations in robotics and automation systems, ensuring precision and performance.

Activity

- Explain that vibration analysis and control are critical in robotics and automation systems to maintain precision, performance, and reliability. Uncontrolled vibrations can affect the accuracy of robotic movements and cause wear on mechanical components.
- Discuss advanced techniques used in vibration control for robotics:
  - Active vibration control (AVC) uses sensors and actuators to detect vibrations and apply counteracting forces in real-time.

AVC is particularly important in robotic arms, where precision is required for tasks like assembly or surgery.

- Feedforward and feedback control: These control strategies use sensor data to adjust robotic movements dynamically, reducing the impact of vibrations on positioning accuracy.
- Vibration isolation platforms are often used in high-precision robotic systems to reduce the impact of environmental vibrations on sensitive tasks, such as semiconductor manufacturing or laboratory automation.

- Provide examples of how vibration control is applied in industrial robots, medical robots, and automated systems to improve performance.

Numerical Example

A robotic arm experiences vibrations at a frequency of $f = 10$ Hz with a displacement amplitude of $A = 0.005$ m. Calculate the maximum velocity and acceleration of the vibrations.

Solution

- Velocity: $v_{max} = 2\pi f A = 2\pi \times 10 \times 0.005 = 0.314$ m/s
- Acceleration: $a_{max} = (2\pi f)^2 A = (2\pi \times 10)^2 \times 0.005 = 19.74 \, \text{m/s}^2$

ChatGPT Integration

- Ask ChatGPT to explore the use of advanced vibration control techniques in cutting-edge robotics applications, such as robotic surgery and autonomous vehicles.

- Use ChatGPT to investigate how industries are incorporating real-time vibration analysis and control into automation systems to improve precision and reduce downtime.

Vibration Wizard Tutor App

This GPT app is designed to complement Part 3 of this book (Vibrations) by providing an interactive and immersive learning experience. Users can explore their preferred chapters and prompts from the book's comprehensive list, enabling them to dive deeper into specific topics of interest. The app features tools for solving numerical examples, step-by-step walkthroughs for complex problems, and instant feedback to enhance understanding. Additionally, it supports interactive simulations, personalized suggestions based on user progress, and detailed explanations to bridge theoretical concepts with practical applications. Whether you are reviewing key concepts, solving practice problems, or seeking clarification, the app serves as a dynamic extension of the book to support a hands-on approach to mastering vibrations.

Users can obtain access and run the Vibration Wizard Tutor app directly through OpenAI GPTs. To get started, when available, navigate to the GPT Apps section within OpenAI's platform and search for "Vibration Wizard Tutor App." Once launched, the app's user-friendly interface allows you to select chapters, prompts, or specific topics of this book. Simply input your questions or problems, and the app will provide tailored guidance, step-by-step solutions, and explanations to enhance your understanding of vibrations.

FIGURE 30.1 Wizard app

Conclusion: The Future of AI-Assisted Engineering Learning

The integration of AI-driven learning tools with engineering education marks a transformative shift in how students, professionals, and educators approach problem-solving, conceptual understanding, and technical mastery. This book, *Mastering Applied Engineering Mechanics with ChatGPT*, is designed as a structured guide to leveraging AI as an interactive companion in engineering studies, focusing on Statics, Dynamics, and Vibrations.

By engaging with **well-structured prompts**, learners can

- strengthen foundational concepts through structured reasoning and step-by-step problem-solving.
- develop problem-solving intuition by exploring numerical examples and real-world applications
- verify and expand their understanding using ChatGPT's interactive capabilities

Beyond serving as a repository of prompts, this book emphasizes the evolution of learning methods, where AI-assisted education fosters deeper engagement by allowing users to ask, explore, and refine their knowledge dynamically.

Engineering Learning in the Age of AI

Traditional textbooks present static content, requiring learners to internalize fixed explanations. By contrast, AI-driven platforms allow for

- personalized learning paths, where users interact with engineering problems at their own pace
- instant feedback loops, helping learners validate their calculations and approaches
- simulations and scenario-based explorations, fostering an applied understanding of theoretical principles

This book bridges traditional engineering knowledge with AI-powered learning, equipping users with the tools to think critically, verify solutions, and enhance their technical skills in real time.

Future Directions in AI-Assisted Engineering

The use of AI in engineering education is still in its early stages, and several exciting possibilities lie ahead, including

- Expanding AI into other engineering disciplines: Future iterations of this book could explore thermodynamics, fluid mechanics, materials science, or control systems, applying the same AI-assisted learning principles.

- AI-driven engineering simulations: AI-enhanced platforms could simulate real-world mechanical behavior, allowing learners to experiment beyond textbook equations.

- Interactive GPT apps for engineering: The Statics, Dynamics, and Vibrations Tutor Apps presented in this book demonstrate how AI can act as a personal tutor, providing step-by-step explanations, numerical walkthroughs, and interactive learning experiences.

- Industry-integrated AI tools: Future engineering curricula may incorporate AI-powered solvers for real-time system analysis, preparing students for modern industry workflows.

As AI evolves, its role in engineering problem-solving and decision-making will only grow, making adaptive learning tools an essential companion for the next generation of engineers.

Final Thoughts: Empowering Engineers with AI

This book serves as a stepping stone toward a more engaging, interactive, and AI-assisted engineering learning environment. Whether you are a student building a strong conceptual foundation, a professional refining your problem-solving skills, or an educator integrating AI tools into your curriculum, the structured prompts in this book will help you

- think critically and systematically about engineering principles
- solve problems interactively, refining your intuition with AI-based feedback
- bridge theoretical knowledge with practical applications, gaining insights that go beyond static explanations

By embracing AI as a learning companion, engineers can develop deeper insights, explore concepts from multiple perspectives, and enhance their technical expertise with an adaptive, intelligent approach to problem-solving.

Engineering education is evolving. AI is here to guide the next generation of learners.

Next Steps

- Explore AI-driven tools further – Use ChatGPT for extended problem-solving, along with MATLAB, COMSOL, or Python-based simulations.

- Explore GPT Apps – Readers are encouraged to explore the various GPT-powered applications available in the OpenAI App Store, many of which are tailored to mechanical engineering topics. These can be accessed via the "Explore GPTs" option in the sidebar menu of ChatGPT.

- Apply AI in professional workflows – AI is increasingly used in engineering for real-time analysis, predictive modeling, and automation.

- Continue expanding AI-assisted learning – Engage with more advanced prompts and explore topics beyond this book's scope.

Latest AI Tools for Engineering Education

The rapid advancement of AI technologies has introduced a variety of tools that enhance engineering education across disciplines, from personalized learning to real-time data analysis and simulations. Below is a curated list of practically useful AI tools, including AI agents, Grok, NotebookLM, and others, with brief descriptions of their applications in engineering education and industry workflows. These tools are selected for their ability to support students, educators, and professionals in mastering complex engineering concepts and applying them in real-world scenarios.

Grok (xAI)

Grok, developed by xAI, is a truth-seeking AI chatbot designed to provide insightful answers with advanced reasoning, coding, and real-time data analysis capabilities. In engineering education, Grok excels in explaining complex concepts in statics, dynamics, and vibrations through step-by-step reasoning, making it ideal for students tackling PhD-level problems or coding tasks. Its DeepSearch mode integrates real-time web and X data, enabling students to stay updated with the latest engineering trends and research. Grok's conversational tone and Think mode allow for in-depth exploration of engineering problems, fostering critical thinking and problem-solving intuition. Available on grok.com, iOS, Android, and X, with higher usage limits via SuperGrok subscriptions.

NotebookLM (Google)

NotebookLM is an AI-powered research assistant that synthesizes, analyzes, and generates content from uploaded sources like PDFs, Google Docs, and web URLs. For engineering students, it creates interactive study guides, summaries, and podcast-style audio overviews, making it easier to grasp complex topics like material science or fluid dynamics. Educators can use it to organize course materials or generate tailored lesson plans. Its ability to handle diverse formats and provide citations ensures accuracy, making it a valuable tool for research and collaborative learning. Available for free with a Google account, with premium features for students via .edu email.

ChatGPT (OpenAI)

ChatGPT, powered by GPT-5, is a versatile AI assistant excelling in creative problem-solving, content generation, and technical explanations. In engineering education, it supports students by generating detailed solutions to mechanics problems, automating content creation for reports, and providing interactive tutoring. Its plugin system, including DALL-E 3 for visualizations, enhances its utility for creating diagrams or simulating engineering scenarios. ChatGPT is widely used for its accessibility and ability to handle nuanced queries, making it a go-to for students and professionals. Free tier available, with paid plans starting at $20/month.

Claude (Anthropic)

Claude, developed by Anthropic, is a conversational AI prioritizing safety and clarity, particularly in coding and technical explanations. Engineering students benefit from its ability to produce clean, well-documented code and explain complex algorithms in plain English, making it ideal for programming-heavy courses like control systems or robotics. Claude's collaborative tone helps learners debug code or explore theoretical concepts, while its reliability minimizes errors. Available with free and premium tiers, Claude is a favorite among developers and educators.

Perplexity AI

Perplexity AI is a conversational search engine that combines AI with real-time web search to provide accurate, referenced answers. For engineering education, it's useful for researching current industry trends, such as advancements in AI-driven simulations or material science. Students can use Perplexity to gather insights for projects or verify theoretical concepts with up-to-date sources. Its ability to reduce hallucinated responses makes it reliable for academic work. Free to use with over 10 million monthly active users.

AI Teaching Assistant (Springs)

AI Teaching Assistants, like those offered by Springs, automate administrative tasks such as grading, scheduling, and attendance, freeing educators to focus on teaching. In engineering education, these agents provide instant feedback on assignments, assess student progress, and create personalized learning paths within Learning Management Systems (LMS). They enhance student engagement through tailored content and real-time analytics, making them ideal for hybrid or online engineering courses. Integrates with existing LMS platforms via APIs.

Burning Glass (Predictive Job Market Analytics)

Burning Glass provides real-time job market data, helping engineering schools align curricula with industry demands in fields like AI, robotics, and machine learning. For students, it offers insights into career paths and skills in demand, such as proficiency in AI tools or simulation software. Educators can use it to guide students toward relevant opportunities, ensuring their

Further Readings

For readers looking to deepen their understanding, the following textbooks provide detailed explanations, problem-solving techniques, and practical applications in statics, dynamics, and vibrations.

Part 1: Statics

1. *Engineering Mechanics: Statics* – J.L. Meriam and L.G. Kraige

   A classic reference with theory, problem-solving techniques, and real-world applications

2. *Vector Mechanics for Engineers: Statics* – F.P. Beer, E.R. Johnston, and D.F. Mazurek

   Covers vector-based problem-solving methods with step-by-step examples

3. *Engineering Mechanics: Statics* – Andrew Pytel and Jaan Kiusalaas

   A comprehensive and structured textbook that provides clear explanations, vector-based analysis

Part 2: Dynamics

1. *Engineering Mechanics: Dynamics* – J.L. Meriam and L.G. Kraige

   Focuses on motion analysis, Newtonian mechanics, and kinetics with real-world applications

2. *Vector Mechanics for Engineers: Dynamics* – F.P. Beer, E.R. Johnston, and P. Cornwell

   Provides detailed derivations and computational approaches for rigid-body dynamics

3. *Dynamics of Particles and Rigid Bodies: A Systematic Approach* – Anil V. Rao

   A modern perspective on analytical mechanics with computational tools and simulations

Part 3: Mechanical Vibrations

1. *Mechanical Vibrations* – S.S. Rao

   A comprehensive guide covering theory, numerical methods, and engineering applications

2. *Fundamentals of Vibrations* – L. Meirovitch

   Emphasizes mathematical formulations and energy-based methods for vibration analysis

3. *Theory of Vibrations with Applications* – W.T. Thomson and M.D. Dahleh

   Combines practical applications with detailed mathematical modeling of vibratory systems

Closing Summary

This book brings together the essentials of applied engineering mechanics—statics, dynamics, and vibrations—framed through practical prompts and supported by AI. The goal was not just to cover theory, but to make problem-solving feel natural, structured, and relevant to how real engineering decisions are made.

By using ChatGPT throughout, you have explored a different way of learning, one that encourages asking questions, checking your work, and thinking through problems interactively. The intent is to make mechanics less abstract and more connected to the kinds of challenges engineers face every day.

Mastering engineering mechanics is not a one-time goal: it is an ongoing mindset. Tools like ChatGPT (or your desired AI LLM-tool) can help make that journey more approachable and adaptable. Whether you are working through textbook problems, preparing for design challenges, or simply trying to understand how systems behave, combining solid fundamentals with flexible tools can give you a real advantage.

Thanks for exploring these ideas. Keep learning, keep solving, and make the tools work for you.

INDEX